FUNDAMENTAL SOIL SCIENCE

FUNDAMENTAL SOIL SCIENCE

MARK S. COYNE
University of Kentucky

JAMES A. THOMPSON
West Virginia University

THOMSON

DELMAR LEARNING

Australia Canada Mexico Singapore Spain United Kingdom United States

THOMSON

DELMAR LEARNING

Fundamental Soil Science

Mark S. Coyne and James A. Thompson

Vice President, Career Education Strategic Business Unit:
Dawn Gerrain

Director of Editorial:
Sherry Gomoll

Acquisitions Editor:
David Rosenbaum

Developmental Editor:
Gerald O'Malley

Editorial Assistant:
Christina Gifford

Director of Production:
Wendy A. Troeger

Production Manager:
J.P. Henkel

Production Editor:
Betty L. Dickson

Project Editor:
Maureen Grealish

Director of Marketing:
Wendy Mapstone

Marketing Specialist:
Gerald McAvey

Cover Design:
TDB Publishing Services

Library of Congress Cataloging-in-Publication Data

Coyne, Mark S., 1960–
 Fundamental soil science /
 by M. S. Coyne and J. A. Thompson.
   p. cm.
 Includes bibliographical references
 and index.
 ISBN 0-7668-4266-5
 1. Soil science. 2. Soils. I. Thompson,
 J. A. (James Allen) II. Title.
 S591.C77 2006
 631.4—dc22

                   2005020802

NOTICE TO THE READER

Publisher does not warrant or guarantee any of the products described herein or perform any independent analysis in connection with any of the product information contained herein. Publisher does not assume, and expressly disclaims, any obligation to obtain and include information other than that provided to it by the manufacturer.

The reader is expressly warned to consider and adopt all safety precautions that might be indicated by the activities herein and to avoid all potential hazards. By following the instructions contained herein, the reader willingly assumes all risks in connection with such instructions.

The Publisher makes no representation or warranties of any kind, including but not limited to, the warranties of fitness for particular purpose or merchantability, nor are any such representations implied with respect to the material set forth herein, and the publisher takes no responsibility with respect to such material. The Publisher shall not be liable for any special, consequential, or exemplary damages resulting, in whole or part, from the readers' use of, or reliance upon, this material.

CONTENTS

PREFACE

This book is designed to introduce you to fundamental concepts of soil science. It will help you understand the properties and processes that are basic to the use and management of soils. Because soils are gaining recognition as important ecological systems upon which life on Earth depends, your ability to understand and appreciate the soil resource will make you a better agricultural/environmental scientist, and will allow you to make better decisions regarding land use and management.

To meet this goal, this book is divided into six sections and sixteen chapters appropriate for a one-semester course in soil science. Section 1 describes soils in general from a landscape perspective and introduces how soils develop, are characterized, and are classified. The next several sections describe how soil is put together and organized in terms of partitioning soil into its major constituents.

Section 2 examines soil physical properties associated with the solid phase of soil and how the architecture of the solid phase influences soil porosity. Porosity subsequently influences important features of the water- and gas-holding properties of soil. Section 3 introduces critical chemical and physical properties of the solid phase of soil. Section 4 describes the biology and biochemistry of the organic component.

Section 5 examines various ways by which the simple soil illustrated in Figure 1-1 is managed, and the consequences of that management. Management can affect soil's physical properties, hydrologic properties, and chemical properties (soil fertility is applied soil chemistry). Finally, in Section 6, you will examine perceptions of soil use and its value as a natural resource.

After reading this book, you should be able to:

❖ Describe the important role of soil in the environment and its contributions to agricultural and nonagricultural systems.

❖ Communicate clearly using the common terms used by soil scientists.

❖ Describe the fundamental physical, chemical, mineralogical, and biological properties of soils, interactions among these properties, and their effects on plant growth, soil behavior, and soil management.

❖ Explain the role of soils in the landscape, particularly as related to the soil's participation in the major cycles of matter and energy (hydrologic cycle, carbon cycle, nitrogen cycle, etc.).

❖ Infer basic differences among soils formed under the influence of differences in climate, organisms, relief, parent materials, or time.

❖ Use soil information (e.g., from soil survey reports or collected from field studies) and understand the benefits and limitations of these data for various land use purposes.

EXTENSIVE TEACHING/LEARNING PACKAGE

The complete supplement package was developed to achieve two goals:

1. To assist students in learning the essential information needed to continue their exploration into the exciting field of Soil Science.

2. To assist instructors in planning and implementing their instructional program for the most efficient use of their time and resources.

Instructor's Manual to Accompany Fundamental Soil Science: The Instructor's Manual provides answers to end of chapter questions and in-chapter "Focus-On..." questions.

Lab Manual (Order #0-7668-4270-3): This comprehensive lab manual reinforces the text content. It is recommended that students complete each lab to confirm understanding of essential content.

Lab Manual Instructor's Manual: The Instructor's Manual provides answers to exercises found in the student lab manual, and additional guidance for instructors.

Math For Soil Scientists (Order #0-7668-4268-1): This unique resource is written for students and practitioners in the field of soil science who must learn or review basic mathematical operations faced when studying or working with soils. Math for Soil Scientists explains the importance of each concept explored, discusses the theory behind each concept, and presents the method for solving each problem.

ACKNOWLEDGMENTS

Special thanks goes to our friends and colleagues at the American Geological Institute. With their help, we were able to include some of the most outstanding photographs available.

SECTION 1

DESCRIBING SOIL

Chapter 1

THE SOIL IS THE NATION

"The soil is the nation. To benefit one is to aid the other."
Anon.

*"The soil is the great connector of our lives, the source
and destination of all."*
Wendell Berry, *The Unsettling of America,* 1977

OVERVIEW

Sometimes it's difficult to understand how vast a subject soil science truly
is, let alone understand why anyone would want to be a soil scientist. The
former question is the topic of this book, while the latter question is the
topic of this chapter. How can anyone become enthused by a topic as com-
monplace as dirt? Well, dirt is soil out of place and admittedly is a nuisance. But soil . . . soil is something you can explore with your senses—you
can see it, you can feel it, you can smell it, and you can even taste the differences in soil. Soil is accessible—it's no farther than your own backyard,
and if you don't have a backyard, no farther than the nearest patch of
earth. Without soil we wouldn't have plants. Without plants we wouldn't
have people. And without people we wouldn't have nations. So, to study
soil is to study the stuff by which societies are built and thrive. And that's
something to be enthused about.

OBJECTIVES

After reading this chapter, you should be able to:

- ✔ Identify two stories that soils tell through observation and
  measurement.
- ✔ Name three reasons why studying soils is important.
- ✔ Indicate factors that account for soil variability.
- ✔ Name five soil-forming factors.
- ✔ Name four distinct ped or aggregate shapes.
- ✔ Understand what a pedon is.

KEY TERMS

atmosphere	hydrosphere	peds
biosphere	lithosphere	productivity
fertility	pedon	weathering
horizon	pedosphere	

WHY STUDY SOIL?

Soil and its properties influence most of your daily life.

As you begin studying soil science you may ask yourself: What makes soil so important? And if *you* don't ask yourself, some friend or family member probably will ask you. You can confidently answer that soil and its properties influence most of your daily life: from where and how to build a house to why the grass won't grow where your neighbor drives off the driveway and over your lawn; from which crops will yield best on your land to why the United States has attained great economic and political stability while Russia's leaders struggle to maintain legitimacy and Africa struggles to feed itself. You'll see why Gilbert Wooding Robinson said, "The affairs of the soil may not have the strange magnificence of the outer universe or the curiosity of the inner recesses of the atom; but they touch our daily lives most intimately" (Robinson, 1937).

Close examination of soil tells two stories: how it came to be and how it can best be used.

Your successful use of soil information will depend on your ability to interpret the messages written in soil. Soil tells you two stories (if you are sufficiently perceptive). First, soil properties you observe or measure, such as color, horizontal layering, organic matter content, mineralogy, texture, pH, or nutrient content, can be used (alone or together with other environmental factors) to infer how that soil came to be. Second, and often more important, the soil properties you observe can be used to infer how a piece of land can be best used, including the management strategies that might be most successful.

THE IMPORTANCE OF SOIL

Soil has many important roles in natural and managed environments.

Soil is a critical component to most agricultural and environmental systems—it is not just a medium for plant growth and support. Whether the objective is crop production, urban development, livestock management, turfgrass maintenance, or environmental protection, understanding the role of soil in physical, chemical, and biological processes cannot be ignored. Soil performs many roles in natural and managed environments, including facilitating plant growth, regulating water supplies, recycling materials, hosting soil organisms, and providing physical support. In most, if not all, cases, it is a combination of the physical, chemical, and biological properties of soil that determine the soil's ability to properly function for these different roles.

Soil Is a Medium for Plant Growth

Soil's upper limit is air or shallow water, and its lower limit extends to the rooting depth of native plants or bedrock. One formal definition of soil is "the unconsolidated mineral matter on the immediate surface of the earth that serves as a natural medium for the growth of land plants (Soil Science Society of America [SSSA], 2004).

Most plants rely on soil to supply many of their growth requirements. At one time roots were actually thought to eat soil (Tisdale and Nelson, 1975). From the standpoint of plant production, soil scientists often focus on the ability of a soil to supply nutrients to growing plants, or soil **fertility.** A soil scientist's understanding of soil fertility is based on his knowledge of the mineralogical (Chapter 7) and chemical (Chapter 8) properties of soils, as well as nutrient cycling (Chapter 10). These principles guide him in his decisions about efficient and effective nutrient management practices (Chapter 14). Anyone farming and fertilizing is therefore practicing soil science.

Soil fertility is only one aspect of soil **productivity.** In addition to nutrients (Chapter 14), plants rely on soil as a source of water, a source of oxygen for roots, and as a means of physical support. Your full appreciation of soil productivity will also require you to appreciate the physical properties of soil (Chapters 4 and 5) and the interactions between soil and water (Chapter 6). These principles guide us in our decisions about both the physical management of soil (for example, tillage and erosion; Chapter 12) and the management of soil water (for example, drainage and irrigation; Chapter 13).

> Soil productivity is more than just soil fertility and nutrient availability.

Soil Is a Natural Body

Another definition of soil is "the unconsolidated material on the surface of the earth that has been subjected to and influenced by genetic and environmental factors of *climate* and *organisms* acting upon *parent material,* as conditioned by *relief,* over periods of *time*" (SSSA, 2004). This view of the soil as a natural body highlights your need to appreciate the factors and processes that have produced the soils seen today (Chapter 2), the methods and tools used to document the spatial variability of soils (Chapter 3), and how this variability influences a person's decisions about the proper use and management of soils (Chapter 15).

> There are five basic soil-forming factors: climate, organisms, relief (or topography), parent material, and time (CL·O·R·P·T).

Soil Is a Natural Resource

Neither of the previous definitions fully recognizes the role of soil in the environment. Soils are the skin of the Earth, and they are intimately connected to the rocks and other sediments below (the **lithosphere**), the plants and animals within and above (the **biosphere**), and the water and air above, below, and throughout (the **hydrosphere** and **atmosphere**) (Figure 1-1). In fact, the **pedosphere** (a technical term for the soil environment) is often the strongest link between these other spheres (Chapter 16).

> Soil (the pedosphere) links the other spheres of Earth's environment.

The soil is an ecosystem that is the habitat for many organisms (Chapter 9). These organisms facilitate many of the functions of soil. You can similarly view soil as a part of the human habitat because of its role as an engineering material supporting roads, buildings, and other human structures. Soil is also where materials accumulate naturally (for example, nutrients and organic debris) and artificially (for example, land-applied wastes and chemical spills). Therefore, soil can be seen as a sink for compounds like carbon, a recycler for nutrients like nitrogen and phosphorus, and a filter for water and waste materials. To understand the role of soil in these environmental processes, you need to thoroughly understand soil's physical (Chapters 4–6), chemical (Chapters 7–8), and biological (Chapters 9–11) processes.

> Soil is an important construction material for human and animal habitats.

So, whether your interest is plant production, environmental protection, land-use planning, civil engineering, or water resource management;

FIGURE 1-1 The pedosphere links the other spheres of the Earth. These other spheres—atmosphere, biosphere, hydrosphere, and lithosphere—represent air, life, water, and rock, respectively.

whether you're a concerned citizen, home gardener, farmer, forester, land developer, or environmental scientist—you need to be knowledgeable in the "affairs of the soil."

What Soil Is

How you define soil depends on its use.

The definition of *soil* changes depending on its intended use or function (real or perceived). Each user may have a different view of the soil resource; each user may be accurate. Before you begin your in-depth examination of soils and their properties, it will be useful to highlight some of the important concepts of soil that will be important throughout your study.

Soil Is a Mixture of Solids, Liquids, and Gases

Soil is a complex mixture of mineral and organic solids, liquids, and gases.

Soil is not just broken-up rocks. It is not just decayed organic materials. Soil is a complex mixture of solids (mineral and organic), liquids, and gases. A typical surface soil will be about 50 percent pores and 50 percent solids, with most of the solids being mineral material (Figure 1-2). If the soil is moist (but not saturated) the pore space will be about 50 percent filled with water (25 percent of the total volume). However, the amounts of water and air in soil pores change rapidly and regularly. Immediately after rainfall the pores may be mostly or completely filled with water. If you fail to water your houseplants for several weeks, the soil pores may only hold 15 percent water. Though not as dynamic as the pore space, the split between mineral and organic material in the solids also is not constant, particularly with depth. Deeper in the soil the solid portion will be mostly mineral material.

Soil Is Formed from Parent Materials

A soil is not just broken-up rocks, though all mineral soils began as a rock: sedimentary rock, igneous rock, or metamorphic rock (Figure 1-3). Many

Organic
Material
5%
Liquid
Phase
15–100%
SOLID PHASE
Mineral
Material
45%
Gas/Air
Phase
0–85%
PORE SPACE

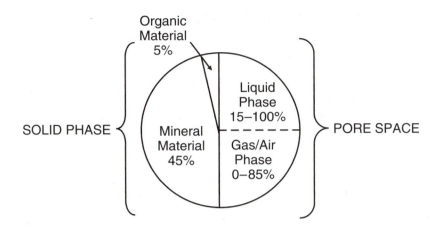

FIGURE 1-2 The idealized soil system in a mineral soil. The total volume is composed of about 50 percent solid material and 50 percent pore space. The pore space can be completely filled with water (saturated) or consist of about 85 percent air when the soil is air dry.

soils form directly from solid rock. Soils also form from unconsolidated parent materials that have been produced during **weathering**, the chemical, physical, and biological decomposition of rocks into smaller-sized particles.

All mineral soils originally came from rock.

Soil Is Colorful

As with people, you can sometimes tell much about a soil based on first impressions. Soils are red, brown, black, white, gray, and mottled. Soils can be olive, green, blue, and even shades of purple. Soil color is one of the

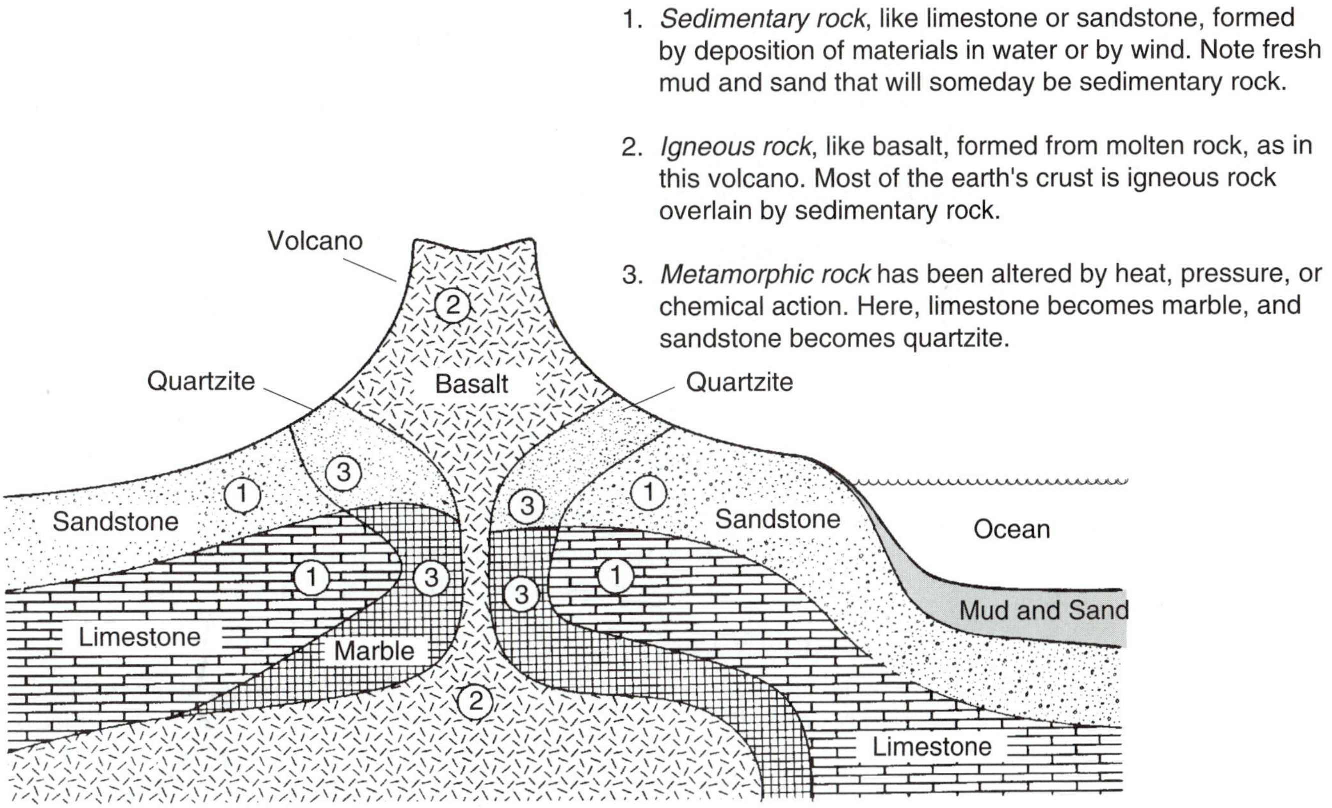

1. *Sedimentary rock*, like limestone or sandstone, formed by deposition of materials in water or by wind. Note fresh mud and sand that will someday be sedimentary rock.
2. *Igneous rock*, like basalt, formed from molten rock, as in this volcano. Most of the earth's crust is igneous rock overlain by sedimentary rock.
3. *Metamorphic rock* has been altered by heat, pressure, or chemical action. Here, limestone becomes marble, and sandstone becomes quartzite.
Volcano
Basalt
Quartzite
Quartzite
Sandstone
Sandstone
Ocean
Limestone
Marble
Mud and Sand
Limestone

FIGURE 1-3 The three basic types of rock: sedimentary (e.g., sandstone, limestone), igneous (e.g., basalt), and metamorphic (e.g., granite, marble). (*Plaster, 1997, p. 18*)

FIGURE 1-4 What are different types of soil structure? There are several different types of structure that you should recognize: spherical or granular, block-like, prismatic, and platey.

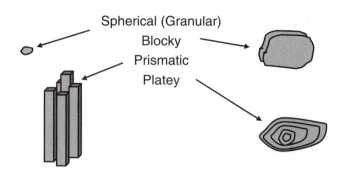

Spherical (Granular)
Blocky
Prismatic
Platey

Soil color can be a diagnostic feature.

most important diagnostic criteria you can use to determine such things as whether a soil is wet or dry, aerobic (oxidized) or anaerobic (reduced), and high or low in organic matter.

Soil Is Structured

There are four basic ped shapes: granular, blocky, prismatic, and platey.

Soil particles are formed into idealized soil structural units called **peds** (synonymous with *aggregates*). The peds can have various shapes depending on the particles from which they formed and the environment in which they occur. Some of the basic shapes of peds are granular, blocky, prismatic, and platey (Figure 1-4). The arrangement of the peds plays a significant role in water, air, and root movement through soil.

Soil Is Layered

Soil profiles show that soils can be layered, like a cake.

If you dig a trench in a field you will likely see that the soil is layered. These layers are the soil **horizons.** A soil profile is an exposed, two-dimensional slice of soil that includes the different soil layers (horizons) from the natural organic layers on the surface to the parent material or other layers beneath the solum (true soil). A soil horizon is a layer of soil approximately parallel to the surface having similar characteristics produced by the soil-forming factors (time, climate, relief, parent material, and vegetation, as well as human influence; Figure 1-5). Horizons can be distinguished partly on the basis of field observations, but sometimes lab data is needed.

Soil Is a Three-Dimensional Body

Pedons are three-dimensional bodies of soil with all the characteristics of a specific soil.

Another organizational structure of the soil is the **pedon,** a three-dimensional body of soil between 1 and 8 m^3 with the minimum volume possessing all the characteristics of that specific soil. Pedons are organized into polypedons, polypedons are organized into map units, and map units are organized into associations as one moves from the level of soil features on an individual basis to soil features across a landscape. For example, differences in organic matter content can be seen across a landscape, with lighter-colored soils on the knolls and darker-colored soils in the swales. The differences in color are due to weathering, the accumulation of organic matter, and erosion, all of which depend on where that soil is in a landscape.

Soil Is Highly Variable

Soils may be similar but are never identical. Soils of a single hillslope will differ mainly due to topography. The soils of Kentucky differ from those of Florida mainly due to parent materials, and differ from those of Kansas mainly due to vegetation, and differ from those of Alaska mainly due to climate.

(a)   (b)

FIGURE 1-5   (a) The Crider soil is the state soil of Kentucky. Well-developed horizons are obvious in these soils, which originally developed beneath forested environments. (b) The Woodson soil series is found in Kansas, Missouri, and Oklahoma. The Woodson series consists of deep, somewhat poorly drained, very slowly permeable soils that formed in silty and clayey sediments. These soils are on uplands. *(Photograph by Jim Fortner, courtesy of USDA-NRCS)*

The soils of the United States reflect the unique contributions of climate, vegetation, and time that help define physiographic regions (for example, the Great Plains compared to the Appalachians; Figure 1-6). The soils of the world differ for those same reasons (Figure 1-7). The factors of parent material and topography, climate and vegetation, and time and human intervention ensure that soils are distinct wherever you go.

Soil Is Complex

Compare soil science to a ball game. Individually, the basic concepts of most ball games are simple: you throw the ball, you catch the ball, you hit or kick the ball. Likewise, the basic individual concepts of soil science are not difficult. For example, soil texture, cation exchange capacity, and soil pH are fundamentally simple concepts. The challenge of soil science, and the foremost difficulty for students in their initial exposure to soil science, is in the complex relationships among the individual topics. For example, soil texture, specifically the amount of clay particles, directly influences the cation exchange capacity of a soil. Cation exchange capacity, in turn,

> Basic concepts in soil science are simple, but their interactions are not.

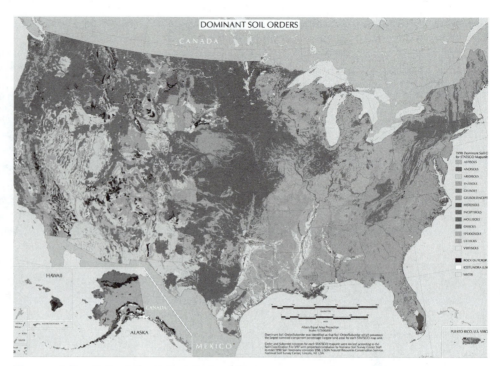

FIGURE 1-6   Soils of the United States. *(Map courtesy of USDA-NRCS)*

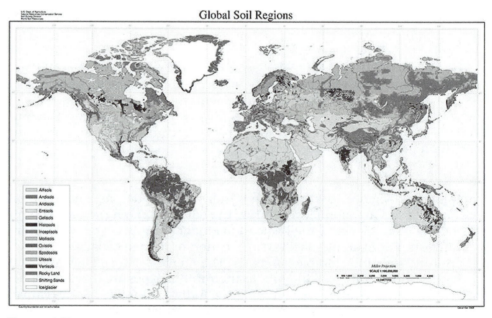

FIGURE 1-7   World soil resources. *(Map courtesy of USDA-NRCS)*

influences the buffer capacity of a soil and its resistance to changes in pH. Indirectly, the amount of clay affects the amount of organic matter in a soil. The amount of organic matter also influences the cation exchange capacity and a soil's resistance to changes in pH. As a further complication, a soil's pH affects the cation exchange capacity of a soil.

These relationships are complex. More important, these relationships are critical to understanding proper soil use and management. Soil texture, for example, a simple and easily determined soil property, will have a complex and profound effect on the lime requirement of a soil (how much alkaline material needs to be added to increase the soil pH), a basic management issue (Chapter 4). Other similar relationships will be described later in this book.

SUMMARY

There is great value to studying soil science beyond the obvious agricultural and engineering implications. Soil science touches our lives daily. In addition to being a medium for plant growth, soils regulate water supplies, recycle materials, host many organisms, and provide physical support. How one defines soil depends on how one uses soil. Although you care about a soil's fertility—its ability to supply plant nutrients—you are also concerned with its productivity, soil's ability to serve other environmentally significant functions.

Soils are continuously evolving and develop as a function of parent material, relief/topography, climate, organisms, and time. Human influences will also affect soils. Consequently, soils will be very different in different locations. The differences can be related to the distribution of minerals, organic matter, and pore space. The differences can also be related to the way the individual peds in soil are organized. Differences in soil can be observed in their color and horizons. Differences in soils can also be observed in their location within a landscape.

Many of the individual concepts in soil science are quite simple. They draw from other basic sciences such as chemistry, physics, and biology. The challenge of soil science is to understand all the complex relationships that occur in soil, because changes in one soil property can have significant effects on other soil properties.

END OF CHAPTER QUESTIONS

1. What are two stories that soil can tell you?
2. In what three ways besides plant growth is soil important?
3. What does soil fertility mean, and how is it different from soil productivity?
4. What are the five basic soil-forming factors?
5. What is the pedosphere, and how does it relate to the other spheres on Earth, such as the lithosphere and atmosphere?

6. What is a simple definition for the limits or boundaries of a true soil?

7. What is the basic parent material of a mineral soil?

8. What kinds of colors will you find in soils, and what will they tell you?

9. What are peds, and what four kinds of basic structures do they have?

10. What is a soil horizon, and how does it relate to a soil profile?

FURTHER READING

Berry, W. 1996. *The unsettling of America: Culture and agriculture.* San Francisco: Sierra Club Books.

(Wendell Berry, author and farmer in central Kentucky, is a renowned exponent of preserving the land resource. The Unsettling of America is an interesting philosophical look at the relationship of soil resources to society.)

Limbrey, S. 1975. *Soil science and archaeology.* New York: Academic Press.

(A useful reference if you want to see how soil science applies to other disciplines.)

Olson, L. 1943. Columella and the beginnings of soil science. *Agricultural History* 17: 65–72.

Olson, L. 1944. Pietro de Crescenzi: The founder of modern agronomy. *Agricultural History* 18: 35–40.

Olson, L. 1945. Cato's views on a farmer's obligation to the land. *Agricultural History* 19: 129–133.

(These three articles by Olson should give you a historical perspective on how the ancient and medieval worlds viewed soils.)

REFERENCES

Robinson, G. W. 1937. *Mother Earth or letters on soil.* London: T. Murby Publisher.

Soil Science Society of America. 2004. *Glossary of soil science terms.* Madison, WI: Author. (http://www.soils.org).

Tisdale, S. L., and W. L. Nelson. 1975. *Soil fertility and fertilizers,* 3rd ed. New York: MacMillan Publishing Co., Inc., p. 12.

Chapter 2

SOIL GEOGRAPHY AND GENESIS

*"Be it deep or shallow, red or black, sand or clay, the soil is the link
between the rock core of the earth and the living things on its surface.
It is the foothold for the plants we grow. Therein lies the main reason
for our interest in soils."*

Roy W. Simonson, *USDA Yearbook of Agriculture,* 1957

OVERVIEW

Look at the big picture first. Say you're standing in a field—any field—and
begin to survey your surroundings (Figure 2-1). If you're a soil scientist,
you start looking for several things. Where are you standing in the landscape? At the bottom of a slope or on a ridgetop? What kind of plants are
around? Is it mostly grasses, or are there a lot of trees? Is there any evidence that people have been present, like fences, houses, or roads? Is it
hot? Is it wet? Do you see any exposed rocks? What kind are they?

A soil scientist asks these questions for a reason. Each question relates
to some aspect of the way a soil was formed. Soil scientists call them the
soil-forming factors. For the past sixty years they have formed the basis for
how soil scientists look at the way soils develop in different environments.
So, before you take an in-depth look at what soil is made of and how it can
be used, you first have to look at how soils form in general, and why the
factors influencing soil formation have a geographical basis.

> The soil-forming factors
> form the basis for how soil
> scientists look at soils in
> different environments.

OBJECTIVES

After reading this chapter, you should be able to:

✔ Identify who Hans Jenny was and why his ideas are important in soil
science.

✔ Describe the factors of soil formation and explain when they are
important.

What is the influence of climate?
What is the role of vegetation and other organisms?
What is relief, and when does it matter?
What are different kinds of parent material?

What role does time play in soil formation?

How does weathering integrate the activities of several of these factors?

✔ Show how a typical soil horizon is characterized.

KEY TERMS

alluvial	igneous	profile
clastic	lacustrine	residual soil
colluvium	leaching	sedimentary
eluvial	loess	solum
eolian	metamorphic	subsoil
floodplain	parent material	terrace
glaciation	pedogenesis	topsoil
horizon	plow layer	weathering
hydrolysis		

SOIL-FORMING FACTORS

Hans Jenny is credited with developing the concept of the soil-forming factors.

Although the basic ideas had been circulating for many years, Hans Jenny (1899–1992) formalized an equation to describe the basic factors that drive geographic variability and formation of soil:

$$\text{Soil} = \text{a function of } \{\text{Cl·O·R·P·T}\}$$

FIGURE 2-1   Studying soil science gives you new perspectives on how to look at landscapes. This view is from a research farm in Lexington, KY. Your view is from the top of a ridge facing west. Approximately 1-2 meters below the soil surface the underlying bedrock is Ordovician-aged limestone.

where: Cl = climate (temperature and moisture)

O = organisms (biological processes including living and dead organisms, plants and soil microorganisms being the main players)

R = relief (topography, slope, landscape)

P = parent material (texture, primary mineralogy)

T = time (time since last disturbance)

Climate reflects the effects of wind and rain, freezing and thawing, weathering and transport of parent materials, plant and microbial growth (both the extent and duration on a seasonal basis), and **leaching** of soluble materials. Water itself is the great solubilizer, as you will see in later chapters.

Organisms reflect the above- and below-ground microflora and macroflora and fauna—the vegetation, insects, and microorganisms. As you saw in Chapter 1, soil is a mixture of organic and inorganic materials.

Relief or *topography* refers to the lay of the land; is it level or hilly? If hilly, how steep are the slopes, and what aspect do they display? (What direction do they face and how steep are they?)

Beneath all soils and within all soils is the **parent material** from which they originally formed. Even organic soils have some mineral matter.

Soils, like people or plants, have their dawn of life, maturity, and senescence. Unlike people or plants, the span of years in the life of soil numbers in the thousands to tens of thousands and in some ancient landscapes, such as the soils of Australia and Africa, perhaps to the hundreds of thousands of years. Early in the development of soil fertility increases, but as time goes on, weathering proceeds, and ultimately fertility decreases.

PARENT MATERIALS

Rock—The Primary Starting Material

The parent material of soil is inherited rather than acquired. It affects important chemical factors such as pH and fertility (compare soils derived from limestone in Kentucky to those derived from granite in New York) as well as texture. Three basic kinds of rock form parent material: igneous, sedimentary, and metamorphic (Table 2-1).

Igneous, sedimentary, and metamorphic rock form most parent material.

TABLE 2-1 Common types of rock. *(Adapted from Plaster, 1997)*

Sedimentary	Igneous	Metamorphic	Main Components
Sandstone		Quartzite	Quartz sand
Limestone		Marble	Calcite
Dolomite		Marble	Dolomite
Shale		Slate	Feldspar, clays
	Granite	Gneiss	Quartz, mica, feldspar
	Basalt	Schist	Feldspar, mica, olvine
	Rhyolite		
	Obsidian		
	Gabbro		

HANS JENNY AND THE SOIL FORMING FACTORS

Hans Jenny was born in Basel, Switzerland on February 7, 1899. After a brief period from 1926–1927 as a Rockefeller Fellow with Selman Waksman at Rutgers, Jenny took a position at the University of Missouri, Columbia in 1927. He subsequently joined the faculty of the University of California, Berkeley in 1936 as an Associate Professor of Soil Chemistry and Morphology.

Jenny gained international recognition for his 1941 publication *Factors of Soil Formation,* the basic ideas of which have diffused so widely that they form part of the soil science school curriculum everywhere. It has been said that Jenny's unification of driving forces for soil formation was no less fundamental than those of Lyell in geology and Darwin in biology.

Part of Jenny's success lay in his ability to bring the ideas of his predecessors such as E. W. Hilgard to fruition by combining field studies and abstract formalism in physical chemistry. He was in many ways a Renaissance man, capable of publishing surveys of soil productivity potential in California as well as lecturing to the Vatican Academy of Science on "The Image of Soil in Landscape Art, Old and New."

After his retirement, Jenny was instrumental in conducting research at the Pygmy Forest Reserve in Mendocino Co., California, a 70-acre (28 ha) tract of wave-cut terrace overlying four progressively younger (by 100,000 years) terraces representing novel soil and biotic environments. The reserve, run jointly by the Nature Conservancy and the University of California, now bears his name.

Igneous Rock

Igneous rock was originally molten rock. The size of the crystals in the rock reflects the speed at which cooling occurred—large crystals for slow-cooling rock and small crystals for fast-cooling rock. About one quarter of the Earth's land is underlain by igneous rock such as granite, gabbro, and basalt. The texture of soils formed from igneous rock depends on the texture of the igneous rock itself: coarse for granites, fine for basalts. Soils derived from granite have a lot of quartz, which resists weathering, and tend to be sandy (Figure 2-2). Soils derived from the finer-textured igneous rock contain little or no quartz, so they tend to be darker and more clayey.

FIGURE 2-2 Weathering of quartz from granite results in sandy soils as can be seen at the base of this granite outcrop. *(Copyright © Larry Fellows, Arizona Geological Survey.)*

(a)

(b)

FIGURE 2-3 Sedimentary rock, like sandstone, is frequently layered.(a) A weathering sandstone outcrop in Kentucky. (b) A sandstone fragment from a surface mine in Kentucky showing banded deposits.

Sedimentary Rock

Sedimentary rock underlies the developing soil in three-quarters of the Earth's land. Sandstone, shale, and limestone make up 99 percent of all sedimentary rock. These rocks originated as soft layers of sediment that were deposited in ancient shores, rivers, and oceans. Particle size was important in determining where the sediments were deposited and what type of sedimentary rock formed. The smallest particles are carried furthest during sedimentation, and sandstone, shale, and limestone have the largest to smallest particles, respectively. Because the sedimentary deposits were often the result of periodic deposition, sedimentary rock will often be layered (Figure 2-3)

One of the most common types of limestone is a clastic variety. **Clastic** means broken or fragmental. Clastic limestone is composed of the fragments of fossils transported from their source and deposited. Conglomerate is another type of sedimentary rock in which small stones of various types are cemented together.

> Sedimentary rock makes up most parent material.

Metamorphic Rock

Metamorphic rocks have been modified by heat and/or pressure activity upon rocks. Gneiss and marble are two examples of metamorphic rock (Figure 2-4). In most cases the metamorphosis creates a more resistant material. For example, marble is a more resistant rock than its starting material, limestone. Consequently, marble is a very suitable, though expensive, building material. Once weathering occurs, however, the metamorphic rocks typically degrade into the same minerals as their parent materials.

> Metamorphosis typically forms more resistant material.

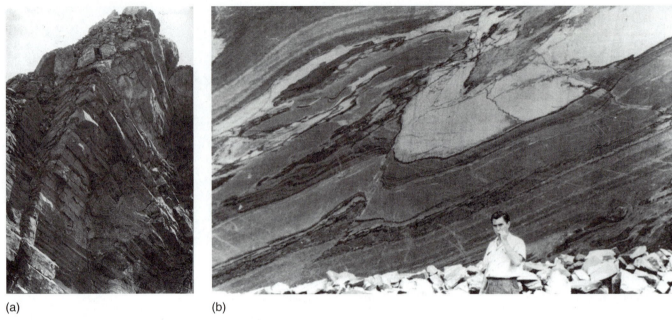

(a)   (b)

FIGURE 2-4   Gneiss and marble are two types of metamorphic parent material. (a) Banded gneiss near Zirkel Mountain, Jackson County, CO, 1911. (b) Reverse fault in dolomitic marble at the Jazida do Urubu in Minas Gerais, Brazil, 1957. The drag folds near the foot of the face, which are well defined along the fault plane, give the marble a distinctive pattern and illustrate the tremendous metamorphic properties within the Earth that led to its formation. *(Photograph courtesy of the USGS)*

FOCUS ON . . . PARENT MATERIALS

1. What are the three basic types of rock that form soil parent material?
2. What best characterizes the way igneous rock is formed?
3. What are examples of igneous rock?
4. How does the texture of igneous rock influence the texture of the soils that form from it?
5. What are examples of sedimentary rock?
6. Why does the weathering of sedimentary rocks such as sandstone create different textured soils?
7. Why are sedimentary rocks often layered?
8. What characterizes the formation of metamorphic rock?
9. What are examples of metamorphic rock?

Classification of Parent Materials

Soils can form from materials weathered in place (residuum) or transported from elsewhere (colluvium, alluvium).

In addition to classifying parent material as to whether it is igneous, metamorphic, or sedimentary, you can also classify parent materials as to whether they have formed soils in place (residual soils or residuum) or formed soils after being deposited elsewhere (transported soils). An example of how different types of residual and transported soils are distributed across the United States is given in Figure 2-5.

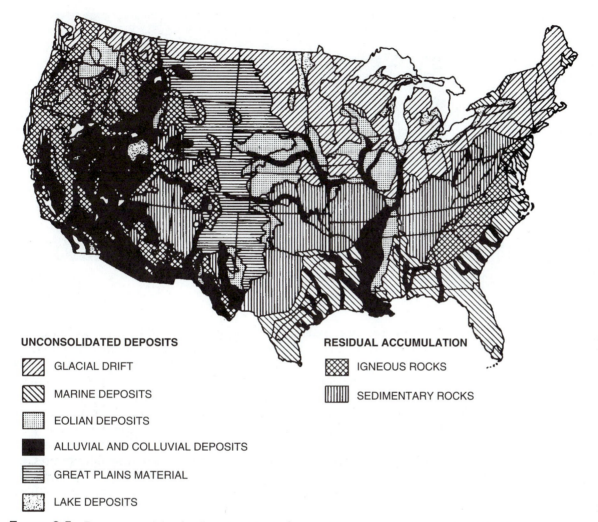

UNCONSOLIDATED DEPOSITS

[///] GLACIAL DRIFT

[\\\] MARINE DEPOSITS

[:::] EOLIAN DEPOSITS

[■] ALLUVIAL AND COLLUVIAL DEPOSITS

[≡] GREAT PLAINS MATERIAL

[:::] LAKE DEPOSITS

RESIDUAL ACCUMULATION

[⊞] IGNEOUS ROCKS

[|||] SEDIMENTARY ROCKS

FIGURE 2-5   Parent materials of soils in the United States. *(Map courtesy of the USDA-NRCS)*

Residual Soils

Residual soils weather in place without much disruption and without much off-site transport. The rate of soil formation is often slow, and the soil appears to form directly on bedrock. Consequently, these soils are often shallow. Texture of these soils is directly related to the texture of the parent material.

Water-Transported Parent Material

There are several kinds of water-transported soils. **Lacustrine** soils are soils deposited in freshwater lakes. Marine soils are soils deposited in oceans. **Alluvial** or fluvial soils are soils that were transported by running water in streams or rivers. There are three types of alluvial deposits: floodplains, alluvial fans, and deltas.

There are many types of alluvial soils.

FIGURE 2-6 Flooding of agricultural land within a floodplain. (Photograph courtesy of the American Geological Institute Photo Gallery, Copyright © Bruce Molnia, Terra Photographics)

FIGURE 2-6 Flooding of agricultural land within a floodplain. *(Photograph courtesy of the American Geological Institute Photo Gallery, Copyright © Bruce Molnia, Terra Photographics)*

Floodplains are that part of a river valley periodically inundated by flooding (Figure 2-6). Prior to the Mississippi River being contained by dikes and levees, for example, vast stretches surrounding the river were floodplains, and much of this was recovered by the Mississippi when massive flooding in 1993 caused the dikes and levees to fail.

Floodplains record each flooding event as a separate layer. Coarse materials like sands are deposited close to the river channel while fine silts and clays are deposited farther away. As streams erode downwards they leave **terraces** that record the previous floodplain. Because floodplains are periodically enriched with new sediment, they can be very deep and very fertile. Some of the earliest great civilizations in Egypt, India, and China developed on floodplains.

Alluvial fans form below hills and mountain ranges as water flowing down these slopes carries eroded material (Figure 2-7). As with marine deposits and floodplains, coarse material will be deposited first at the base of the mountain or hill, with finer material being deposited progressively further away. Deltas, such as the Mississippi Delta and the Nile Delta, deposit fine-grained material at the river mouth where river meets ocean. Deltas, in a sense, are a continuation of the floodplain.

Gravity-Transported Parent Material

Colluvium moves by the force of gravity.

Gravity-transported material is called **colluvium**. It is material that slides or rolls down slopes due to the force of gravity. At the base of the slope, it accumulates in material called *talus* (Figure 2-8). Colluvium is unsorted and unstable, and subject to landslides.

Wind-Transported Parent Material

Eolian material is transported by wind.

Another term for wind-transported material is **eolian** deposit. Two types of wind-transported material are **loess** and sand. Cover sands are common in arid and subhumid areas where sandstone parent material has weathered and been shifted by wind. The Sand Hills of western Nebraska are a good example of this process. Another good place to see dunes occurs along the eastern shoreline of Lake Michigan (Figure 2-9)

FIGURE 2-8   Talus at the foot of a hill (mesa). *(USDA-NRCS)*

FIGURE 2-7   Aerial view south of Salem showing pre-Lake Bonneville alluvial fan at the foot of the Wasatch Mountains cut by shoreline terraces of Lake Bonneville. Utah County, Utah, ca. 1940. *(Photograph courtesy of the USGS)*

FIGURE 2-9 Sand dunes at Sleeping Bear Dunes National Park on the eastern shore of Lake Michigan. These dunes represent aeolian deposits. Vegetation has a tenuous grasp in dunes, which makes them extremely sensitive environments.

Loess is silt-sized material typically blown from floodplains and glacial outwash. Loess soils are very fertile and can be found in Iowa along the Missouri River, in the north of China, and in the Danube Valley of Europe. Loess deposits can be extremely deep. Loess deposits also have the property of having greater stability if cuts through them are as vertical as possible.

In volcanic regions such as the western United States, the Andes, and Indonesia, an important type of windborne deposit is volcanic ash. These materials weather rapidly into a material called *allophane*. Allophane has the unique property of considerable anion exchange capacity.

Ice-Transported Parent Material

Glaciation has had a tremendous influence on soil formation in some regions.

Much of North America was **glaciated** during the last ice age, which ended some 10,000 years ago. As glaciers moved back and forth they acted as large bulldozers scraping away the developing soils and depositing them as unconsolidated terminal moraines at their boundaries. Other types of glacial debris are termed *eskers* and *drumlins.* Melting water from the glaciers also carried unconsolidated material of various materials into outwash plains.

FOCUS ON . . . TRANSPORTED PARENT MATERIALS

1. By what methods can parent material be transported?

2. How does water transport tend to sort parent materials of different texture?

3. What is the origin of terraces?

4. Why did some of the first great civilizations develop in floodplains?

5. How do colluvium and alluvium differ?

WEATHERING

Weathering results in both destruction and formation of soil material.

All parent material must weather before it can become soil. **Weathering** is a biochemical process that results in physical and chemical disintegration as well as new synthesis of material. One definition of weathering is *the response of materials that were once in equilibrium with the Earth's crust to new conditions at or near contact with water, air, and living matter.* Weathering integrates the functions of climate, vegetation, time, and even topography. Climate plays a critical role in weathering; so does vegetation, and vegetation is influenced by climate. Likewise, the nature of the starting material and its resistance to weathering determines how fast weathering will proceed.

Geologists and soil scientists are fond of using grave markers in cemeteries to illustrate weathering (Figure 2-10). The markers sit in relatively undisturbed environments and are dated so one has a useful and relatively precise gauge of how much weathering has occurred in a given period. Nature is full of examples in which parent material of different resistance to weathering has created unusual designs in the landscape (Figure 2-11).

The parent rock from which grave markers are cut weathers differentially partly because of the primary minerals from which they were formed. Figures 2-12 and 2-13 illustrate this relationship for primary and secondary minerals. Slate, a metamorphic rock composed primarily of quartz and feldspars, is consequently much more resistant to weathering than limestone. In highly weathered soil, all that remains is a mixture of highly resistant iron-aluminum-oxides and quartz.

(a)

(b)

(c)

FIGURE 2-10 Differential weathering of grave markers in African Cemetery No. 2, Lexington, KY. (a) Note how much more weathered is the earlier inscription on this limestone monument. (b) This fragment carved into slate is still legible after nearly 200 years. (c) In contrast, weathering of this limestone has made the inscription almost illegible after less than 100 years.

FIGURE 2-11   Rocks shaped by the wind and time sit like statues in the southeastern Utah landscape. *(Courtesy of USDA-NRCS)*

FIGURE 2-12   Sequence of weathering in primary materials in humid temperature regions. Feldspars and olvine weather rapidly in contrast to quartz. *(Adapted from Brady and Weil, 2002; Leet and Judson, 1954; redrawn by M. S. Coyne)*

Weathering occurs by physical (mechanical), chemical, and biological means (Table 2-2).

Physical/Mechanical Weathering

Physical weathering is most pronounced in cold and dry climates and climates that undergo extreme temperature variation. As the surface of a rock warms and cools, the minerals begin to expand and contract, which

Most Resistant
Least Resistant
Goethite (FeOOH)
Hematite (Fe_2O_3)
Gibbsite ($Al_2O_3 \cdot 3H_2O$)
Quartz (SiO_2)
Clay Minerals (e.g., Kaolinite, Montmorillonite)
Olivine ($[Mg,Fe]_2SiO_4$)
Dolomite ($Ca,MgCO_3$)
Calcite ($CaCO_3$)
Gypsum ($CaSO_4 \cdot 2H_2O$)

FIGURE 2-13 Sequence of weathering in secondary materials in humid temperate regions primary minerals quartz and olivine are added for comparison. Young soils with little weathering (which can also occur in arid regions) will retain gypsum and calcite. Moderately weathered soils will be dominated by silicate clays such as kaolinite and montmorillonite. Highly weathered soils will accumulate resistant mineral fractions dominated by iron and aluminum minerals. *(Adapted from Brady and Weil, 2002; redrawn by M. S. Coyne)*

TABLE 2-2 Types of weathering in soil.

Physical	Chemical	Biological
Freezing/thawing	Hydration	Acidification
Abrasion	Hydrolysis	Chelation
	Dissolution	Root growth
	Carbonation/acid reactions	
	Oxidation-reduction	
	Complexation	

creates differential stress within the rock. The stresses can lead to a process called *exfoliation* in which the outer layers of a rock peel away like the layers of an onion. Physical weathering can be accelerated if water gets in between the layers of rock or into crevices. As water freezes it expands, and this expansion can create enormous force—about 1465 Mg m^{-2}.

Abrasion by wind, water, and ice is another important physical weathering phenomenon. Windblown abrasive sediments can wear away rock and produce such phenomenon as natural rock arches (Figure 2-14). Water-borne sediment abrasion is an even more important process, as made evident by such natural phenomena as the Grand Canyon—which was almost entirely formed by the erosive action of the Colorado River—and other canyons and gorges. Abrasion can turn the roughest stones smooth with time, which makes river stone a valuable landscaping commodity.

Yosemite Valley in California was carved out by glacial movement, and in glaciated areas you can still see places where ice-borne rocks scoured the underlying parent material.

FIGURE 2-14   A natural arch at Mackinac Island, Michigan formed by the abrasive action of wind-borne sediment and water cutting through the parent limestone material.

For the most part, physical and mechanical weathering causes the rocks to disintegrate, and they become smaller, which increases the surface area and chemical reactivity. But the rocks maintain their overall chemical composition.

Chemical Weathering

Chemical weathering is prominent in hot and humid climates.

Chemical weathering is the dominant type of weathering in most hot and humid environments. Temperature is important because for every 10°C rise in temperature, most chemical reaction rates double. Water is important because water is involved in all of the most basic kinds of chemical weathering processes. The effect of physical weathering can create small cracks in the surface of rocks, which will subsequently allow for chemical weathering as solutions percolate within. In some material a weathering rind develops. The weathering rind is simply a layer around rocks, penetrating no more than a few millimeters into the rock, that displays the effects of physical and chemical weathering.

Hydration

In hydration, water is incorporated into the chemical structure of the mineral, which changes the mineral properties.

Hydrolysis

In **hydrolysis,** water molecules are split into hydrogen (H^+) and hydroxyl (OH^-). This splitting allows the H^+ to replace a cation, such as potassium (K^+), in the mineral structure. (Mineral structure will be discussed more fully in Chapters 7 and 8.) The K^+ can be adsorbed onto the cation exchange sites of soil. Likewise, any silicic acid (H_4SiO_4) that is formed can slowly leach or also react to form secondary minerals in soil. Remember that weather can create as well as destroy compounds in soil.

An interesting artifact of weathering frequently found in limestone-rich soils is the geode. Geodes have a hard outer shell and clear or colored quartz crystals projecting into the interior of a hollow center. They apparently form as dissolved solutes leach into and precipitate in voids left by dissolution of the limestone rock. Some geodes are quite large and are valuable art objects. Figure 2-15a shows a geode that has almost completely filled.

> Hydration, hydrolysis, and dissolution are three important water-related weathering processes.

Dissolution

Dissolution occurs in many minerals when water hydrates the cations and anions until they begin to dissociate from one another and become surrounded by water molecules.

Carbonation and Other Acid Reactions

Acids increase the hydrogen activity, which of course increases the potential for hydrolysis. When carbon dioxide (CO_2) dissolves in soil water it makes a small amount of carbonic acid (H_2CO_3), which increases the hydrolysis of mineralizable compounds like dolomite, gypsum, and limestone.

> Acidification often starts with oxides: CO2, NO2, and SO2.

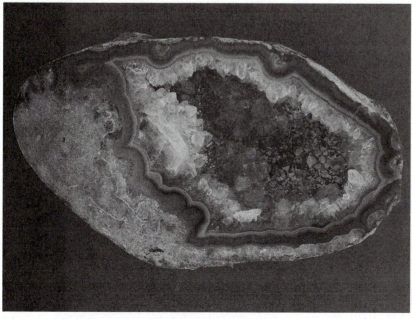

(a)

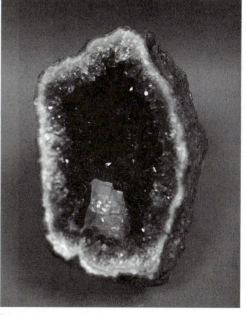

(b)

FIGURE 2-15 Geodes form when silica (quartz) precipitates inside voids within limestone-derived soils. (a) Geodes are much-prized collector's items because of their intrinsic beauty. (b) Note how quartz crystals have filled the interior of this geode. *(Copyright © Stone Trust, Inc.)*

Likewise, when metals such as ferric iron (Fe^{3+}) hydrolyze in soil they can release acidity. Sulfur dioxide (SO_2) and nitrogen dioxide (NO_2) released by coal combustion can oxidize in the atmosphere and precipitate as nitric and sulfuric acid.

Oxidation-Reduction

When compounds such as iron, manganese, and sulfur change their oxidation state (become either more oxidized or more reduced) they can influence the solubility of the minerals that contain them. So, oxidation of ferrous iron (Fe^{2+}) can destabilize minerals and likewise, because ferrous iron is more soluble, reduction can also destabilize minerals.

Biological Weathering

Living organisms contribute to physical and chemical weathering.

Although it is somewhat artificial to separate biological weathering from either physical or chemical weathering, it is useful to do so if only to illustrate that without the added contribution of plant and biological life, soil weathering would be much slower. Plants and microorganisms contribute to weathering primarily through root growth, acidification reactions, and complexation reactions.

Physical Weathering

If you doubt the capacity of plant roots to disintegrate rock, you need look no further than the example of the ancient Egyptians who used the expansion of dry wood inserted into carved slots to help cleave the massive limestone blocks used to build the pyramids. As Figure 2-16 illustrates, the sapling springing from this rock will eventually help to pry open the crevice it has exploited.

Acidification

As they respire, plant roots and microorganisms release carbon dioxide that becomes a weak carbonic acid solution in soil. Likewise, growing plant roots can release hydrogen ions (protons) that will acidify their environment and dissolve soil minerals. In addition, through cation uptake, plant

FIGURE 2-16 Tree roots contributing to the physical disintegration of parent material from a site in northern Michigan.

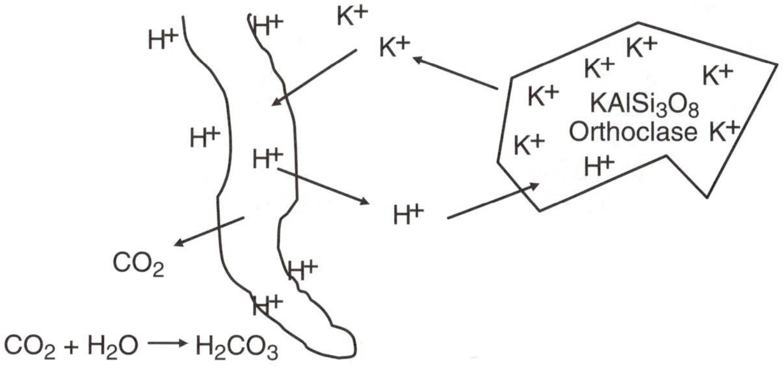

$CO_2 + H_2O \longrightarrow H_2CO_3$

$KAlSi_3O_8$
Orthoclase

FIGURE 2-17 Contribution of root growth to soil acidification and transformation of orthoclase. Absorption of soluble K^+ causes a diffusion gradient that makes more K^+ go into solution from orthoclase. Simultaneous release of H^+ by the plant root to maintain a charge balance can lead to further replacement of K^+ on orthoclase by H^+.

roots can create small concentration gradients that will stimulate mineral dissolution, and some of the protons they release can be used in dissolution reactions to release nutrients like K^+ from primary minerals such as orthoclase (Figure 2-17).

Lichens, a symbiotic association between fungi and cyanobacteria or green algae, are essential agents in the biochemical weathering of rock, in part through the carbonic acid they secrete. Lichens not only dissolve the rock, but the small cavities they leave also store water that can contribute to further weathering through freezing and thawing (Figure 2-18).

In addition to the carbon dioxide they produce during respiration, soil microbes are also important in acidifying soil through various oxidative processes they conduct, such as nitrification, iron and manganese oxidation, and sulfur oxidation. You'll learn more about these processes in Chapters 9 and 10.

FIGURE 2-18 Lichen play an important role in the initial biochemical weathering of rocks. The lichen release CO_2 during respiration, which forms carbonic acid (H_2CO_3) when it dissolves in water. The carbonic acid is not a very strong acid, but it is strong enough to gradually dissolve the underlying rock. *(Copyright © Larry Fellows, Arizona Geological Survey)*

Complexation Reactions

Plant roots and soil microbes release a variety of organic acids such as oxalic, citric, and tartaric acids that act as complexing and chelating agents. The carboxyl groups (−COOH) on these molecules form soluble complexes with metals such as iron and aluminum. This destabilizes the original materials, and it also allows the metals to leach through the soil profile. Weathering will be discussed more fully in Chapter 7.

> Complexation destabilizes soil minerals and promotes leaching.

FOCUS ON . . . WEATHERING AND SOIL FORMATION

1. Do different soil parent materials weather at the same rate? Which types persist the longest?

2. How does climate play a role in weathering?

3. What are the three major types of weathering processes?

4. What are five ways that water is involved in weathering?

5. What role does soil respiration play in weathering?

6. In what ways is weathering not a destructive process?

7. How does chemical weathering fundamentally differ from physical weathering?

RELIEF

> Relief is also called the "lay of the land."

Relief, or topography, is more commonly referred to as the "lay of the land." Why is topography important? It affects soil properties such as:

1. Water movement and regime
2. Soil formation and weathering
3. Soil depth
4. Vegetation
5. Material movement and type
   a. Erosion and deposition
   b. Organic matter content
   c. Thickness of soil layers
   d. Soil color
   e. Soil reaction
   f. Base saturation (fertility)
   g. Soil temperature
   h. Degree of impermeable layer development

For these reasons, soils developing in different parts of a landscape will have very different characteristics.

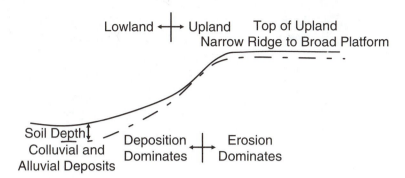

FIGURE 2-19 Slope and curvature influence soil depth. The shallowest soils are usually on the slope; the deepest are in the foot slope, which accumulates colluvial and alluvial material.

Topographic Position and Soil Depth

One of the most important influences of topography is on the depth of soil that develops in a given landscape. Soils tend to be thickest at the bottom of slopes, where deposits from erosion have been dropped, and where there is more water to support plant growth. Soils tend to be thinnest on the shoulders of slopes where the erosion is greatest and the retention of water least (Figure 2-19).

> The position of a soil in a landscape greatly influences the soil depth.

The Influence of Topography on Soil

Relief affects the amount of water that will infiltrate the soil. Because relief influences the soil depth in a given landscape position, it also influences the total water-holding capacity of a given parcel of land.

Aspect

Aspect (the orientation of a surface) is important because it influences the amount of solar radiation that a soil receives, and therefore its temperature. The greater the angle or slope of a soil, the greater the surface area over which solar radiation is distributed. In the northern hemisphere a north-facing slope will also typically be cooler than a south-facing slope.

Slope

Slope is the rise versus the run of a landscape, usually measured in percent (for example, ft per 100 ft or m per 100 m). Steep slopes encourage runoff and erosion and consequently receive less effective rainfall than flatter areas. Low-lying areas will receive runoff and become wetter than their surroundings. Different locations on a slope are given different names to help characterize their positions, because soil properties will differ at each position. So, whether a soil forms at a summit, back slope, or toe slope influences the soil depth, development, and water content (Figure 2-20).

> Slope is usually described in terms of percent (rise divided by run 3 100).

Curvature

Curvature, whether a slope is linear, convex, or concave, influences water transport, with convex slopes shedding water and concave slopes channeling water (Figure 2-21).

FIGURE 2-20 Slope positions and names.

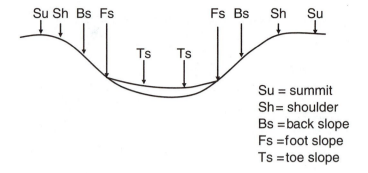

Su = summit
Sh = shoulder
Bs = back slope
Fs = foot slope
Ts = toe slope

FIGURE 2-21 Types of slope curvature. Arrows mark the direction of water flow. The consequence of curvature is that it can lead to channeling of water in concave systems, which can accelerate erosion.

L = Linear
V = Convex
C = Concave

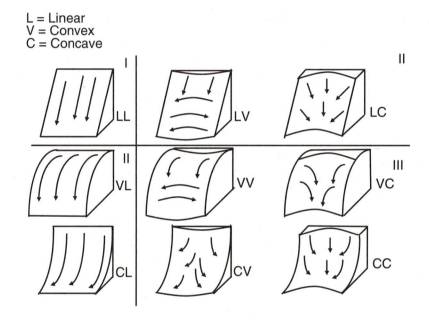

FOCUS ON . . . TOPOGRAPHY AND SOIL FORMATION

1. What does topography refer to?

2. How does topography influence solar radiation?

3. How does topography influence water retention?

4. What are three key components of topography?

5. Why might a concave slope be more prone to water erosion than a convex slope?

6. Why might a convex slope have less vegetation than a concave slope?

7. What is the slope of a landscape that rises 3 m for every 200 m of length?

8. What is the slope of a hill that rises 10 m for every 75 m of distance?

9. How much will your elevation change if you travel 25 m down a 30 percent slope?

CLIMATE AND ORGANISMS (PLANTS, MICROBES, AND ANIMALS)

Climate is often considered the most important soil-forming process because of the major role it plays in weathering. Temperature helps determine the rate of chemical reactions. Temperature also influences the amount of water, which controls leaching, pH, and aeration in addition to profoundly influencing vegetation. Wind, water, and ice help control the transport of materials.

How do temperature and rainfall affect soil genesis? If you were to make a transect from the arctic to the equator you might observe a change in weathering that looks something like Figure 2-22.

The climate also helps determine the type of vegetation that develops. Table 2-3 illustrates how different ecosystem types develop in response to actual and potential evapotranspiration, and how the different vegetation types that develop subsequently contribute to the formation of different soil types.

> Climate helps to control the type of vegetation that develops.

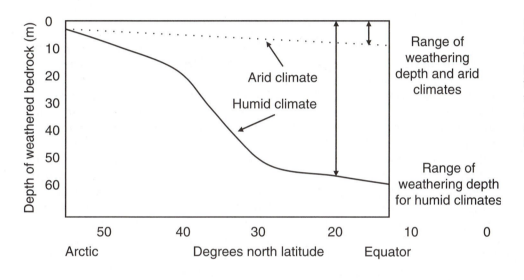

FIGURE 2-22 Effect of temperature and precipitation on depth of weathering. The weathering depth increases from the arctic to the equator. The depth of weathering is much less in arid environments. *(Adapted from Brady and Weil, 2002; redrawn by M. S. Coyne)*

TABLE 2-3 Ecosystem types and soil orders in relation to actual and potential evapotranspiration. *(Adapted from http://culter.colorado.edu:1030/,tims/jan27.html)*

Ecosystem Type	AET	PET	AET/PET	Dominant Soil Order
Tundra	150	300	0.50	Histosol
Boreal Forest	300	450	0.67	Spodosol/Entisol
Temperate Grassland	300	420	0.71	Mollisol
Temperate Deciduous Forest	690	700	0.99	Alfisol
Temperate Shrublands	300	400	0.75	Aridisol
Temperate Coniferous Forest	550	600	0.92	Spodosol
Temperate Rain Forest	700	700	1.00	Spodosol
Tropical Grassland	400	800	0.50	Mollisol
Tropical Shrubland	400	800	0.50	Aridisol
Tropical Seasonal Forest	1000	1400	0.71	Ultisol
Tropical Rain Forest	1600	1600	1.00	Oxisol

AET = actual evapotranspiration (mm)

PET = potential evapotranspiration (mm)

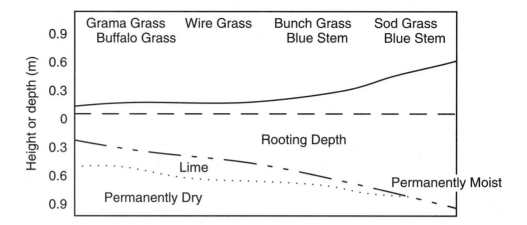

FIGURE 2-23 Vegetative growth, soil moisture, and soil depth in a west-east transect across the Great Plains. The height of plant and rooting depth increase as one moves to the east and precipitation increases. Residual lime in the soil profile only occurs at deeper and deeper depths because it is being more rapidly dissolved in the eastern side of the transect. *(Adapted from Shantz, 1923; redrawn by M. S. Coyne)*

Figure 2-23 illustrates how both the depth of soil and the type of vegetation change as one crosses the Great Plains from west to east—from drier to wetter environments. The change in water and temperature therefore influences vegetation. In turn this influences the amount of organic matter that is deposited. But as temperature increases, so does the rate at which decomposition can occur; thus, these two factors exist in equilibrium with one another.

One of the reasons that the production and loss of organic matter from growing vegetation is in equilibrium is that there is an entire community of microorganisms and animals (insects, earthworms, mites, etc.) that grow at the expense of this vegetation. So, as the vegetation thrives in a particular climate the microbial and animal community thrives as well. The larger animals contribute to *bioturbation*, physical movement and mixing of soil by burrowing and channeling activities. There will be much more discussion of specific activities of soil organisms in Chapter 9.

THE INTERACTING EFFECTS OF TEMPERATURE AND RAINFALL ON SOIL ORGANIC MATTER

The United States is a good example of how climate can influence soil through the interaction of rainfall, temperature, and vegetation. If you traverse the United States from south to north the temperature becomes cooler (discounting changes in elevation for the moment), and as you traverse from west to east there is generally more rainfall. So what happens if you look at the interaction of temperature and rainfall?

Rainfall is, of course, going to influence the type and amount of vegetation that forms. As more rain is available, the vegetation changes to grassland in the Great Plains running from Montana to Texas. Approximately east of the Missouri and Missisippi the native vegetation changes to forest—coniferous in northern states like Maine, Michigan, and Wisconsin and deciduous in states farther south.

However, although rainfall permits an environment to generate more organic matter, greater rainfall also leads to greater leaching, and higher temperatures also lead to greater rates of soil organic matter decomposition. More rain and cooler temperatures lead to soil organic matter accumulation. Although plant growth may be slow because of the cooler temperatures, decomposition rates are slow as well. High rainfall and high temperatures lead to soil organic matter degradation despite fast plant growth (Figure 2-24).

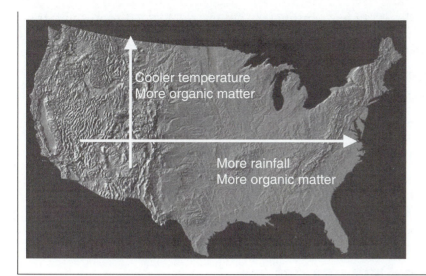

FIGURE 2-24 The influence of climate on soil organic matter. *(Topographic map courtesy of USGS. Thelin, G. P., and Pike, R. J., Landforms of the conterminous United States; A digital shaded-relief portrayal: U.S. Geological Survey Miscellaneous Investigations Map I-2206 scale 1:3,500,000)*

Cooler temperature
More organic matter

More rainfall
More organic matter

FOCUS ON . . . CLIMATE AND ORGANISMS

1. Why can climate be considered one of the major driving forces behind soil genesis?
2. How does temperature influence the depth to which soils weather?
3. What role does rainfall play in determining vegetation?
4. Why does soil development and depth appear to be greater in some cold, wet environments compared to hot, arid environments?

TIME

A chronosequence is a series of soils that have developed in the same topography, climate, and parent material but differ in the amount of time that each has been exposed to weathering. Alluvial terraces can sometimes be used to construct chronosequences because the uppermost terraces begin weathering before any lower terraces. As time progressed you would be able to see more and more layers (horizons) develop in the soil as plants grew, organic matter was deposited, elements leached, and soil minerals weathered (Figure 2-25). With the exception of organic matter accumulation, many changes in a soil profile are so subtle that they require a chemical analysis to demonstrate that true differences have occurred.

> Chronosequences are soils differing only in the amount of time for their development.

Soil depth

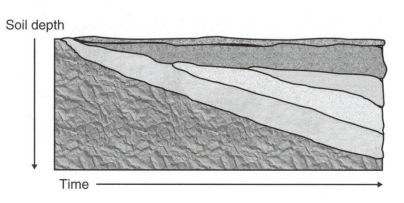

Time

FIGURE 2-25 The presence or absence of soil-horizons (layers) depends on the soil-forming factors, particularly time and place. The longer a soil develops in one spot, the more layers or horizons it is likely to contain.

As a general rule, every few centimeters of new soil formed represents the combined action of the soil-forming factors over periods of several hundreds to thousands of years.

FOCUS ON . . . TIME

1. What is a chronosequence?
2. Where might you find good examples of chronosequences?
3. Why might all soils not develop at the same rate?

ANTHROPOGENESIS (HUMAN INFLUENCE)

Humans have influenced soil formation in multiple ways.

Anthropogenesis refers to the creation of new soil by human intervention. You can think of this as a sixth soil-forming factor because the soil is being manipulated in ways that would not normally have occurred for its place and time. (Jenny was fully aware that humans influence soil formation, but preferred to think of genesis in terms of development after soil disturbance had occurred.) Some examples of ways in which humans have affected soil genesis are by (Figure 2-26):

❖ Burning of prairie grassland to maintain grazing areas
❖ Agricultural activity
❖ Urban landscaping
❖ Strip mining
❖ Deforestation
❖ Irrigation
❖ Drainage

FIGURE 2-26 Native Americans commonly burned prairie systems to maintain grazing areas for wild game and to prevent the succession of other vegetation. Similar management continues today in some prairie remnants such as the Konza Prairie in Kansas. Prairie burns will stimulate new growth and recycle nutrients. *(Photograph courtesy of Jeff Vanuga, USDA-NRCS; Jeff Vanuga, Konza Biological Research Station, Manhattan, KS)*

FIGURE 2-27 Installing perforated drainage tubing into a wet area in northcentral Iowa in the 1980s. By changing drainage patterns and increasing aeration, this practice will alter the way in which a soil develops compared to its surroundings. *(Photograph courtesy of Lynn Betts, USDA-NRCS)*

Agricultural activity, for example, has a dramatic influence on soil properties. The Morrow plots at the University of Illinois have been continuously cropped with corn since 1876. During that time the nitrogen content of the soil has declined from approximately 3.0 to 0.1 percent. Draining wetlands to increase cropland was a common practice until recently when the importance of wetlands in regulating the hydrological cycle in soil was fully recognized (Figure 2-27).

Urban development is another human activity that completely alters soil. The Washington Mall, for example, was originally marshland adjacent to the Potomac River. Over time the mall area has been raised over 6 m (18 feet) by fill from surrounding excavations, the most recent having occurred in the 1920s to bring the Mall up to its present level. Because of the periodicity of filling, the soil in the Mall had a chance to develop for brief periods, which has left an unusual pattern of buried layers.

> The soil profile of the Washington Mall tells a story of repeated human disturbance.

Current strip-mining laws, particularly those that occur on "prime farmland," require that the soils above the minerals be removed by layer, stockpiled, and then replaced in their original order (i.e., replaced by horizon) so that the original land use can be restored once the strip mining is completed. Occasionally, insufficient soil is available to complete the task and more has to be brought in. Thus, although the soil may have the original material, the process of removing, stockpiling, and replacing will have severely disrupted any original structure that it had.

> Sometimes the replacement soil in strip mines is better than the original material.

FOCUS ON . . . HUMAN INFLUENCES

1. Why can human disturbance be regarded as a soil-forming process?
2. Is human disturbance of soil any different from natural disturbance such as landslides or wind throw?
3. What are examples of human disturbance to soil environments?
4. How does managed burning of prairies affect soil formation?
5. How will installing tile in wet soils affect soil formation?

THE SOIL PROFILE

So far you have taken a very broad look at the factors that form soil, and there have been hints that these factors lead to soils of different types and appearance. Now it's time to examine more closely how these soil-forming factors integrate to form the material that is recognizable as soil.

Here is another definition of soil:

> *Soil is the collection of natural bodies on the Earth's surface, in places modified or even made by man, of earthy material containing living matter and supporting or capable of supporting plants out of doors. Its upper limit is air or shallow water. Its lower limit extends to the depth of the rooting of native plants or the bottom of the genetic soil (i.e., bedrock).*

The process by which soils are made is referred to as **pedogenesis.** There are four basic processes involved in pedogenesis:

❖ Transformations

❖ Translocations

❖ Additions

❖ Losses

As these processes occur, under the influence of the soil-forming factors, they either build, destroy, change, or move the constituents of soil. Because these processes will differ from location to location, soil that evolves in each location will be slightly different, although soils will share some similarities because they may have some of the original soil-forming factors in common.

Processes of Soil Formation

Transformations occur when soil constituents are physically and chemically changed. These changes can be destructive, or they can generate new compounds. The decomposition of primary minerals into silicate clays is a good example of a chemical transformation. Another good example is the dissolution of primary minerals into smaller pieces. Decomposition of plant residues on the soil surface, formation of organic matter, synthesis of organic acids, and precipitation of dissolved minerals are all examples of transformation processes. Changes in oxidation status of oxidized ferric iron (Fe^{3+}) to reduced ferrous iron (Fe^{2+}) and back again is a transforma-

> There are four basic processes in pedogenesis: transformations, translocations, additions, and losses.

> Transformations are physical and chemical changes.

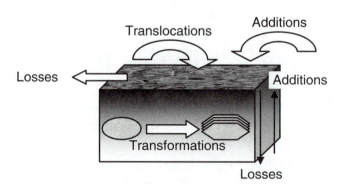

FIGURE 2-28 Basic processes of soil formation (pedogenesis). The basic processes are surface and subsurface loss, additions, and translocations.

tion in soil driven by aeration status and leads to a distinctive soil characteristic known as mottling.

Translocations involve the movement of organic and inorganic soil components into and out of the soil environment. The burrowing activity of insects, earthworms, and crayfish, for example, translocates soil material from within the soil to the soil surface. Likewise, these burrowing animals help to bring organic matter into the soil profile. Plants translocate O_2 from the atmosphere to plant roots, and translocate nutrients, such as potassium (K^+), from the soil. Plants also translocate fixed carbon from photosynthesis in the plant leaves down into the root environment. Water percolating through soil will carry dissolved nutrients and chelated or complexed compounds deeper into the soil profile. Capillary water movement from the water table below a soil can carry dissolved salts, such as nitrate (NO_3^-) and sodium (Na^+), to the soil surface where they will form crusts.

| Translocations involve movement. |

Developing soils can receive deposition from windborne (eolian), gravity-borne (colluvial), or waterborne (alluvial) sediments, as well as salts carried in aerosols from ocean environments. Other additions are the residues from plant growth (Figure 2-29) and human additions such as animal manures, fertilizers, pesticides, and toxic chemicals such as oil and coal slurry.

| Additions add materials to soils. |

Soils can lose materials. Soils can be mined for industrial materials such as diatomaceous earth (large deposits of one-celled marine organisms called diatoms) and kaolin (a type of clay). Soils can lose nutrients through leaching and plant uptake. One way of looking at agriculture is as a form of mining the soil for essential nutrients for human growth. Clay minerals can also be leached through soil and move from the **topsoil** into the **subsoil**. Organic matter in soil can be lost through leaching and decomposition. Soils can lose water through evaporation, transpiration, and leaching. Most importantly, soils can lose the developing surface horizons by wind and water erosion.

| Losses remove materials from soils. |

FIGURE 2-29 No-till agriculture is a practice that adds abundant residues to the soil surface. These young corn plants are growing in the residue of a previous wheat crop. *(Courtesy of USDA-NRCS)*

Master Horizons

Given enough time for these soil-forming processes to occur, the soil will develop distinctive layers or **horizons.** As noted, a soil horizon is a layer of soil approximately parallel to the surface with characteristics produced by the soil-forming factors (time, climate, relief, parent material, organisms— and human influence if you include it). Soil horizons are readily apparent as cross-sections in pits or road cuts of the soil called soil **profiles.**

The unique circumstances of each soil's development ensure that within close vicinity the soil profiles can differ, and they certainly differ when there are major differences in the soil-forming factors such as climate and vegetation.

Five master soil horizons are used to describe and characterize soil profiles: O, A, E, B, and C (Figure 2-30). The underlying bedrock is sometimes designated R (for regolith). Within each master horizon are subclassifications (for example, a, e, i) that are used for further characterization (Figure 2-31). These subclassifications are further described in Table 2-4.

> Roadcuts are ideal places to view soil horizons.

> The master horizons are designated O, A, E, B, and C.

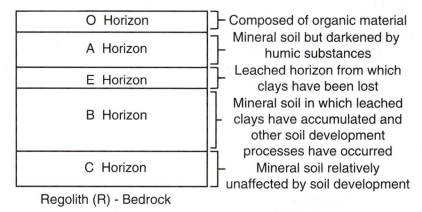

FIGURE 2-30 Master horizons. The master horizons are designated O, A, E, B, and C. In agricultural environments, where a litter layer does not typically occur, there is no O horizon.

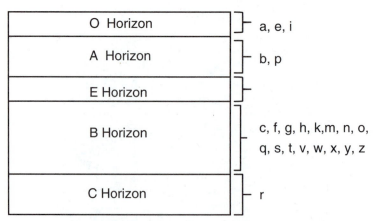

FIGURE 2-31 Diagnostic horizon features. Each horizon may have some specific distinguishing features that affect its chemical or physical properties. See the text for further details about the meaning of each code.

TABLE 2-4 Important subhorizon designations.

Letter Designation	Characteristics
a, e, i	Refers to highly (sapric), intermediate (hemic), or slightly (fibric) decomposed material
b	Used with any horizon that has been buried by overlying deposits
c	Concretions present
g	Strongly gleyed (referring to the amount of reduced iron present)
h	Illuvial (from leaching above), accumulation of organic material
m	90% cemented
p	Mixed by plowing
r	Weathered bedrock that can be dug with a spade
t	Accumulation of clay
x	Fragipan present
z	Accumulation of soluble salts

O Horizons

The O horizon forms above the mineral soil and is comprised of organic material. Typically the O horizon would be found in undisturbed or uncultivated environments such as forests. Much of the organic material is in the early stages of decomposition and can still be recognized. The *i, e,* and *a* subclassifications refer to *fibric, hemic,* and *sapric* materials, respectively. Fibric materials are the least decomposed and may form fibers if rolled between the fingers. Hemic material is more decomposed; the organic material is finer and hardly recognizable, although it may still form fibers. Sapric material is highly decomposed and nonrecognizable.

A Horizons

The A horizon is what most individuals would recognize as topsoil, and is the mineral layer where most of the decomposing organic matter added to soil accumulates. The organic matter often gives the A horizon a dark color. This layer is subject to leaching of clays and other materials. It is also the soil layer most exposed to climatic extremes and human intervention, and so is the most weathered layer.

> The A horizon is what is typically recognized as topsoil.

The *b* and *p* subclassifications refer to a buried horizon and a **plow layer,** respectively. Because of deposition by wind, water, and landfill, the A horizon may be buried beneath new soil that resets the clock for soil formation at the new soil surface. The A horizon thus represents a discontinuity in the soil profile development. An A_p horizon reflects the intervention of humans by plowing. Since plows rarely go below 40 cm depth (approximately 15 inches), the A_p horizon is limited in depth.

E Horizons

The E or **eluvial** horizon is a layer below the A horizon from which soluble materials have left (think *E* as in *e*migrated). These layers have lost organic matter, clay, iron, and aluminum oxides, and have left behind resistant materials like quartz. The E horizon is usually lighter in color than the overlying A horizon and usually occurs in forest soils with lots of rainfall.

> The E horizon is a zone of obvious translocation.

B Horizons

The B horizons of soils are zones of accumulation or "illuviation" (think of *i* as in *i*mmigrated). These horizons are the zones of maximum clay, iron, and aluminum oxide accumulation. Calcium carbonate (limestone) can also accumulate in the B horizon. Although the B horizon is often considered to be synonymous with the subsoil, from a technical perspective it is not. In shallow soils B horizon material might be mixed in with A horizon material during plowing, and in deep soils only the upper part of the A horizon may actually be the topsoil. The A, E, and B horizons are collectively called the **solum** and represent the layers of soil where most plant roots grow.

C Horizons

The C horizon represents unconsolidated material that is in the process of weathering, but more nearly resembles the parent material. That parent material, of course, may or may not represent the same parent material as the overlying layers depending on what sort of deposition process has occurred. There is very little organic matter and very little biological activity in C horizons.

Transitional Horizons and Other Distinctions

In addition to the five master horizons, there are additional horizons that have intermediate characteristics—they may be dominated by the characteristics of one horizon, but they still demonstrate the characteristics of a distinctly different horizon. These horizons are given combination names such as AE or BE, with the first letter representing the horizon that most contributes to the properties of the transition horizon (Figure 2-32).

It is also important to remember that every soil may not have every master horizon or subclass. For example, a well-developed soil in an undisturbed environment may have most of the horizons, but if uncultivated it will lack an A$_p$ horizon. Many soil profiles are truncated; that is, they have lost one or more horizons due to erosion. Severe erosion, for example, may

> Every soil does not have every horizon.

FIGURE 2-32 Transitional horizons. Most soils will have transitional horizons that reflect some but not all of the properties of overlying and underlying layers.

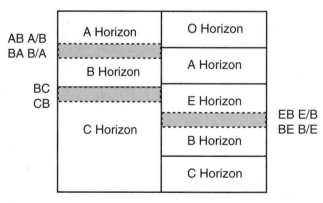

AB = More like A than B
A/B = A but has < 50% B-type material
BA = More like B than A
B/A = B but has <50% A-type material

EB = More like E than B
E/B = E but has < 50% B-type material
BE = More like B than E
B/E = B but has <50% E-type material

completely remove the A horizon of a soil. However, in soils continually receiving fresh deposits there may be too little time for the weathering required to develop a true B horizon. As a consequence, one can collect soil profiles from many different sites and observe that both the number and depth of profiles differs dramatically.

FOCUS ON . . . SOIL HORIZONS

1. What is the difference between a horizon and a profile?
2. What are the five master horizons?
3. How is the level of decomposition in the O horizon designated?
4. What happens when a soil profile is truncated?
5. Why do the depths of soil horizons differ?

SUMMARY

In this chapter you examined how geography and factors of soil formation influence the types of soils that form in different landscapes. Soils represent the unique environments in which they formed as a result of the five soil-forming factors: climate, organisms (vegetation), relief, parent material, and time.

Climate is the major force influencing soil formation through weathering of the soil material, through transport of the soil parent material by wind and precipitation, and through its influence on the type and amount of vegetation that will be present in a soil environment. Weathering will, by mechanical and chemical means, physically destroy the parent material of soils, chemically alter it, and in some cases produce completely new materials.

Vegetation plays a significant role in physical and chemical weathering of soil. Soil animals and microorganisms also play a role in developing soil in an environment. Without the influence of this soil-forming factor, the development of soil formation would be greatly slowed.

Relief (topography), or the "lay of the land," influences the formation of soil through aspect, slope, and curvature, which collectively influence the degree of heating a soil may receive, the amount of water it can hold, and how likely erosion is to occur. Steep slopes lead to greater erosion, which is why the shoulders of slopes, which have some of the steepest slopes, also have some of the shallowest soils.

Beneath all soil is parent material, which is composed of igneous, sedimentary, or metamorphic rock. Sedimentary rock is the most significant soil-forming parent material and generally consists of sandstone, limestone, or shale. Few soils are residual soils, soils that formed directly in place from their parent material. Most soils are eolian, colluvial, or alluvial, meaning they have been transported from one location to another by wind, gravity, or water, respectively.

Soil formation is a long process. Without enough time very little soil formation can occur. Decades are required to form an appreciable amount of a distinctive soil, and this can be rapidly lost if erosion occurs. In many cases human activity has restarted the clock for soil formation because of alterations in either the type of parent material or removal of the soil itself.

The interaction of all the soil-forming factors results in pedogenesis, or soil formation, which can be described by processes such as translocation, transformation, addition, and loss of materials in soil. These processes therefore result in distinct soil horizons depending on the environment of a soil, which are made obvious when the horizons are exposed in soil profiles.

In the next chapter (Chapter 3) you will examine how some sense and order is given to the variability of soil through the application of soil classification.

END OF CHAPTER QUESTIONS

1. How can weathering be both *constructive* and *destructive?*
2. Give an example of each of the following:

   Hydrolysis

   Hydration

   Dissolution

   Complexation
3. What are the five soil-forming factors?
4. If you compared a desert in Arizona with a prairie in Kansas, how would each of the soil-forming factors differ?
5. What distinguishes the way in which soil parent materials are moved by colluvial, alluvial, and eolian transport?
6. What is loess and where would you be likely to find it?
7. Draw a sketch of the soil profile of an alluvial plain just after decades of annual flooding events and draw another profile to illustrate the appearance of the soil 1000 years after flooding has stopped.
8. What are two examples of physical weathering?
9. Sandstone, limestone, and shale are transformed into what kinds of rock during metamorphosis?
10. Where are you most likely to find glacial soils in the United States?
11. Does a BC horizon most nearly resemble a B horizon or a C horizon?
12. Where are you most likely to find residual soils?
13. What is one of the most likely materials remaining in a highly weathered soil?

14. What factors might explain greater soil depth in an alluvial soil but greater soil development in a residual soil?

15. All things being equal, where in the United States are you most likely to find soils with the highest organic matter content, and why?

16. Why is the B horizon not always the subsoil?

17. What is the difference between eluviation and illuviation?

18. Where are you most likely to find talus?

19. What are the four processes involved in pedogenesis? Give an example of each.

20. Is soil development constant with time? Why or why not?

21. What are three examples of anthropogenic influences on soil formation?

22. If you discovered an A_b horizon, what would that mean and how might it have gotten there?

23. What is the significance of Hans Jenny to the development of soil science?

24. What characteristics of alluvial soils make them ideal for agriculture?

25. How do wind and water contribute to soil formation?

26. If southern exposures tend to receive more sunlight than northern exposures, what would you expect to be the relationship between soil organic matter and water content in soils developing with either a southern or northern exposure?

27. Where would you be most likely to find a chronosequence, and if you did, how would the various soils differ in terms of weathering?

28. What are some of the ways in which relief affects soil development?

29. What typically happens to soil nitrogen and carbon contents when a soil is brought into cultivation?

30. Would you expect to find an O horizon in a plowed soil? Why or why not?

FURTHER READING

It would hardly do to reflect on the importance of Hans Jenny's role in synthesizing ideas about soil formation without providing a source for the books in which he presented it.

Jenny, Hans. 1941. *Factors of soil formation: A system of quantitative pedology.* New York: Dover Press (originally McGraw Hill).

Jenny, Hans. 1980. *The soil resource—origins and behavior. Ecological studies*, vol. 37. New York: Springer-Verlag.

REFERENCES

Brady, N. C., and R. R. Weil. 2002. *The nature and property of soils.* Upper Saddle River, NJ: Prentice-Hall, Inc.

Leet, L. D. and S. Judson. 1954. *Physical geology.* New York: Prentice-Hall, Inc.

Plaster, E. J. 1997. *Soil science & management.* Clifton Park, NY: Thomson Delmar Learning.

Shantz, H. L. 1923. The natural vegetation of the Great Plains region. *Annals of the Association of American Geography* 13: 81–107.

Troeh, F. R., and L. M. Thompson. 1993. *Soils and soil fertility.* New York: Oxford University Press.

Chapter 3

SOIL TAXONOMY AND CLASSIFICATION

"A rainbow of soil is under our feet; red as a barn and black as a peat. It's yellow as lemon and white as the snow; bluish gray. So many colors below. Hidden in darkness as thick as the night; The only rainbow that can form without light. Dig you a pit, or bore you a hole, you'll find enough colors to well rest your soil."

F. D. Hole, *A Rainbow of Soil Words,* 1985

OVERVIEW

In Chapter 2 you examined how the soil-forming factors interact to create soil, and how the manifestation of that activity in pedogenesis is apparent in the soil profile. Soil profiles develop distinct horizons depending on the processes of addition, loss, translocation, and transformation.

In this chapter you are going to examine the design and operation of a system to take the information evident in a soil profile, as well as other physical and chemical measurements, and use that information to classify soils into identifiable groups. Much of this chapter deals with terminology, and it may sound like a foreign language (in many cases it is). This is unfortunate, but the reward is that once you understand the terminology, and the basis behind why the terminology is used, you will have a much better grasp of the differences between soils, why they developed in different ways, and how that knowledge can be used for management.

OBJECTIVES

After reading this chapter, you should be able to:

✔ Understand the basic terminology used in soil classification and taxonomy.

✔ Outline how soil classification schemes have developed in the United States.

✔ Interpret the approximate meaning of most soil taxonomic names.

✔ Identify the twelve soil orders of *Soil Taxonomy* and provide basic characteristics of each.

    ✔ Identify the basic properties of soils to the suborder level, and identify where they are likely to be found.

    ✔ Use a Munsell color chart to determine the diagnostic color of a soil.

KEY TERMS

Alfisol	Histosol	soil series
Andisol	horizon	*Soil Taxonomy*
Aridisol	hue	Spodosol
chroma	Inceptisol	subgroup
Entisol	Mollisol	suborder
epipedon	Oxisol	Ultisol
family	pedon	value
Gelisol	polypedon	Vertisol
great group	soil order	

WHAT IS CLASSIFICATION?

Classification organizes knowledge in meaningful ways.	Classification organizes knowledge in a meaningful way. On one hand it helps to characterize what's present, and on the other hand it helps to predict what to expect from similar examples. So, classifying soils is a way of cataloging the available soil resources, and at the same time helping land use managers (especially farmers) predict how a soil will look and behave in a given field when they only have specific knowledge about part of that field.
Classification systems are artificial; they are logical for a certain place and time.	Classification systems are completely artificial. There is no "natural" system of soil classification. Rather, there are classification systems that have a certain logic for their place and time. No classification system is better than any other; those that are valuable prove most adequate at meeting the purpose for their development. If there are multiple purposes, it makes sense to have multiple types of classification. The most useful classification system permits the largest number of the most important statements about a given class of objects—in this case different soils.

INDIGENOUS SOIL CLASSIFICATION SYSTEMS—LARI AND COLCA VALLEYS, SOUTHERN PERU

Soil classification systems are not unique to industrial agricultural societies. Many if not all indigenous agricultural cultures have some sort of system that allows their members to assess either the quality or value of land.

The residents of Lari and Colca Valleys in southern Peru have inherited an agricultural tradition at least 1500 years old. The farmers in these communities recognize different types of soil and different types of soil materials, including those that are beneficial after consumption (some of the minerals in these soils may adsorb glycoalkaloid phytotoxins in the potatoes that are an important food in the region; Callahan, 2003), those that are poorly drained, those that are excessively drained,

those that are infertile, those that have different horizons, and those that have impeding layers.

Classification by these farmers reflects their agricultural needs. So in this case, the most important characteristics of soil are texture, drainage, workability, fertility, and location. Soil color plays an important role in distinguishing differences in soil. An example of one farmer's classification system is given in Figure 3-1. In this example, the farmer recognizes four categorical levels of soil, including several types and colors of clay. At the second categorical level the five classes of agricultural soil are arranged from left to right in terms of decreasing productivity. At the far right are soils that are nonagricultural or serve some other purpose besides agriculture.

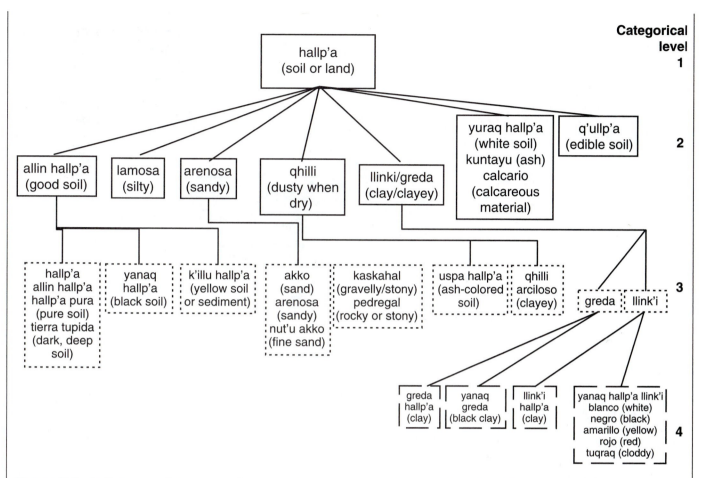

FIGURE 3-1 Indigenous soil taxonomy of farmers of Lari and Colca Valley, Southern Peru. Four levels of classification occur, which includes soils used for agricultural and nonagricultural purposes. (*Adapted from Sandor and Furbee, 1996; redrawn by M.S. Coyne*)

BASIS OF CLASSIFICATION

The basis of soil classification is the *natural body*, which owes its properties to the five soil-forming factors of climate, organisms (vegetation), relief, parent material, and time. Thus, soil classification begins at the landscape, where all five of those features are manifested. Within each landscape are **polypedons,** collections of soil units that share the same properties. The smallest classifiable unit in each area is called the **pedon** and reflects a surface area of approximately one square meter (Figure 3-2).

Using the pedon as the smallest classifiable natural body of soil, the U.S. system of soil classification outlined in the United States Department of Agriculture (USDA) publication **Soil Taxonomy** uses a hierarchical approach to classify the soil. There are six categories of classification in *Soil Taxonomy* ranging from very broad and encompassing to the field scale: (1) **order,** (2) **suborder,** (3) **great group,** (4) **subgroup,** (5) **family,** (6) **series.** A seventh layer of classification is often considered to be "type." There are 12 orders, 63 suborders, 319 great groups, 2484 subgroups, approximately 8000 families, and approximately 19,000 soil series at present. The relationship between the classification levels in *Soil Taxonomy* and another familiar classification scheme is given in Figure 3-3.

The U.S. system of soil classification is outlined in the USDA publication Soil Taxonomy.

FIGURE 3-2 Pedons are the smallest classifiable unit in soil taxonomy. Polypedons are groups of contiguous pedons that share the same characteristics. The landscape will have multiple polypedons depending on how varied the topography and vegetation are.

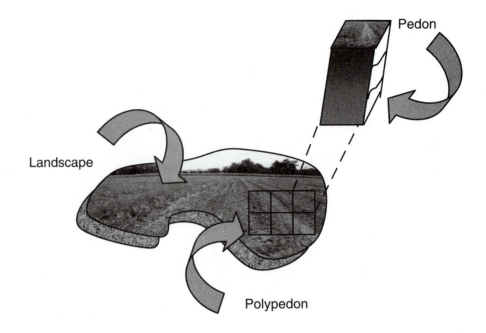

FIGURE 3-3 Comparison of classification in *Soil Taxonomy* to the Linnaean classification system for plants and animals.

Kingdom ⟹ Phylum ⟹ Class ⟹ Order ⟹ Family ⟹ Genus ⟹ Species

Order ⟹ Suborder ⟹ Great group ⟹ Subgroup ⟹ Family ⟹ Series ⟹ Type ⟹ Phase

One of the most important features in *Soil Taxonomy*, developed by C. E. Kellogg and Guy Smith, is that it attempts to classify the soils based on observable and quantifiable soil properties that can be viewed and sampled in an exposed pedon. For example, are certain **horizons** present or not? At what depth do they occur? How thick are they? What are properties of temperature and climate? Thus, the first step in classification in this system relies on interpreting some basic information from diagnostic surface and subsurface horizons.

Classifying soils has been likened to a visit to the doctor's office. The doctor can make some characterizations based on obvious appearance (fat/thin, tall/short, pale/tan, black/white), symptoms (coughing, sneezing, bleeding, rash), and measurements (temperature, heartbeat, raspy lungs), but for a real diagnosis some additional clinical tests have to be made. This is analogous to the surface and subsurface diagnoses that occur in soil taxonomy, which are always followed up by more thorough laboratory analysis before a classification is made.

> Classification is like a visit to a doctor's office—diagnosis is based on observations and tests.

Diagnostic Surface Horizons—The Epipedon

> The surface horizon of the soil is called the epipedon.

In *Soil Taxonomy* the entrance to classification is made by diagnosing the presence and depth of the surface horizon of soil or **epipedon**. There are

GUY SMITH AND THE DEVELOPMENT OF *SOIL TAXONOMY*

The first scientific system of soil classification is credited to the Russian V. V. Dokuchaiev. His work was published in 1879 and introduced to the United States via translation in 1927 by C. F. Marbut, who partially adapted Dokuchaiev's "zonal" concept of soil development and created several "great groups" reflecting soil-forming conditions in the United States. Marbut's classification system was modified by C. E. Kellogg and others for the 1938 *Yearbook of Agriculture* at the bequest of Secretary of Agriculture Henry Wallace. Because less than a year was given to develop a classification system, they borrowed heavily from the original Russian system. The 1938 soil classification system lacked real definitions of any of the great groups and contained mostly broad definitions. In addition, the scientists involved realized that the system was flawed because they were unable to find any single soil property that included all the soils collected into the largest groups.

In response to these limitations, Guy Smith (1902–1981) of the USDA spent his career trying to devise a manageable soil classification system that could be based on specific quantifiable soil properties rather than observational notes about soil color or potential genesis. The first *Soil Taxonomy* was published in 1960 by the Soil Survey Staff and officially adopted by the Soil Conservation Service in 1965. The seventh approximation (and the one officially published as *Soil Taxonomy*) appeared in 1975. The size of *Soil Taxonomy* has grown tremendously since its inception. There are now approximately 19,000 different soil series classified by *Soil Taxonomy*. The most recent soil order added is the Gelisol.

TABLE 3-1 Diagnostic surface horizons (epipedons).

Diagnostic Horizon	Major Features
Anthropic	Human-modified, mollic-like horizon
Histic	Very high in organic matter, periodically wet
Melanic	Thick, black, >6% organic C, common in volcanic soils
Mollic	Thick, dark-colored, well structured, high base saturation
Ochric	Too light-colored, too little organic matter, or too thin to be mollic
Plaggen	Human-made sodlike horizon created by years of manuring
Umbric	Same as mollic except for low base saturation

seven diagnostic epipedons, which are outlined in Table 3-1. Each diagnostic horizon has a rigid definition to which it must be held, but as a general rule, most comparisons are made on the basis of its similarity or dissimilarity to the mollic epipedon.

Mollic Epipedon

The formal definition of a mollic epipedon is a thick, dark-colored upper horizon usually associated with grassland soils. It must contain at least 0.6 percent organic carbon and have at least 50 percent base saturation (at least 50 percent of the cation exchange capacity must be saturated by cations such as Ca^{2+}, Mg^{2+}, and K^+). The minimum thickness ranges from 10 cm over bedrock through one-third of the solum to 25 cm in a deep soil. Mollic epipedons typically have granular structure and easy workability. Hence, the name derives from the Latin word *mollus*, meaning "soft."

Umbric Epipedon

The formal definition of an umbric epipedon is one that in most respects mirrors that of the mollic epipedon except that the base saturation is < 50 percent. The name comes from the Latin word *umbra*, meaning "shade," which refers to the dark color of the horizon. Umbric epipedons are found in environments where more weathering has occurred.

Ochric Epipedon

Ochric epipedons show more weathering than mollic epipedons.

The formal definition of an ochric epipedon describes a horizon that is too thin, too light in color, or has too low a carbon content to be a mollic epipedon. It reflects more soil weathering. The ochric epipedon is the most common type described. The name comes from the Greek word *ochros*, meaning "pale." Typical plowed fields with an ochric epipedon have a grayish or yellowish-brown appearance.

Histic Epipedon

Histic epipedons show the accumulation of organic matter.

The histic epipedon reflects the accumulation of organic matter on the soil surface. When this develops into a layer 20 to 40 cm deep (8 to 16 inches) it is called a histic epipedon. *Histic* comes from the Greek word *histose*, meaning "tissue."

Melanic, Anthropic, and Plaggen Epipedons

Three other less common epipedons are the melanic, anthropic, and plaggen epipedons. The melanic epipedon is a dark upper horizon dominated by volcanic ash and allophane, which has high anion exchange capacity. Anthropic epipedons show the influence of human activities such as fertilization or irrigation. Plaggen epipedons are also artificial epipedons formed through long-term plowing and heavy manure application.

Diagnostic Subsurface Horizons

Diagnostic subsurface horizons focus on distinctive soil properties, some of which have to be measured.

There are eighteen common subsurface diagnostic horizons. Some of these are briefly described in Table 3-2. As before, more formal definitions for each exist in *Soil Taxonomy*. While the epipedons focus on such observable features as soil color, appearance, and fertility, the diagnostic subsurface horizons focus more on properties associated with translocation and transformation: weathering, accumulation of clays, presence of iron and aluminum, bleaching, salt accumulation, and so on. Other features that may be examined are the presence of impermeable layers of various types.

TABLE 3-2 Diagnostic subsurface horizons and features.

Diagnostic Horizon (typical location and designation)	Major Features
Albic (E)	Light-colored, clay and Fe and Al oxides mostly removed
Agric (A or B)	Organic matter and clay accumulation just below the plow layer
Argillic (Bt)	Zone of silicate clay accumulation
Calcic (Bk)	Accumulation of $CaCO_3$ and $MgCO_3$
Cambric (Bw, Bg)	Nonilluvial physical or chemical change
Duripan (Bqm)	Hardpan cemented by silica
Fragipan (Bx)	Dense, brittle, loamy-textured pan
Gypsic (By)	Accumulation of gypsum
Natric (Btn)	Argillic horizon high in sodium
Oxic (Bo)	Highly weathered horizon with a mixture of Fe and Al oxides and silicate clays
Petrocalcic (Ckm)	Cemented calcic horizon
Petrogypsic (Cym) Salic (Bz)	Cemented gypsic horizon
Spodic (Bh, Bs)	Organic matter and Al and Fe oxide accumulation
Sulfuric (Cj)	Highly acidic with Jarosite mottles

Among the most important subsurface diagnostic horizons to remember are the albic, cambic, argillic, spodic, and oxic horizons.

Albic Diagnostic Horizons

Albic horizons are eluvial horizons that form below the A horizon. They are at least 1 cm thick and at least 85 percent of this volume is filled by bleached materials. The bleaching is due to intense leaching, often because of water saturation. Albic comes from the Latin word *albus,* meaning "white."

Cambic Diagnostic Horizons

A cambic horizon formed by weathering of materials within the horizon rather than by gaining materials through illuviation. Cambic horizons can have bright colors, but are too weakly developed to be considered either argillic or spodic horizons. Cambic is from the Latin word *cambriare,* meaning "to change."

> Cambic horizons indicate change within the horizon.

Argillic Diagnostic Horizons

An argillic horizon is an illuvial horizon that is at least 10 percent as thick as the overlying A horizon and contains 3 to 8 percent more clay. There should also be the accumulation of clay films on soil peds, pores, and sand grains. Argillic comes from the Latin word *argilla,* meaning "white clay." One type of argillic horizon is a natric horizon, which contains sodium and causes soil sealing. Another type of argillic horizon is the kandic horizon, which is rich in kaolinite clays that do not retain nutrients well.

> Argillic horizons typically reflect the accumulation of translocated clay.

Spodic Horizon

The spodic horizon represents an accumulation of illuviated organic matter and iron and aluminum oxides lying underneath a bleached horizon. The colors of the horizon can be very bright. The name comes from the Greek word *spodos,* meaning "wood ash."

Oxic Horizons

The oxic diagnostic horizon is an impoverished subsoil layer so highly weathered that almost no minerals other than quartz, kaolinite, and metal oxides persist. Oxic horizons do not appear distinct from the other subsurface horizons in these soils. Most oxic horizons have very low cation exchange capacity.

> Oxic horizons reflect extreme weathering.

FOCUS ON . . . CLASSIFICATION AND DIAGNOSTIC HORIZONS

1. Are soil classification systems unique to modern agricultural societies?

2. Where was the first scientific soil classification system developed?

3. How did *Soil Taxonomy* become the basis of soil classification in the United States?

4. What is a diagnostic horizon?

5. How does a mollic epipedon differ from an umbric epipedon?

6. What does an argillic subsurface horizon refer to?

7. Which two surface epipedons reflect human influence?

SOIL ORDERS

The U.S. soil classification system has twelve soil orders.

Based on information collected from surface and subsurface diagnostic horizons, and some additional information on organic matter content, moisture content, clay type, and parent material identified in Figure 3-4, soils can be placed in one of twelve soil orders, which are the broadest classification in *Soil Taxonomy* (Table 3-3).

Details about the soil orders will be given later in the chapter. From the perspective of weathering and soil development, the soil orders can be arranged as in Table 3-4, which also gives their relative distribution.

Classification into Suborders

Suborders provide information about climatic regime.

To classify soils to the suborder level requires some additional information, mostly about climate. Other descriptive names associated with the nomenclature of suborders are given in Table 3-5.

Soil Moisture Regimes

Soil moisture regimes reflect water availability in the soil and, indirectly, how much water is likely to affect leaching.

Aquic soils are saturated with water for at least part of the year.

Udic soils are in humid climates that provide enough water to meet most plant growth needs. Extremely wet regions are perudic.

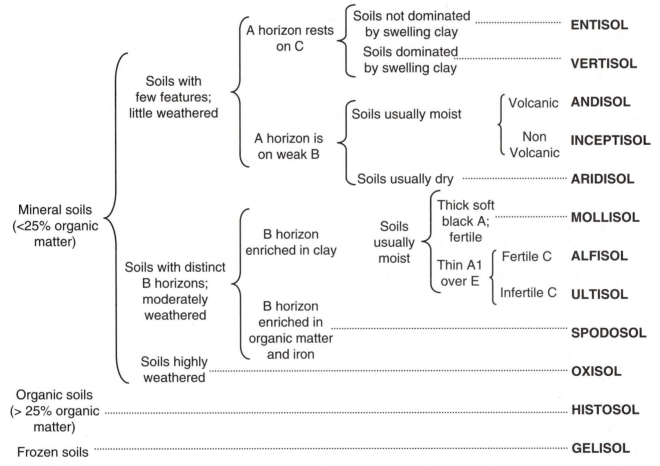

FIGURE 3-4   Key to the classification of soils to the order level based on *Soil Taxonomy.* (Adapted from Harpstead et al., 1997; redrawn by M. S. Coyne)

TABLE 3-3 Names of the soil orders in *Soil Taxonomy* and their major characteristics.

Order	Major Characteristics
Inceptisol	Immature soils with some development Few diagnostic features, ochric or umbic epipedon, cambic horizon
Alfisol	Fertile soils of hardwood forests Medium to high base saturation, some argillic or natric horizons
Mollisol	Fertile grassland soils Dark soils with mollic epipedons and high base saturation, some argillic or natric horizons
Andisol	Volcanic soils Formed from volcanic ejecta, high in allophane or Al-humic materials
Spodosol	Acid soils of coniferous pine forests Spodic horizons with iron and aluminum oxides and humus accumulation
Ultisol	Low fertility forest soils of warm humid regions Argillic horizons present, low base saturation
Aridisol	Desert soils Ochric epipedons, sometimes argillic or natric horizons
Vertisol	Cracking, dark clay soils High in swelling clays
Entisol	Primitive (young) developing soils Little profile development, ochric epipedon common
Histosol	Organic soils Peat or bog soils, > 20% organic matter
Oxisol	Highly weathered tropical soils Oxic epipedons, no argillic horizons
Gelisol	Frozen soils Permafrost and frost churning visible

A useful mnemonic device to remember the names of the 12 soil orders is to take the first letter of each order, as represented here, I AM A SUAVE HOG.

Ustic climates are intermediate between *udic* and *aridic* environments; some periods of drought occur.

Aridic environments are arid and moist for less than ninety consecutive days. *Torric* soils are both hot and dry.

Xeric soils are found in Mediterranean-type climates that have hot dry summers and cool wet winters. Long periods of drought may occur in summer.

Soil Temperature Regimes

Torric soils are both hot and dry. *Cryic* soils, in contrast, are extremely cold. These are the two most common temperature terms used to categorize suborders.

> Torric and cryic refer to hot and cold temperature regimes, respectively.

FOCUS ON . . . SOIL ORDERS AND SUBORDERS

1. What is a simple device for remembering the soil orders?

2. Which soil order is most likely found only in the far north or at high elevation?

3. Which soil order is most prevalent in the United States?

4. Which soil order is least prevalent in the United States?

5. What are the most important factors used to designate the dominant suborders for each soil order?

TABLE 3-4 Global and U.S. distribution of major soil types classified by *Soil Taxonomy*.
(Adapted from Brady and Weil, 2002)

	Order	Environment	Global[a]	United States[b]
			% of Ice-Free Land	
Least Weathered	Entisols	Recently deposited	16.3	12.2
	Inceptisols	Various conditions	9.9	9.1
	Andisols	Mildly weathered ejecta	0.7	1.7
	Histosols	Wet, organic	1.2	1.3
	Gelisols	Very cold	8.6	7.5
	Aridisols	Dry, desert shrubs, grass, alkaline	12.1	8.8
	Vertisols	Wet/dry seasons	2.4	1.7
	Alfisols	Moist, mildly acid	9.6	13.9
	Mollisols	Semiarid to moist	6.9	22.4
	Ultisols	Wet tropical and subtropical	8.5	9.6
	Spodosols	Cool, acid coniferous	2.6	3.3
Most Weathered	Oxisols	Hot, wet tropical	7.6	<0.1
	Regolith/Sand		14.1	7.8

[a]Data from FAO world database.
[b]Data from USDA/NRCS Soil Survey Division.

The advantages of the United States in terms of its soil resources are abundantly clear in this table; almost 40 percent of the ice-free land is classified in soil types that have inherently moderate to high fertility.

TABLE 3-5 Order and suborder names in *Soil Taxonomy*.

Order	Suborder	Characteristic Environment, Climate, or Feature
Alfisols	Aqualfs	Wet
	Cryalfs	Cold climates
	Udalfs	Humid climates
	Ustalfs	Semiarid climates
	Xeralfs	Mediterranean climates
Andisols	Aquands	Wet
	Cryands	Cold climates
	Torrands	Hot and dry climates
	Udands	Humid climates
	Ustands	Semiarid climates
	Vitrands	Volcanic glass
	Xerands	Mediterranean climates
Aridisols	Argids	Clay accumulation
	Calcids	Carbonate accumulation
	Cambids	Typical
	Cryids	Cold climates
	Durids	Duripans present
	Gypsids	Gypsum accumulation
	Salids	Salty

(continued)

TABLE 3-5 Order and suborder names in *Soil Taxonomy (continued)*.

Order	Suborder	Characteristic Environment, Climate, or Feature
Entisols	Aquents	Wet
	Arents	Mixed horizons
	Fluvents	Alluvial deposits
	Orthents	Typical
	Psamments	Sandy
Gelisols	Histels	High organic content
	Orthels	Typical
	Turbels	Cryoturbation
Histosols	Fibrists	Partially decomposed (peats)
	Folists	Undecomposed leaf and twig litter mats
	Hemists	Moderately decomposed
	Saprists	Highly decomposed (mucks)
Inceptisols	Anthrepts	Human-influenced
	Aquepts	Wet
	Cryepts	Cold climates
	Udepts	Humid climates
	Ustepts	Semiarid climates
	Xerepts	Mediterranean climates
Mollisols	Albolls	Albic horizons
	Aquolls	Wet
	Cryolls	Cold climates
	Rendolls	Calcareous
	Udolls	Humid climates
	Ustolls	Semiarid climates
	Xerolls	Mediterranean climates
Oxisols	Aquox	Wet
	Perox	Very humid climates
	Torrox	Very hot and dry climates
	Udox	Humid climates
	Ustox	Semiarid climates
Spodosols	Aquods	Wet
	Cryods	Cold climates
	Humods	Humus accumulation
	Orthods	Typical
Ultisols	Aquults	Wet
	Humults	Humus accumulation
	Udults	Humid climates
	Ustults	Semiarid climates
	Xerults	Mediterranean climates
Vertisols	Aquerts	Wet
	Cryerts	Cold climates
	Uderts	Humid climates
	Usterts	Semiarid climates
	Xererts	Mediterranean climates

NOMENCLATURE AND ORDERING IN *SOIL TAXONOMY*

Constructing soil taxonomic names is systematic.

The sequence of constructing taxonomic names in *Soil Taxonomy* makes a considerable amount of sense once you work with it a little. The naming of the four highest categories (order, suborder, great group, subgroup) is systematic. Each addition to the name adds a little more information about the soil properties of the soil involved. For example, take Woolper silty clay loam, taxonomically described as a Typic Argiudoll. What does that mean?

Order—Mollisol (This is a Mollisol, it has a mollic epipedon and probably formed in grassland.)

Suborder—Udoll (This Mollisol formed in a udic or humid climate regime.)

Great Group—Argiudoll (This Mollisol also has an argillic horizon in the subsurface in addition to the mollic epipedon. Some weathering and translocation of clay particles has occurred.)

Subgroup—Typic Argiudoll (This is a typical soil for this great group.)

So, each new name in the taxonomy, or the formative syllable, tells you something more about the soil when you add it. The hardest thing is learning what all the formative names refer to. Examples of formative names to give to subgroups are given in Table 3-6. Examples of formative names to add for great groups are given in Table 3-7.

Soil names at the family level have adjectives describing important properties.

Even more information can be given at the family level, although the emphasis here, rather than on soil-forming factors, is on descriptive adjectives indicating properties of texture, mineralogy, and fertility, to name a few. A list of some of the common adjectives used is given in Table 3-8.

Ultimately a series name appears. There are approximately 19,000 soil series names. These series names are given to soils with very similar properties. Note that it is *very similar* properties that determine classification, not *identical* properties. Soil series names are derived from a town or community near where the soil was first officially described. Obviously, a similarly named series in an adjacent state will not be precisely alike, but it will be sufficiently close for management purposes. The Woolper series described earlier, for example, was established in Bath Co., Kentucky in 1960 and is distributed in Kentucky, Ohio, and Tennessee. The complete description of the Woolper soil is given in Figure 3-5, which gives you another look at how taxonomic names are constructed.

Soil series names come from a town or community close to where a soil was first described.

FOCUS ON . . . NOMENCLATURE IN *SOIL TAXONOMY*

1. What is the order of classification in *Soil Taxonomy?*

2. What is the smallest level of classification in *Soil Taxonomy?*

3. Does "Cryoll" refer to a great group or suborder?

4. In what kind of environment did an aquent develop?

5. Soils in xeric climate regimes have what type of winters?

6. In the name "typic Hapludoll," which term refers to the great group?

TABLE 3-6 Formative element names in the suborders of *Soil Taxonomy*.

Formative Element	Connotes	Denotes
alb	White	Albic (bleached eluvial) horizon
anthr	Human	Anthropic or plaggen (plowed or tilled) horizon
aqu	Water	Characteristics of wetness
ar	Plowed	Mixed horizons
arg	White clay	Argillic horizon (horizon with illuvial clay)
calc	Lime	Presence of calcic horizon
camb	Change	Presence of cambric horizon
cry	Cold	Characteristics of cold
dur	Hard	Presence of a duripan
fibr	Fibrous	Least decomposed stage of organic materials
fluv	River	Floodplains
fol	Leaf	Mass of leaves
gyps	Gypsum	Presence of gypsic horizon
hem	Half	Intermediate stage of decomposition
hist	Tissue	Presence of histic epipedon
hum	Earth/Humus	Presence of organic matter
orth	True	Common
per	Throughout time	Year-round climates
psamm	Sand	Sand textures
rend	Rendzina	High in carbonate
sal	Salty	Salic (saline) horizon
sapr	Rotton	Most decomposed
torr	Torrid	Hot and dry
turb	Turbulent	Perturbed (such as by frost heaving)
ud	Humid	Of humid climates
ust	Burnt	Of dry climates
vitr	Glass	Resembling glass
xer	Dry	Dry summers, moist winters

THE SOIL ORDERS IN *SOIL TAXONOMY*

What are the characteristics and distribution of the soil orders classified in *Soil Taxonomy?* Each soil order has a specific suite of diagnostic surface and subsurface features that allow you to distinguish one from the other. However, our survey will just take an overview of the distinguishing characteristics of each soil type.

Alfisols

Alfisols are soils that typically occur under deciduous forests. Although they usually have an ochric epipedon because the humus returned to soil by leaf litter is not thick, Alfisols can sometimes have an umbric epipedon. Alfisols are generally fertile, with base saturation greater than 35 percent (Figure 3-6). The subsurface is characterized by an argillic or natric horizon. The mean soil temperature in which Alfisols form is usually > 8°C (47°F). About 13.9 percent of the surface area in the United States consists of Alfisols (Table 3-4). Once cleared of forest they are usually considered to be fertile cropland.

> Alfisols are soils of deciduous forests.

TABLE 3-7 Formative elements for names of great groups in *Soil Taxonomy*.

Formative Element	Connotation	Formative Element	Connotation
acr	Extreme weathering	quartz	High quartz
pale	Old development	fibr	Least decomposed
agr	Agric horizon	fol	Mass of leaves
alb	Albic horizon	hem	Intermediate decomposition
argi	Argillic horizon	hist	Presence of organic material
calc, calci	Calcic horizon	hum	Humus
camb	Cambic horizon	sapr	Most decomposed
fulv	Light melanic horizon	sphagn	Sphagnum moss
gyps	Gypsic horizon	verm	Wormy or mixed by animals
hapl	Minimum horizon	and	Ando-like
natr	Natric horizon	chrom	High chroma
petr	Cemented horizon	lithic	Near stone
plagg	Plaggen horizon	luv, lu	Illuvial
sal	Salic horizon	melan	Melanic epipedon
somb	Dark horizon	molli	Mollic epipedon
al	High Al, low Fe	rhod	Dark red colors
dystr, dys	Low base saturation	umbr	Umbric epipedon
eutr	High base saturation	cry	Cold
ferr	Iron	torr	Usually dry and hot
hal	Salty	ud	Humid climates
kand	Low activity 1:1 silicate clay	ust	Dry climate
plinth	Plinthite	xer	Dry summers, moist winters
sulf	Sulfur	anhy	Anhydrous
vitr	Glass	aqu	Water saturated
dur	Duripan	endo	Fully water saturated
fragi	Fragipan	epi	Perched water table
plac	Thin pan	fluv	Floodplain
psamm	Sand texture	hydr	Water

TABLE 3-8 Characteristics used to distinguish families.

Particle Size Class	Mineralogy Class[a]	CEC Class[b]	Temperature Regime Class (Mean annual temperature, °C)
Ashy	Mixed	Superactive	Hypergelic (<-10)
Fragmental	Micaceous	Active	Pergelic (-10 to -4)
Sandy-skeletal	Silicaceous	Semiactive	Subgelic (-4 to $+1$)
Sandy	Kaolinitic	Subactive	Cryic ($< +8$)
Loamy	Smectitic		Frigid ($< +8$)
Clayey	Gibbsitic		Mesic ($+$ to $+15$)
Fine-silty	Carbonic		Thermic ($+15$ to $+22$)
Fine-loamy	Gypsic		Hyperthermic ($> +22$)

[a]Mixed = mixed mineralogy, Micaceous = dominated by mica; Silicaceous = dominated by silica/quartz; Kaolinitic/Smectitic = dominated by 1:1 or 2:1 silicate clays, respectively; Gibbsitic = dominated by aluminum oxides and hydroxides; Carbonic = carbonate deposits; Gypsic = gypsum deposits.

[b]Reflects the amount of CEC relative to % clay content. Superactive is highest while subactive is lowest.

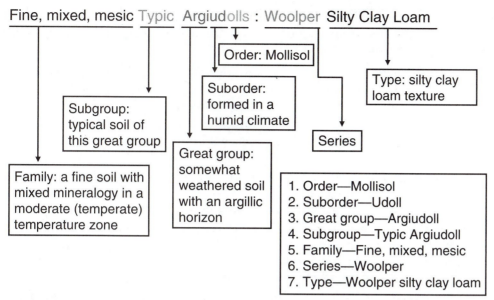

Fine, mixed, mesic Typic Argiudolls : Woolper Silty Clay Loam

Order: Mollisol

Suborder: formed in a humid climate

Type: silty clay loam texture

Subgroup: typical soil of this great group

Great group: somewhat weathered soil with an argillic horizon

Series

Family: a fine soil with mixed mineralogy in a moderate (temperate) temperature zone

1. Order—Mollisol
2. Suborder—Udoll
3. Great group—Argiudoll
4. Subgroup—Typic Argiudoll
5. Family—Fine, mixed, mesic
6. Series—Woolper
7. Type—Woolper silty clay loam

FIGURE 3-5 Key to the classification of soils to the Order level based on *Soil Taxonomy.* *(Adapted from Harpstead et al., 1997)*

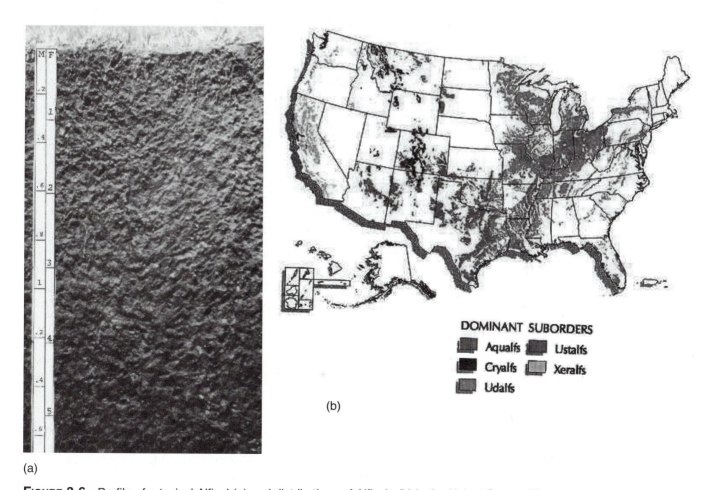

(a)

(b)

DOMINANT SUBORDERS
Aqualfs   Ustalfs
Cryalfs   Xeralfs
Udalfs

FIGURE 3-6 Profile of a typical Alfisol (a) and distributions of Alfisols (b) in the United States. *(Photograph and map courtesy of USDA-NRCS)*

Dominant Suborders of Alfisols

There are five dominant suborders of Alfisols: Aqualfs, Cryalfs, Udalfs, Ustalfs, and Xeralfs (Table 3-5).

❖ Aqualfs form in warm, wet conditions (Figure 3-6). Most Aqualfs are believed to have originally been in forest, and drainage or water control are usually necessary to bring them into cultivation.

❖ Cryalfs are cold Alfisols that occur at high elevations (Figure 3-6). Because these regions have short growing seasons they usually remain in forest.

❖ Udalfs are more extensive than the other suborders of Alfisols (Figure 3-6). They have udic moisture regimes. All are believed to have supported forest growth prior to any cultivation.

❖ Ustalfs, as the name implies, have ustic moisture regimes and form in drier environments than Udalfs (Figure 3-6). They can have pronounced dry seasons. Ustalfs support savanna-type vegetation and grassland and most are used for either grazing or cropping.

❖ Xeralfs are the driest of the Alfisols. Most Xeralfs occur in California (Figure 3-6). The most common cultivated crops are small grains. The original vegetation of Xeralfs was a mixture of grasses, forbs, and woody shrubs, although coniferous forest grew on the cooler and moister Xeralfs.

Andisols

> Andisols are soils of volcanic regions.

Andisols are volcanic soils. They developed in volcanic ejecta of volcanoclastic materials. Andisols represent less than 2 percent of the surface land in the United States (Table 3-4). Andisols are dominated by rapidly weathering materials and volcanic glass, the proportion of which is a measure of how long weathering has proceeded. The layers of volcanic material can be thick. Andisols typically have a low bulk density, high carbon content, and a unique capacity to immobilize phosphorus. They are often very dark and can be extremely fertile (Figure 3-7).

Dominant Suborders of Andisols

The dominant suborders of Andisols are Aquands, Cryands, Torrands, Udands, Ustands, Vitrands, and Xerands (Table 3-5).

❖ Aquands occur in aquic conditions in lower elevations with forest or grass vegetation (Figure 3-7). Some Aquands are drained for agriculture.

❖ Cryands are found in cold climates in the mountains of the Pacific Northwest and in Alaska (Figure 3-7). Most formed beneath coniferous forest and are used as forest.

❖ Torrands are warmer and drier Andisols that formed under grassy or shrub vegetation (Figure 3-7).

❖ Udands, although they can have a high water content, hold it too tightly for plants to use, and so most of these soils remain in the original forest (Figure 3-7).

❖ Ustands are very similar to Udands in terms of water availability, although of much less extent. They formed under savanna or forest-type vegetation and are mostly found in Hawaii (Figure 3-7).

❖ Vitrands can have udic or ustic moisture regimes and are coarse-textured Andisols. They typically formed under coniferous forest in Oregon and Washington (Figure 3-7). Most Vitrands are still used as forest land but they can be cropped.

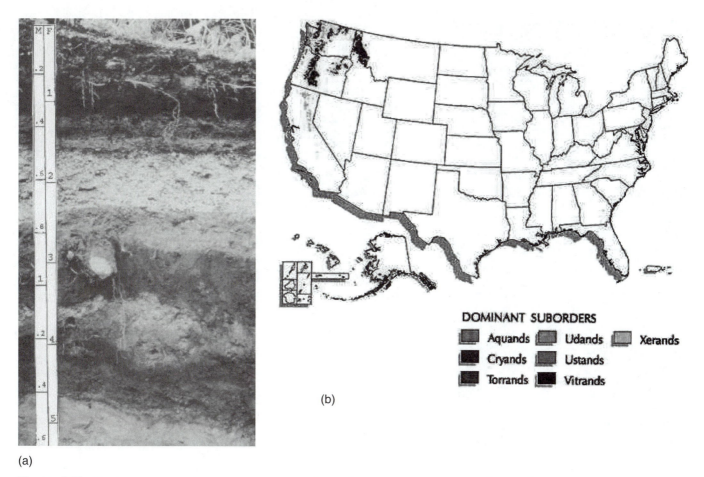

(a)

(b)

DOMINANT SUBORDERS
- Aquands
- Cryands
- Torrands
- Udands
- Ustands
- Vitrands
- Xerands

FIGURE 3-7   Profile of a typical Andisol (a) and distribution of Andisols (b) in the United States. *(Photograph and map courtesy of USDA-NRCS)*

❖ Xerands have xeric moisture regimes and temperature regimes that range from frigid to thermic. Xerands with frigid and mesic temperature regimes typically formed beneath coniferous forest, and those with thermic temperature regimes formed under grass and shrub vegetation (Figure 3-7).

Aridisols

Aridisols are desert soils and represent about 9 percent of the surface soils in the United States (Table 3-4). Some but not all Aridisols are salty, and most contain lime (Figure 3-8). Aridisols will have one or more of the following features within 100 cm of the soil surface: a calcic, cambic, gypsic, natric, petrocalcic, petrogypsic, or a salic horizon, or a duripan or an argillic horizon. In some Aridisols, for example, the lime has cemented in the subsoil to form a petrocalcic horizon. One common name for this phenomenon is *caliche.* In other cases the soil has cemented with silica to form a duripan.

Black alkali soils are Aridisols that have so much sodium that any organic matter has dispersed over the soil surface and colored it dark. White alkali soils lack the sodium and remain light, but they are also salty. Aridisols are highly productive if given water, but extremely fragile in their natural state. In their native state they are too dry for mesophytic vegetation to survive. Overgrazing of desert soils has been a major environmental problem.

> Aridisols are desert soils.

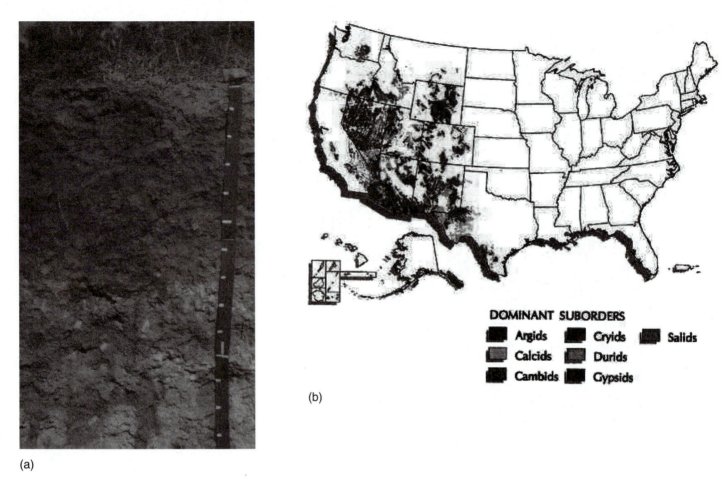

(a)

(b)

FIGURE 3-8   Profile of a typical Aridisol (a) and distribution of Aridisols (b) in the United States. *(Photograph and map courtesy of USDA-NRCS)*

Dominant Suborders of Aridisols

Aridisols are extensive throughout the American southwest and throughout Nevada, where there is no period of plant-available water longer than three months (Figure 3-8). The dominant suborders are Argids, Calcids, Cambids, Cryids, Durids, Gypsids, and Salids (Table 3-5).

❖ Argids have an argillic or natric horizon, but not a duripan or a gypsic, petrocalcic, petrogypsic, or salic horizon.

❖ Calcids have calcic or petrocalcic horizons and have calcium carbonate in the layers above. The parent materials have a high carbonate content and the lack of precipitation prevents carbonates from leaching through the soil profile to any great extent.

❖ Cambids are Aridisols with the least degree of soil development.

❖ Cryids are cold desert soils such as those found in the soils at high elevations in mountain valleys and basins in Idaho.

❖ Durids are Aridisols with a duripan; they are typically found in Idaho and Nevada.

❖ Gypsids are Aridisols with a gypsic or petrogypsic horizon.

❖ Salids are salty Aridisols commonly found in depressions (playa). They are unsuitable for agriculture unless the salts are leached.

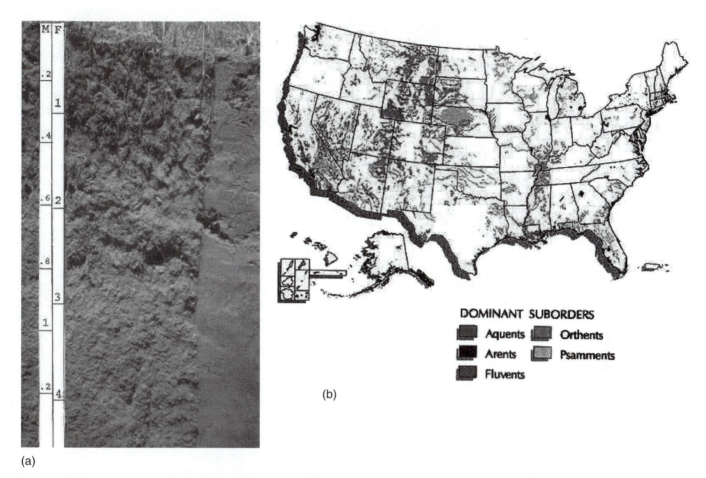

(a)

(b)

DOMINANT SUBORDERS
Aquents   Orthents
Arents   Psamments
Fluvents

FIGURE 3-9   Profile of a typical Entisol (a) and distribution of Entisols (b) in the United States. *(Photograph and map courtesy of USDA-NRCS)*

Entisols

Entisols are primitive, early soils with little or no evidence that pedogenic horizons have developed. This may be because the parent material is very resistant to weathering and soil formation is very slow, or because the newly developing soil is either eroded or buried by new deposits before distinctive horizons have a chance to develop.

> Entisols are very young soils that are just developing.

Many Entisols have an ochric epipedon and a few have an anthropic epipedon. Many are sandy or very shallow. Entisols have no diagnostic horizons within one meter (39 in.) of the soil surface. They typically occur on steep slopes that are actively eroding and in floodplains or glacial outwash plains (Figure 3-9). About 12 percent of the surface land in the United States consists of Entisols (Table 3-4). Entisols are widely distributed throughout the United States because the processes that form them are also widely distributed (Figure 3-9).

Dominant Suborders of Entisols

The dominant suborders of Entisols are Aquents, Arents, Fluvents, Orthents, and Psamments (Table 3-5).

❖ Aquents are wet Entisols that form in coastal regions and floodplains from recent sediments. They support vegetation that tolerates temporary flooding. Cropping occurs, but most Aquents are used as pasture or wildlife habitat.

❖ Arents are anthropogenic soils. They do not have diagnostic features because they have been plowed, spaded, or unearthed by human activity. With irrigation they are valuable cropland, particularly in California.

❖ Fluvents are freely draining Entisols typically found in floodplains, alluvial fans, and deltas. They show evidence of stratification. Some can be used as cropland if protected from flooding by dikes and levees.

❖ Orthents reflect recent erosional surfaces.

❖ Psamments are sandy in all layers. They can be productive rangeland, but bare psamments are subject to significant wind erosion and drifting.

Gelisols

Gelisols are soils of cold regions.

Gelisols are frozen soils of the north (although these soils can also be found in the southern hemisphere and at high elevation). Although Gelisols make up 7.5 percent of the ice-free surface land in the United States and 8.6 percent globally, those areas are almost exclusively in Alaska, Canada, Scandinavia, and Siberia. Gelisols are characterized by permafrost (permanently frozen soil) within 100 cm of the soil surface (Figure 3-10). Gelisols will have evidence of cryoturbation (frost heaving and churning) in the soil above the permafrost. The permafrost acts as a barrier to soil development by preventing the downward movement of solutes.

FIGURE 3-10 Profile of a typical Gelisol. *(Photograph courtesy of USDA-NRCS)*

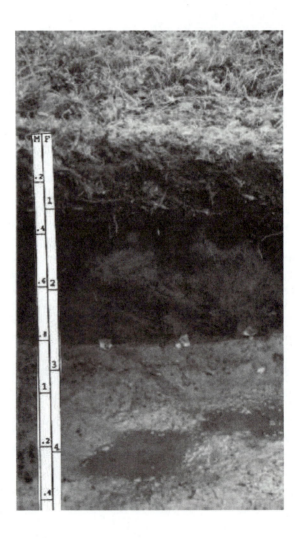

Dominant Suborders of Gelisols

The dominant suborders of Gelisols are Histels, Orthels, and Turbels (Table 3-5).

❖ Histels are Gelisols with large amounts of organic carbon. They are found in Alaska. The organic carbon comes from the tundra vegetation in this region, which consists of mosses, sedges, and shrubs. The short growing season precludes agricultural activity, so these soils mostly sustain wildlife.

❖ Orthels show little or no evidence of cryoturbation. In addition to mosses, sedges, and shrubs they also contain some black and white spruce.

❖ Turbels show cryoturbation, which appears as irregular, broken, or distorted horizon boundaries, ice wedges, or oriented rock fragments on top of the permafrost. Turbels occur on slopes that receive more sunlight or areas where fire and land clearing have altered the thermal properties of the soil so that the permafrost can thaw periodically.

Histosols

Histosols are soils that are dominantly organic. They make up a small (1.3 percent) but important part of the soil resources in the United States from the north to the Everglades and in some mountain regions (Figure 3-11). They are commonly called bogs, moors, peats, or mucks. The organic matter

> Histosols are dominantly organic.

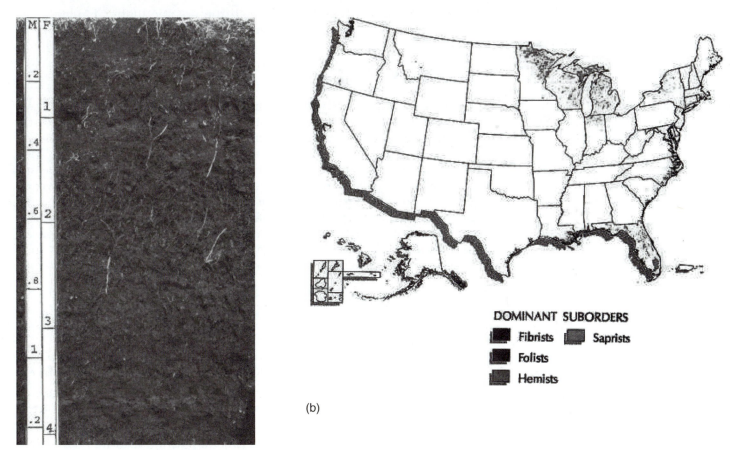

(a)

(b)

DOMINANT SUBORDERS
- Fibrists
- Folists
- Hemists
- Saprists

FIGURE 3-11 Profile of a typical Histosol (a) and distribution of Histosols (b) in the United States. *(Photograph and map courtesy of USDA-NRCS)*

accumulates because these soils are too wet or cold to allow organic matter decomposition to keep up with organic matter deposition. A soil is classified as a Histosol if it has more than 25 percent organic matter and does not have permafrost (Figure 3-11). A general rule is that if half or more of the upper 80 cm of soil is organic, the soil is classified as a Histosol.

Many Histosols form in old glacial lakes that have filled with dead plants. Although Histosols have great agricultural potential, drainage causes some unique problems. When Histosols are drained and aerated, decomposition can accelerate, and the soil can drop (subside) as much as 30 cm (about a foot) in ten years.

Dominant Suborders of Histosols

The dominant suborders of Histosols reflect the extent of organic matter decomposition that has occurred within them and consist of Fibrists, Folists, Hemists, and Saprists (Table 3-5).

* ❖ Fibrists are wet, slightly decomposed Histosols often referred to as peat. Most are found in southern Alaska. There is a long tradition of harvesting peat for fuel in countries such as Ireland.

* ❖ Folists are freely drained Histosols with horizons consisting of leaf litter, twigs, and branches resting on bedrock or rock fragments. They mostly occur in Alaska and Hawaii. The high organic matter content prevents them from being classified as Entisols.

* ❖ Hemists are wet Histosols in which the organic matter is slightly decomposed.

* ❖ Saprists are wet Histosols in which the organic matter is highly decomposed. These are also called muck soils and may appear very black.

Inceptisols

Inceptisols are immature, developing soils.

Inceptisols are immature soils of humid and subhumid regions that have altered horizons with evidence of chemical and physical change, but still have some weatherable material and lack a clear illuviated zone (Figure 3-12). Most Inceptisols have an ochric A horizon and a cambic B horizon. They can have other diagnostic horizons, but because they are early in development they lack argillic, natric, kandic, spodic, or oxic horizons that would be indicative of weathering and solute movement.

Inceptisols are widely distributed across the United States (Figure 3-12) and make up about 10 percent of the surface soils. They are found in all climate zones except the desert. Common locations for Inceptisols in a landscape are depressions.

Dominant Suborders of Inceptisols

The dominant suborders of Inceptisols are Anthrepts, Aquepts, Cryepts, Udepts, Ustepts, and Xerepts (Table 3-5).

* ❖ Anthrepts are freely drained Inceptisols that have an anthropic or plaggen epipedon. They are not known to occur in the United States.

* ❖ Aquepts are wet Inceptisols with poor drainage and groundwater at or near the soil surface at least some parts of the year. They can have any type of vegetation. Water is a limitation for agricultural use.

* ❖ Cryepts are cold Inceptisols occurring in mountains or high latitudes. The native vegetation is often conifers or mixed conifers and hardwood trees.

* ❖ Udepts are freely draining Inceptisols with udic moisture regimes. Most are used as forest but many have been cleared for cropland.

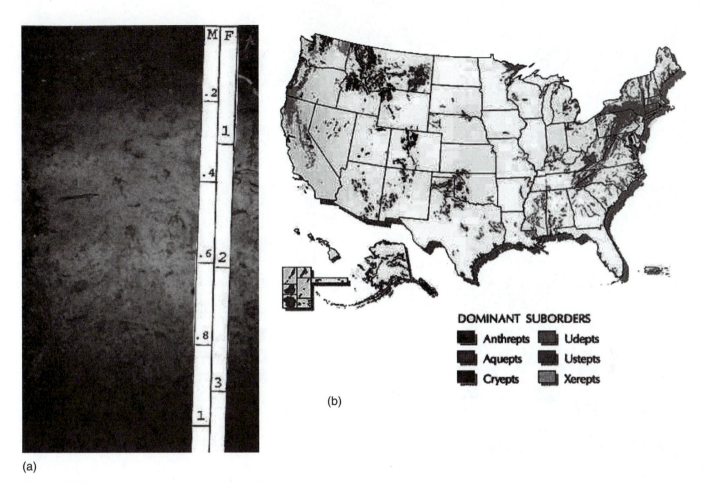

(a)

(b)

DOMINANT SUBORDERS

Anthrepts   Udepts
Aquepts   Ustepts
Cryepts   Xerepts

FIGURE 3-12   Profile of a typical Inceptisol (a) and distribution of Inceptisols (b) in the United States. *(Photograph and map courtesy of USDA-NRCS)*

❖ Ustepts have ustic moisture regimes. They are drier Inceptisols and are common in the Great Plains (Figure 3-12). The native vegetation was originally mostly grass.

❖ Xerepts have xeric moisture regimes and are found in the western United States. The temperature regime can range from frigid to thermic.

Mollisols

Mollisols are grassland soils extensively distributed throughout the United States, but primarily west of the Mississippi River (Figure 3-13). The dominant feature of Mollisol is a deep, dark-colored A horizon that is base rich and extremely fertile. Other diagnostic horizons are argillic, natric, cambic, or calcic horizons. Mollisols are among the most productive agricultural soils in the world. Small grains are grown in drier regions and maize (corn) and soybeans in the warmer, humid regions. Mollisols were among the first soils scientifically described in Russia, where they were called Chernozems (black earth).

> Mollisols are grassland soils and among the most productive soils in the world.

Dominant Suborders of Mollisols

In the United States the dominant suborders of Mollisols appear as sequential bands across the country reflecting the moisture regimes in which each Mollisol developed: Albolls, Aquolls, Cryolls, Rendolls, Udolls, Ustolls, and Xerolls (Table 3-5; Figure 3-13).

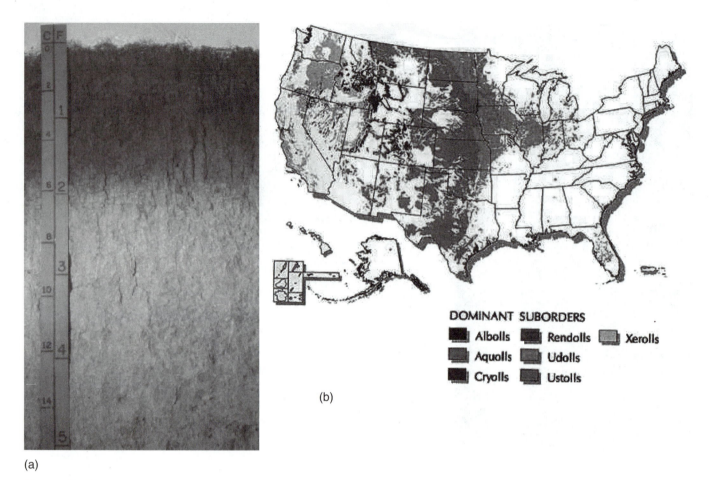

(a)

(b)

FIGURE 3-13 Profile of a typical Mollisol (a) and distribution of Mollisols (b) in the United States. *(Photograph and map courtesy of USDA-NRCS)*

❖ Albolls have an albic horizon and fluctuating groundwater table. They generally occur on gentle slopes.

❖ Aquolls are wet Mollisols, most of which have been drained for agricultural activity.

❖ Cryolls are cool or cold Mollisols, generally freely draining, that occur in high elevations or Alaska. Their agricultural use is limited by short growing seasons.

❖ Rendolls are Mollisols that formed in humid regions in highly calcareous parent materials such as limestone, chalk, and drift in places such as Florida, tropical islands, and a few mountainous regions in the western United States.

❖ Udolls are freely draining Mollisols in udic (humid) moisture regimes in the eastern Great Plains. The native vegetation was originally tall grass prairie.

❖ Ustolls are Mollisols formed in subhumid to semiarid climates in the western Great Plains. The original vegetation was grass, but most of these soils have been converted to cropland. Rainfall can be periodic, and drought severe, and in the absence of permanent soil cover wind erosion can be significant.

❖ Xerolls are Mollisols that form in an environment with a Mediterranean climate (hot, dry summers; cool, wet winters). The original vegetation was bunch grasses, shrubs, and trees in environments with mesic or frigid temperatures and savanna with perennial oak and fir in areas with thermic temperature regimes. Many Xerolls are irrigated, and many are used as rangeland.

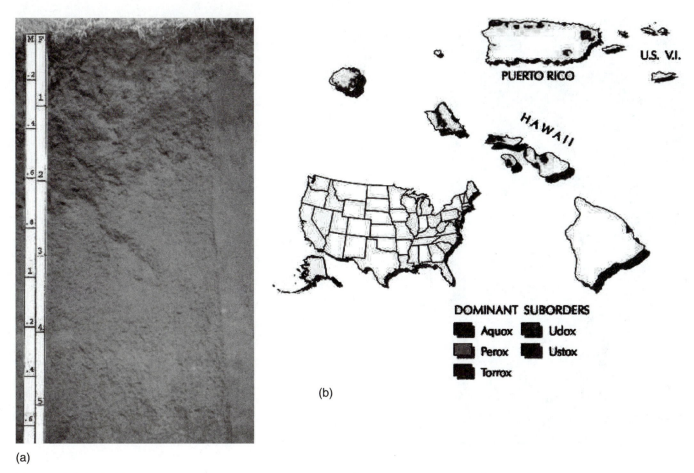

(a)

(b)

FIGURE 3-14 Profile of a typical Oxisol (a) and distribution of Oxisols (b) in the United States. *(Photograph and map courtesy of USDA-NRCS)*

Oxisols

Oxisols are highly weathered soils of great age and are virtually absent from the continental United States (Figure 3-14). They form in tropical and subtropical environments. Oxisols are characterized by a high content of quartz, inert clays, and oxides of iron and aluminum that give these soils their distinctive red color (Figure 3-14). The only type of layered clay remaining in these soils is kaolinite. Horizon designations are usually arbitrary.

> Oxisols are highly weathered soils.

Although the structure of Oxisols is granular and good for root growth, leaching is severe and nutrient retention is low. Traditional global management for Oxisols has been slash-and-burn, wherein after a short period of cultivation trees and shrubs are allowed to return to recycle nutrients from lower profiles and form at least some organic residue near the soil surface that can be exploited after the overlying vegetation is burned. More intensive agricultural use requires substantial inputs.

Dominant Suborders of Oxisols

The dominant suborders of Oxisols are Aquox, Torrox, Udox, and Ustox (Table 3-5).

❖ Aquox are wet Oxisols occurring only in Puerto Rico and Hawaii.

❖ Torrox are Oxisols of arid regions.

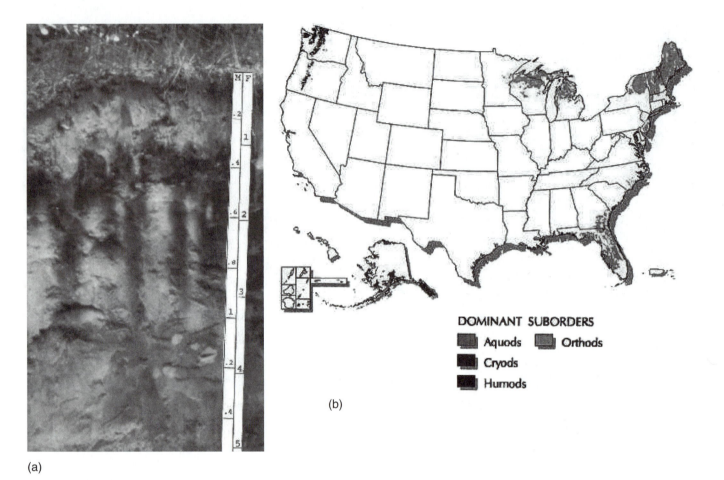

(a)

(b)

FIGURE 3-15 Profile of a typical Spodosol (a) and distribution of Spodosols (b) in the United States. *(Photograph and map courtesy of USDA-NRCS)*

❖ Udox are well-drained Oxisols with a udic moisture regime. They have a year-round growing season and generally sufficient precipitation to support nonirrigated agriculture year-round.

❖ Ustox are Oxisols that have a ustic moisture regime and can support one rain-fed crop. Otherwise there is a period of at least ninety days in most years when rainfall is inadequate to support crop growth.

Spodosols

| Spodosols are soils of acid coniferous regions. |

Spodosols have very discrete distribution within the United States (Figure 3-15) and make up about 3.3 percent of the land area (Table 3-4). Spodosols are soils of acid sandy pine land occurring in either the north, south, or far northwest. The original Russian term for these soils was *Podzol,* meaning "ash beneath," because they have a distinctive, nearly white, bleached horizon (Figure 3-15). The spodic horizon, immediately beneath this bleached zone, consists of accumulated organic matter and amorphous aluminum and/or iron. Many Spodosols originally harvested for their abundant timber proved to be poor agricultural land and have since been returned to timber production as their major use.

Dominant Suborders of Spodosols

The dominant suborders of Spodosols are Aquods, Cryods, Humods, and Orthods (Table 3-5).

❖ Aquods are wet Spodosols characterized by a shallow, fluctuating water table, particularly in Florida and along the Atlantic coast. They support water-loving plants. Although the native fertility is low, they can be managed successfully for agriculture.

❖ Cryods are cold Spodosols at high elevations or high latitudes. They consist primarily of coniferous forest.

❖ Humods are freely drained Spodosols that have accumulated significant organic matter in the spodic horizon. They mostly occur in the Pacific northwest (Figure 3-15).

❖ Orthods are Spodosols that are freely draining and have accumulated a modest amount of organic matter in the spodic horizon. They are found extensively in the northeast and Great Lakes states (Minnesota, Michigan, Wisconsin). These regions were extensively harvested for timber.

Ultisols

Ultisols are low-base-status forest soils of warm, humid regions. Base (e.g., calcium, magnesium, potassium) cycling occurs in the forest cover but is not as great as in Alfisols because the parent material usually has less limestone and the weathering and leaching has been more severe. Ultisols are usually older than Alfisols by tens of thousands of years. Because of the iron and aluminum content, which is unmasked by lower organic matter levels, Ultisols tend to be colorful soils ranging from red and yellow to gray, depending on drainage. Native fertility is limited, but the growing season can be long. The history of Ultisols in the southern United States is one of lost fertility and erosion. But current agricultural practices have made many of these soils extremely productive.

> Ultisols are soils of warm, humid regions.

Ultisols are extensive in the United States (Figure 3-16), primarily in the southeast (about 10 percent of the surface area, Table 3-4). Ultisols have B horizons that contain considerable amounts of translocated clay (argillic or kandic horizons), base saturation < 35 percent, and an ochric epipedon (Figure 3-16). The base saturation decreases with depth. The original vegetation was mixed conifers and hardwoods.

Dominant Suborders of Ultisols

The dominant suborders of Ultisols are Aquults, Humults, Udults, Ustults, and Xerults (Table 3-5).

❖ Aquults are Ultisols forming in wet areas in which groundwater is close to the surface during part of the year (usually winter and spring). Most Aquults are on coastal plains with gentle slopes.

❖ Humults are freely drained, humus-rich Ultisols. They receive considerable rainfall, but can be periodically dry.

❖ Udults form in humid environments. They are freely drained but humus-poor.

❖ Ustults form in ustic moisture regimes. They have low organic matter content.

❖ Xerults form in Mediterranean climates with xeric moisture regimes. The vegetation is typically coniferous forest.

Vertisols

Vertisols are rich in clay. They make up about 2 percent of the land surface in the United States with spotty distribution (Figure 3-17, Table 3-4).

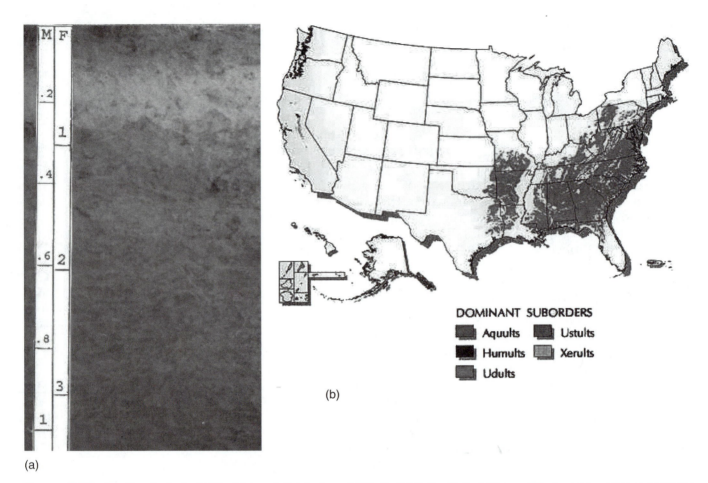

(a)

(b)

DOMINANT SUBORDERS
- Aquults
- Humults
- Udults
- Ustults
- Xerults

FIGURE 3-16 Profile of a typical Ultisol (a) and distribution of Ultisols (b) in the United States. *(Photograph and map courtesy of USDA-NRCS)*

Vertisols are shrinking and swelling soils.

Vertisols are described as cracking, dark clay soils; they shrink when dry and swell when wet. They occur in warm temperate and tropical environments, but do not appear leached because they tend to have a lime content because shrinking and swelling of the indigenous clays churns up fresh limey material from the C horizon to replenish that which was lost or leached (Figure 3-17).

The shrinking and swelling action causes Vertisols to seal and shed water in wet seasons. In summer the shrinking soils form lens-shaped blocks. These blocks can slide past one another causing polished surfaces called *slickensides.* The shrinking and swelling also causes buckling, which shapes the landscape into mounds and hollows called *gilgai.* Although these soil properties do not prevent agricultural use, as long as adequate precipitation and irrigation exists to maintain relatively constant moisture content, they pose some significant engineering problems.

Dominant Suborders of Vertisols

The dominant suborders of Vertisols are Aquerts, Cryerts, Torrerts, Uderts, Usterts, and Xererts (Table 3-5).

❖ Aquerts are wet Vertisols, although cracks can open during some parts of the year. Drainage on these soils is a problem because the saturated hydraulic conductivity is low.

❖ Cryerts do not typically occur in the United States. They exist in cold environments where there is just enough of a summer thaw to allow cracking to occur.

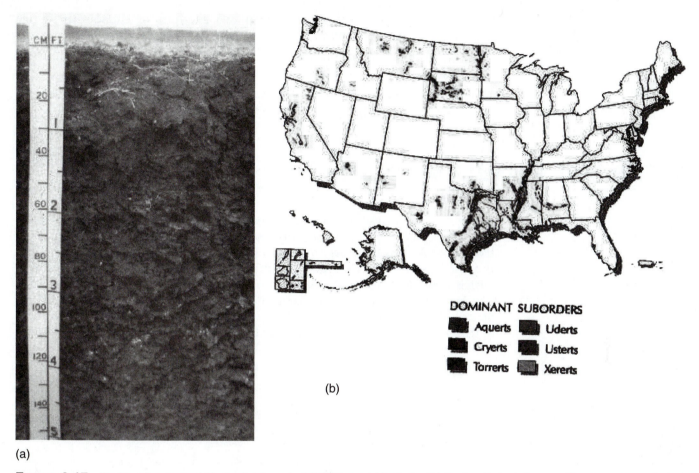

(a)

(b)

DOMINANT SUBORDERS

- Aquerts
- Cryerts
- Torrerts
- Uderts
- Usterts
- Xererts

FIGURE 3-17 Profile of a typical Vertisol (a) and distribution of Vertisols (b) in the United States. *(Photograph and map courtesy of USDA-NRCS)*

❖ Torrerts are Vertisols of arid environments. The cracks formed stay open for most of the year, but periodically close in winter. Saturated hydraulic conductivity is low despite the soil cracking, and this can lead to salt accumulation at the soil surface.

❖ Uderts form in udic moisture regimes. Soil cracks open and close in response to precipitation. Subsurface irrigation is usually used to convert these soils to agricultural use.

❖ Usterts pose many of the same problems as Uderts, and for the same reasons, although the opening and closing of cracks is more limited because of the greater periodicity of rainfall they experience.

❖ Xererts form in Mediterranean climates that have cool wet winters and warm dry summers. Consequently, these soils have cracks that regularly open and close, which can damage overlying structures such as roads. Their main use is for rangeland.

FOCUS ON . . . SOIL ORDERS

1. Why aren't there any Andisols in the eastern United States?

2. What other parts of the world would you predict would have Andisols?

3. What soil orders often have duripans?

4. There is a curious gap between two sections of Ultisols in Figure 3-17. Can you explain it?

5. Why does the suborder classification of Mollisols continuously change as you go from east to west?

6. Why is designating horizon boundaries in Oxisols mostly arbitrary?

OTHER SOIL CLASSIFICATION SYSTEMS

There are, of course, many classification systems throughout the world. Canada, for example, has a unique system for classifying its soils. The FAO (Food and Agriculture Organization) has developed a worldwide soil classification system. The FAO, for example, has twenty-six distinct soil orders, which are roughly described in Table 3-9. Although some of the names overlap with similar designations in *Soil Taxonomy,* many do not, and instead reflect the much greater diversity of soils than can be found outside of the United States when global-soil forming environments are considered.

TABLE 3-9 Soil order names in the FAO system of classification.

Order	Characteristic
Histosol	Organic soils
Lithosol	Hard rock within 10 cm of soil surface
Vertisol	Clay cracking soils
Fluvisol	Recent alluvial deposits without development
Solonchaks	Saline soils
Gleysols	Hydromorphic features within 50 cm of the soil surface
Andosols	Volcanic soils
Arenosols	Coarse-textured soils with albic material
Regosols	Developing soils without diagnostic horizons—ochric epipedon
Rankers	Developing soils without diagnostic horizons—umbric epipedon
Rendzinas	Developing soils without diagnostic horizons—mollic epipedon
Podzols	Spodic B horizon
Ferralsols	Oxic B horizon
Planosols	Impermeable horizons such as a fragipan
Solonetz	Natric (high sodium) B horizon
Greyzems	Mollic A epipedon with bleaching
Chernozems	Deep, dark mollic epipedon
Kastanozems	Mollic epipedon
Phaeozems	Mollic epipedon
Podzoluvisols	Irregular broken horizons
Xerosols	Weak A and aridic moisture regime
Vermosols	Weak ochric A and aridic moisture regime
Nitosols	Having an argillic B horizon
Acrisols	Having an argillic B horizon
Luvisols	Having an argillic B horizon
Cambisols	Having a cambric B or umbric A horizon 25 cm thick

SOIL COLOR

This chapter started with a poem about soil color, and so the last section appropriately ends with that topic. Soil color is a manifestation of physical and chemical properties. The presence of brown and black is indicative of organic matter content. The presence of reds and yellows is indicative of oxidized iron. The presence of blues, grays, and greens is indicative of reducing environments. The absence of color or presence of albic (white) layers is indicative of leaching.

Soil color, however, also plays a diagnostic role, which you can hardly doubt if you read a list of some of the names in the 1938 classification system that preceded *Soil Taxonomy:* brown soils, chestnut soils, gray-brown podzolics, red desert soils, reddish brown soils, and red-yellow podzolics, for example. The Russian terms from Dokuchaiev's original classification often refer to color: Sierozem (gray soil), Brunizem (brown soil), Chernozem (black soil). Soil colors are among the most obvious and consistent properties related to climatic and vegetation factors affecting soil.

Remember that *Soil Taxonomy* aims to quantify soil properties to assist in classification. How do you quantify soil color so that it can be used as a diagnostic tool? Soil scientists use three quantifiable soil properties—hue, value, and chroma—and a series of standardized color chips to assist them in this process (Figure 3-18).

> Soil color is one of soil's most diagnostic features.

Hue

Hue reflects the dominant spectral (rainbow) color in terms of the five cardinal colors (blue, green, yellow, red, and purple). In most soils in the United States the hue is yellow-red (YR), which reflects the dominance of iron oxides in determining soil color. The range for each hue goes from

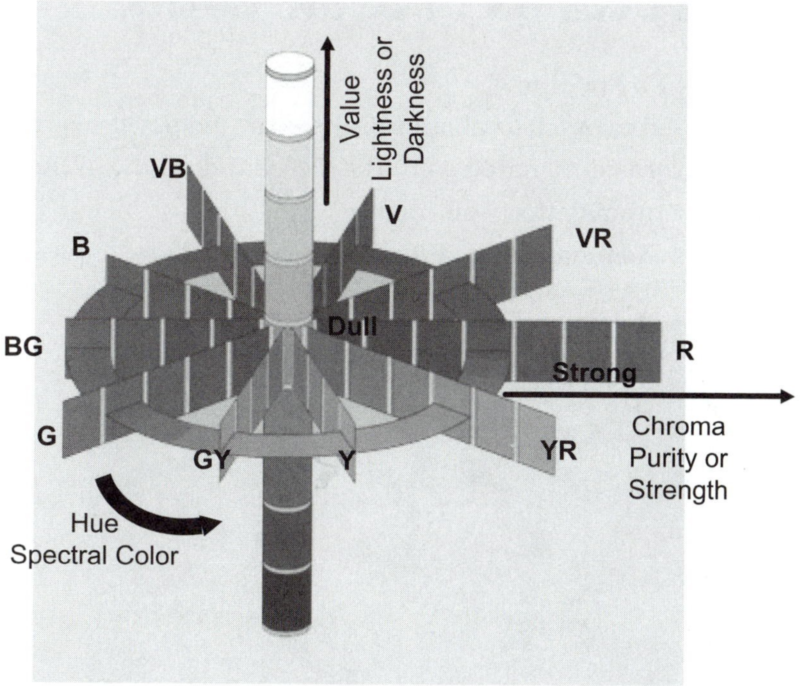

FIGURE 3-18 Diagram illustrating the relationship between hue, value, and chroma, which are used to quantitatively describe soil color. *(Courtesy of USDA-NRCS)*

0 to 10. Within the YR hue, for example, the hue becomes more yellow and less red as the numbers increase.

Value

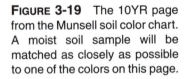
Hue, value, and chroma are used to quantify soil color.

Value is a measure of the lightness or darkness of a material and ranges from 0 (black) to 10 (white).

Chroma

Chroma reflects the strength or dullness of a color. A pure, brilliant color would have a chroma of about 20. The chroma in most soils is < 8. Actual soil colors can be made by mixing pure colors with neutral gray colors, which have a chroma of 0.

How to Determine Soil Color

The Munsell soil color chart is used to help describe soil color.

The Munsell (2000) nomenclature for soil color combines hue, value, and chroma in a standard set of symbols such as 10YR 5/4 (Hue Value/Chroma). The Munsell soil color chart contains 322 chips that reflect the typical colors found in soil (Figure 3-19). A moist soil sample is prepared and compared to the hue page that most closely matches the color. Then the soil is moved up and down (to determine value) and left to right (to determine chroma) until as close a match as possible is made. The usual practice is to give the color name and the appropriate Munsell designation (Troeh and Thompson, 1993). Gleyed soils, soils that have been saturated, and in which the iron is reduced, have a grayish-green color, and their color is described by a second set of color charts.

FOCUS ON . . . SOIL COLOR

1. What does hue refer to?
2. What can soil color tell us about soil chemical and physical properties?
3. If a soil color is designated as 5YR 4/3, what does the 4/3 refer to?
4. How does chroma affect soil color?
5. What is the dominant hue of most soils in the United States?

FIGURE 3-19 The 10YR page from the Munsell soil color chart. A moist soil sample will be matched as closely as possible to one of the colors on this page.

SUMMARY

In this chapter we learned the rationale for soil classification systems and reviewed how the soil classification system evolved in the United States. The pedon is the smallest classifiable unit of soil that is recognized. The guidelines in *Soil Taxonomy,* which was adopted in 1965 as the standard method for classifying soils in the United States, uses a diagnostic and quantitative approach to classifying soils. The first step in classifying soils in this system is to identify the diagnostic epipedon (surface horizon) and diagnostic subsurface horizons. Thereafter, the soil can be placed into one of twelve soil orders. The soils can be further classified to suborder, great group, subgroup, family, and soil series by progressively adding more information to the soil description, such as climatic regime, distinctive horizons, texture, and mineralogy. The addition of distinctive syllables to the soil order gives each soil its unique taxonomic name. Soil series are typically named after a nearby community but can be located in many other areas. There are over 19,000 described soil series.

Although soil color reflects physical and chemical properties, it has been made quantifiable so that it can be used for diagnostic purposes in classification. Soil color is one of the first differences you notice between representative soil profiles of each of the soil orders.

In the next chapter we enter the soil profile and learn how the soil is actually put together.

END OF CHAPTER QUESTIONS

1. What is a unit of natural soil called in classification?

2. What was V. V. Dokuchaiev's great contribution to soil science?

3. What is the basis for soil classification in *Soil Taxonomy* and how does it differ from earlier classification systems?

4. Why is the Mollic epipedon a standard against which other epipedons are measured?

5. What are the levels of classification in *Soil Taxonomy?*

6. Is there a "natural" system of soil classification? Why or why not?

7. Which two soil orders most likely developed under forest vegetation?

8. What most distinguishes Entisols from Inceptisols?

9. What are the most weathered soil orders you are likely to find in the United States? What color will they have? Why?

10. Are duripans and fragipans desirable soil characteristics?

11. Why do Vertisols represent unique engineering problems?

12. What conditions cause Entisols to form?

13. When keying out soil orders, which are the easiest to identify?

14. Where are you most likely to find a spodic horizon? What chemical would be found in it?

15. What is base saturation, and does a low or high base saturation indicate high fertility?

16. Why have many Ultisols suffered from severe erosion?

17. What is one problem you might face in draining a Histosol and building a house on it?

18. A soil that has a hue of 2.5YR will appear to be what color?

19. What are the characteristics of a fine-loamy, mixed-mesic, typic Haludoll?

20. What is the taxonomic name of your state soil?

FURTHER READING

The ideal gateway into soil taxonomy and classification is through the USDA-NRCS Web site at http://www.nrcs.usda.gov. There you will be able to open an electronic version of *Soil Taxonomy* as well as explore the properties of the over 19,000 soil series currently classified in the United States. There are also transcribed interviews with Guy Smith detailing the history and philosophy of how soil taxonomy and classification evolved in the United States from 1938 to the present.

REFERENCES

Brady, N. C., and R. R. Weil. 2002. *The nature and property of soils*, 13th ed. Upper Saddle River, NJ: Prentice Hall.

Callahan, G. N. 2003. Eating dirt. *Emerging and Infectious Diseases* 9: 1016–1021.

Harpstead, M. I., T. J. Sauer, and W. F. Bennett. 1997. *Soil science simplified*, 3rd edition. Ames: Iowa State University Press.

Munsell ®Color. 2000. Munsell soil color charts. New Windsor, NY: Gretag Macbeth.

Sandor, J. A., and L. Furbee. 1996. Indigenous knowledge and classification of soils in the Andes of southern Peru. *Soil Science Society of America Journal* 60: 1502–1512.

Troeh, F. R., and L. M. Thompson. 1993. *Soils and soil fertility*, 5th ed. New York: Oxford University Press, New York.

SECTION 2

SOIL PHYSICAL PROPERTIES

Chapter 4

SOIL SOLIDS: PARTICLE SIZE AND TEXTURE

"To see a world in a grain of sand and a heaven in a wildflower."

William Blake, *Auguries of Innocence*

OVERVIEW

It has been said that the whole is greater than the sum of its parts. So too is the soil greater than the sum of its constituent parts. However, to understand the nature of the soil, the processes that occur in the soil, and the unique role of the soil in both natural and managed systems, it is necessary to appreciate the special properties of the mineral and organic material in the soil and how these materials interact with each other and the water and air in the soil.

In this chapter you will begin examining the physical properties of soil. You will examine the mineral solids in the soil and answer several key questions about them:

✔ What are the primary particles that, on a weight basis, dominate the composition of most soils?

✔ How does texture influence other soil characteristics that influence the use and management of soils, such as permeability, water-holding capacity, and nutrient-holding capacity?

OBJECTIVES

After reading this chapter, you should be able to:

✔ Name and define the different particle size separates and list the corresponding differences in physical (size, shape, surface area) and chemical (mineralogy, surface charge, retention/exchange capacity) soil properties.

✔ Use the textural triangle to determine the soil textural class of a soil sample with known sand, silt, and clay percentages.

✔ Estimate soil textural class using the feel method.

KEY TERMS

aeration
clay
coarse fragments
crusting
erodibility
infiltration
loam

percolation
pore size distribution
porosity
primary minerals
quartz
sand

secondary minerals
silt
soil separates
soil textural classes
textural triangle
water-holding capacity

PRIMARY SOIL PARTICLES

The basic building blocks of most soils are the primary particles of the mineral fraction. The relative amount of these different particles influences many other soil properties. Furthermore, aside from massive earth-moving operations, the mineral composition of the soil cannot be changed with (normal) management.

Soil Separates

> The classification of a soil particle depends on the particle diameter.

Soil particles, or **soil separates,** are divided into three general classes based on the diameter of the particle. The names of the particle size classes are familiar terms: sand, silt, and clay. You have likely experienced the grittiness of **sand,** the largest soil particles, or the stickiness of **clay,** the smallest soil particles. Intermediate to these is **silt,** which has the smooth feel of flour. In the United States, the most commonly used size ranges for sand, silt, and clay are those adopted by the U.S. Department of Agriculture (USDA; Table 4-1). Particles with a diameter greater than 2.0 mm are considered; **coarse fragments** (for example, gravel, cobble, stone, boulder).

TABLE 4-1 Soil particle size separates and their associated physical properties.

Soil Separate (including major subdivisions)	Diameter Range[a]	Particles Per Gram[b,c]	Surface Area Per Gram[c]
	mm		cm^2
Sand	2.00–0.05	90	11.32
Very coarse sand	2.00–1.00	90	11.32
Coarse sand	1.00–0.50	721	22.64
Medium sand	0.50–0.25	5,770	45.28
Fine sand	0.25–0.10	46,005	90.57
Very fine sand	0.10–0.05	721,000	226.42
Silt	0.05–0.002	5.77×10^6	452.83
Coarse silt	0.05–0.02	5.77×10^6	452.83
Fine silt	0.02–0.002	9.01×10^7	1,132.08
Clay	<0.002	9.01×10^{10}	11,320.75

[a]According to the U.S. Department of Agriculture classification system.

[b]Assuming a particle density of 2.65 g cm^{-3}.

[c]Treating all particles as perfect spheres, and using the maximum diameter for a given soil separate.

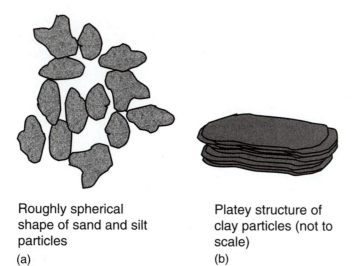

FIGURE 4-1 The idealized shape of soil particles.

Roughly spherical shape of sand and silt particles

(a)

Platey structure of clay particles (not to scale)

(b)

Characteristics of Soil Separates

Physical size is what defines a soil particle as one of the various soil separates, but other differences exist among the separates. These differences contribute to the physical and chemical behavior of the different particles.

Physical Characteristics

Though not perfect spheres, sand and silt particles tend to be rounded or blocky in shape, depending on how weathered they are (Figure 4-1a). Conversely, clay particles are flat in shape, with multiple platelike clay particles layered together (Figure 4-1b). These differences in both the size and shape among sand, silt, and clay particles strongly influence the specific surface area of the particles (the surface area per gram of soil). You can see the effect of size alone by taking a large block, a rock for example, and repeatedly dividing it into multiple smaller blocks (Figure 4-2). Without increasing the amount of material, with each successive division the exposed surface area increases.

This analogy easily explains the increase in specific surface area as particle size decreases from coarse sand to fine silt (Table 4-1) because these particles are relatively block-like. However, the sizeable increase in specific surface area from silt to clay is further attributed to the platelike shape of the clay particles. The effect of particle size on surface area is particularly important in soil science because many of the important reactions that occur in the soil (weathering, water retention, nutrient exchange) occur at the surfaces of the particles where the particles contact the surrounding air or water.

> Surface area is an important property of the different soil particles.

Mineralogical Characteristics

In addition to basic differences in the physical characteristics of the different soil separates, there are differences in the types of minerals from which these particles are made. You learned about these minerals in Chapter 2. Sand and silt particles are predominantly **primary minerals.** Sand particles are mostly grains of **quartz,** which is highly resistant to weathering. Silt particles are also typically grains of quartz, with other resistant primary silicate minerals (such as feldspar) also being common. The clay particles are dominated by **secondary minerals.** These are formed from the weathering of the less resistant primary minerals (such as Olivine, see

> Sand and silt particles primarily contain primary minerals; clays are primarily composed of secondary minerals.

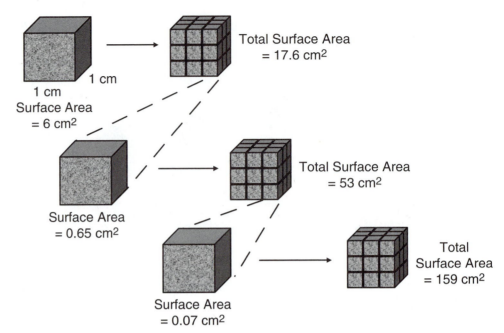

FIGURE 4-2 Surface area increases as a block is subdivided and particle size decreases. The block starts with a surface area (SA) of 6 cm². If we divide each side by thirds we create 27 new blocks with a combined surface area of 18 cm². If we take one of those new blocks and divide it by thirds the new surface area of each block is only 0.07 cm², but there are 729 such blocks, so the combined surface area is 53 cm². If we do this again, each new block only has a surface area of 0.008 cm², but because there are 19,683 of these sized blocks in the original block, the total surface area is now a whopping 159 cm².

Figure 2-12). These secondary silicate minerals have a profound effect on the chemical characteristics of soils, and are discussed to a greater extent in Chapter 7.

Chemical Characteristics

The primary minerals that dominate the sand and silt particles are for the most part inert and have little influence on the chemical characteristics of the soil. The secondary silicate minerals of the clay particles are the source of much of the chemical activity of the soils. These minerals carry a charge (usually negative), which gives them the ability to adsorb water, nutrient ions, and certain contaminants.

FOCUS ON . . . SOIL SEPARATES

1. What are the three soil separates, their size limits (in the USDA system), and an adjective useful in describing how each "feels" (e.g., when rubbed between your fingers)?

2. How does the shape of clay particles differ from that of sand and silt?

3. What is the dominant mineral type in the sand and silt separates? Is this mineral type considered a primary or a secondary mineral?

4. Which soil separate is most likely to adsorb the greatest amount of water per gram of soil? Why?

5. Which soil separate is most likely to adsorb the most nutrient cations (such as Ca^{2+} or K^+) per gram of soil? Why?

SOIL TEXTURE

Soil texture describes the relative proportions of sand, silt, and clay particles in a soil. These amounts, expressed as percentages, are divided into twelve different **soil textural classes.** The textural classes describe whether the soil properties are dominated by one or more of the three particle size separates (for example, sandy, silty, or clayey). If the soil is characterized as a **loam** or loamy soil this suggests that it has a mixture of the three separates such that the soil's properties exhibit an equal *influence* of sand, silt, and clay.

There are twelve soil textural classes.

The soil textural classes are readily illustrated on a **textural triangle** (Figure 4-3). From the textural triangle, you see that to be considered a

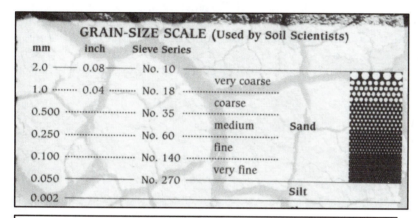

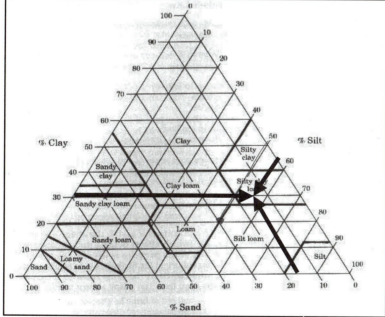

FIGURE 4-3 The textural triangle. The silt percent increases down the right axis, the clay percent increases up the left axis, and the sand content increases from right to left along the bottom axis. An example of its use is shown for a sample with 15 percent sand, 55 percent silt, and 30 percent clay: Find 30 percent on the clay axis and draw a line to the right parallel to the sand axis; find 55 percent on the silt axis and draw a line down and to the left, parallel to the clay axis; then find 15 percent on the sand axis and draw a line up and to the left, parallel to the silt axis. Where these three lines intersect identifies the textural class—in this case as silty clay loam. A grain size scale used to define the size of particles is also present. Anything smaller than 0.002 mm is defined as clay. *(Courtesy of USDA-NRCS)*

Clay has a disproportionate effect on the physical and chemical properties of soil textural classes.

sand textural class (loamy sand or sand), a soil must have > 70 percent sand particles and < 15 percent clay particles. To be considered a silt textural class, a soil must have > 80 percent silt particles and have < 12 percent clay particles. However, to be considered a clay textural class (sandy clay, silty clay, or clay) a sample must have only > 40 percent clay particles. This further emphasizes the pronounced effect of clay particles on the physical, chemical, and biological properties of a soil. A little clay goes a long way in influencing a soil's characteristics, and therefore how a soil is used or managed. Likewise, the loam textural class is not in the center of the triangle, and only has a maximum of 27 percent clay. A sample with ⅓ sand, ⅓ silt, and ⅓ clay would fall into the clay loam textural class.

Why Is Texture an Important Diagnostic Property of the Soil?

Texture may be the most important diagnostic characteristic of a soil, influencing many of a soil's physical, chemical, and biological properties. Consequently, knowing the texture of a soil can allow you to make many inferences about important characteristics of a soil (Table 4-2).

TABLE 4-2 Generalized influence of soil separates on some properties and behaviors of soils.[a]

	Rating Associated with Soil Separates		
Property/Behavior	Sand	Silt	Clay
Porosity	Low	Moderate	High
Percent large pores	High	Moderate	Low
Total water-holding capacity	Low	Moderate to high	High
Available water-holding capacity	Low	Moderate to high	Moderate
Infiltration rate	High	Moderate	Low (unless cracked)
Drainage (percolation) rate	High to very high	Slow to moderate	Very slow
Aeration	Good	Moderate	Poor
Cation exchange capacity	Low	Low to moderate	High
Ability to store plant nutrients	Very low	Low to moderate	High
Resistance to pH change	Very low	Low to moderate	High
Soil organic matter level	Low	Moderate to high	High to moderate
Decomposition of organic matter	Rapid	Moderate	Slow
Shrink-swell potential	Very low	Low	Moderate to very high
Compactability	Low	Moderate	High
Root penetration	Easy	Moderate	Moderate to difficult
Ease of cultivation	Easy	Moderate	Difficult
Suitability for tillage after rain	Good	Moderate	Poor
Warm-up in spring	Rapid	Moderate	Slow
Detachability of particles	Easy	Moderate	Very difficult
Transportability of particles	Very difficult	Moderate	Easy
Susceptibility to wind erosion	Moderate	High	Low
Susceptibility to water erosion	Low	High	Low if aggregated, high if not
Sealing of ponds, dams, and landfills	Poor	Poor	Good
Pollutant leaching potential	High	Moderate	Low (unless cracked)

[a]Exceptions to these generalizations do occur, especially as a result of soil structure, clay mineralogy, and/or organic matter content.

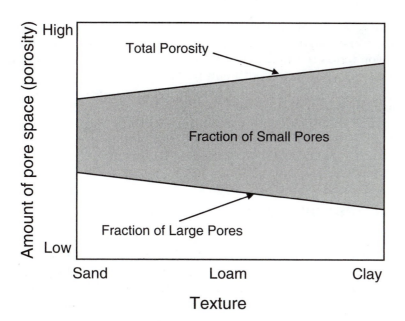

FIGURE 4-4 As particle size decreases from sand to clay, total porosity increases even as pore size declines. So, as a soil acquires more clay, it will have more pores, but the pores will be smaller. *(Adapted from Plaster, 1997)*

Physical Properties

Many of a soil's physical properties are influenced by soil texture, including **porosity, pore size distribution, water-holding capacity,** and permeability. Sandy soils are more permeable and have higher **infiltration** and **percolation** rates than clayey soils. Sandy soils also tend to have better **aeration.** In general, a sandy soil has a lower water-holding capacity than a silty or clayey soil. The reason for these differences is because sandy soils have much larger pore size distribution, or more big pores and few small pores. Conversely, silty or clayey soils have many small pores and fewer large pores. However, the porosity, or the sum of all the pore spaces, of sandy soils is often considerably less than that of clayey soils. Even though they are smaller in size, the total volume of pore space in a clayey soil is greater than that of a sandy soil (Figure 4-4).

> Soil texture influences porosity, pore size distribution, water-holding capacity, and permeability.

Chemical and Biological Properties

The amount of clay particles, which are dominated by the secondary silicate minerals, will dictate much of the chemical characteristics of a soil. As will be discussed in Chapters 7 and 8, these minerals carry a charge (usually negative). The presence of this charge enables the soil to adsorb water, nutrient ions (Table 4-3), and certain contaminants. Therefore, a sandy soil will have a lower water- or nutrient-holding capacity than a silty or clayey soil.

TABLE 4-3 Soil particle size separates and their associated nutrient content.

Separate	Total P (%)	Total K (%)	Total Ca (%)
Sand	0.05	1.4	2.5
Silt	0.10	2.0	3.4
Clay	0.30	2.5	3.4

Particle size also affects buffering capacity in soil.

The charge associated with clay particles also promotes associations between the clay particles and organic matter particles, such that a silty or clayey soil tends to have a greater organic matter content than a sandy soil. The charge associated with soil particles similarly affects a soil's resistance to pH change. Silty and clayey soils require more lime to effectively raise soil pH to desired levels. Biologically, the large surface area of the clay particles provides ample locations for microorganisms to attach and colonize. This (along with greater nutrient and water availability) contributes to silty and clayey soils tending to have more biological activity than sandy soils.

Use and Management

Soil texture affects a soil's load-bearing capacity, compaction, and shrink-swell potential.

These differences in the properties of the different soil separates strongly influence the use and management of different soils (Table 4-2). From an engineering standpoint, soil texture influences the soil's load-bearing capacity, compaction, and shrink-swell potential. The higher porosity of fine-textured soils increases their susceptibility to compaction because this pore space is readily collapsed. The flat shape of the clay particles also allows for closer packing than is possible in coarse-textured soils. Clayey soils are more susceptible to shrinking and swelling with alternating drying and wetting cycles because of their ability to adsorb water and expand the small pore spaces between clay particles.

CHOOSING THE TEXTURE: ARTIFICIAL POTTING AND ROOTING MEDIA

For most settings, it is not practical to alter soil texture. This is mainly because the only way to modify soil texture is to add sand or clay, especially over large areas. Even in small gardens this would require many cubic meters of material. For high-return operations (nurseries, greenhouses, sports fields, for example) using artificial potting or rooting media is more common. For these intensively managed situations, a media made from field soil would be too finely textured to promote adequate aeration and water movement once the natural soil structure has been disturbed. Artificial soil mixtures contain lots of organic matter, such as sphagnum peat moss, composted bark, or leaves. The mixtures also contain stable, porous mineral materials such as pumice, vermiculite, or perlite. This combination provides a ready source of available plant nutrients and favorable water-holding capabilities. These soils must retain some water, but also must release enough water to allow oxygen to reach roots. For plants that require or can tolerate drier conditions, such as cacti, sand can be added to the mixture. Similarly, sports turf (including golf tees and greens) is commonly established in a sandy rooting media that promotes rapid drainage and greater trafficability.

Ease of agricultural management is similarly influenced by texture. In general, soils of medium texture—sandy loam to loam—are usually easiest to manage. This is because they have intermediate levels of many of the soil properties influenced by texture (Table 4-2). As a result, they retain water, but they remain sufficiently drained and aerated; they are suitable for tillage or traffic relatively soon after rainfall; and they warm up readily in spring.

HOW IS TEXTURE DETERMINED?

Considering the great importance of soil texture, especially for management concerns, it is valuable to know how to measure soil texture. There are two general ways to do this: one method can be done quickly in the field without need for special equipment (texture-by-feel); one method is done in a laboratory, which is more accurate but requires more time and specialized equipment (sieving and sedimentation).

In the Field: The "Feel" Method

Estimating the texture of a soil sample in the field is a useful skill for most on-site investigations and consultations. The goal is to sense the relative amounts of sand, silt, and clay by considering how each feels when rubbed between your fingers. The typical method involves only a small handful of soil and some water. Before adding water, use your fingers to break up any large aggregates. It helps to have the soil as crushed as possible before adding water. Then add enough water to moisten the sample, without making a slurry (Figure 4-5). Knead the soil in your hand to uniformly mix the sample. The sample should have the consistency of cookie dough. If it is too dry, add more water; if it is too wet, add more dry soil. Once properly moistened, continue to knead the soil in your hand, noting the grittiness or smoothness of the soil, and making balls and ribbons, which you will use to assess the stickiness and plasticity of the sample.

At the extreme, a soil with a sand textural class will not even stick together to form a ball when squeezed in your hand. A soil that is high in sand is not sticky and does not form a good ribbon. As the clay content increases, the length and strength of the ribbon increase. A common way to determine texture by feel, especially for a beginner, is to follow a flowchart that highlights the necessary decisions in the process (Figure 4-6). However, the only way to develop an accurate sense of texture by feel is to practice on samples of various textures to "calibrate" yourself.

In the Lab: Sieves and Sedimentation

A more quantitative and accurate means of determining the textural class of a soil sample than texture-by-feel is to use a laboratory procedure of particle size analysis. The general steps are to break up any soil aggregates and disperse the soil particles, then separate them by sieving, sedimentation, or both. Sieving is generally only effective down to a particle size of 0.05 mm, so it works best at segregating sand-sized fractions between 2.0 and 0.05 mm, or separating the sand particles from the silt and clay. The basic premise, as with any other sieving operation, is that the holes in the sieve (usually a wire mesh) allow smaller particles to pass through while

> Texture can be determined in the field by feel or in the laboratory by sieving and sedimentation.

> Texture-by-feel is easy to do, but requires practice and experience to do well.

FIGURE 4-5 One feature of determining texture-by-feel is estimating the length of soil ribbon that can be made. *(Photograph courtesy of the USDA-NRCS)*

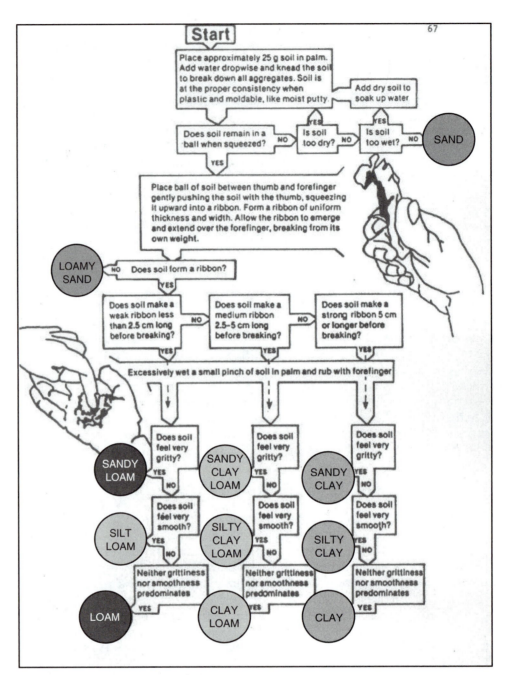

FIGURE 4-6   Flowcharts are a useful tool when determining texture-by-feel. *(Adapted from Thien, 1979)*

Sieving can be done wet
or dry.

holding larger particles (Figure 4-7). The sieving can be done using dry soil and shaking vigorously to encourage the smaller particles to fall through the sieve, or it can be done using wet soil and a stream of water to push the smaller particles through the openings in the mesh. By starting with a known mass of soil, the percentage of a particular particle size separate is calculated by dividing the mass of the material that remained in the sieve by the total initial mass.

Determining particle
fraction by
sedimentation is based
on Stoke's law.

Sedimentation can be used to separate particles of different sizes because the time it takes for a particle to settle from a liquid is mainly a function of the diameter of the particle. Mathematically, this is described by Stoke's law:

$$v = \frac{h}{t} = \frac{d^2 g(\rho_s - \rho_l)}{18\eta}$$

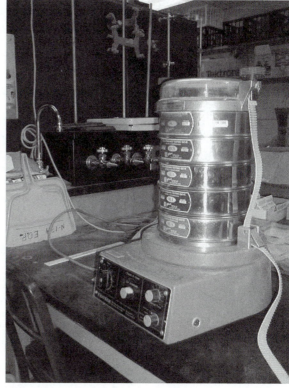

FIGURE 4-7 (a) Metal sieves are used to help determine particle size. (b) Multiple particle size fractions can be obtained by sieving dry soil in a nested set of sieves with decreasing opening from bottom to top.

(a)

(b)

where v is velocity (mm/sec), h is the distance the falling particle travels (m), t is the time that the particles are allowed to fall (sec), d is the effective diameter of particle (m), g is the gravitational force (9.81 Newtons/kg, or 9.81×10^{-3} Newtons/g), ρ_s is the density of the particle (2.65 g/cm^3, or 2.65×10^3 kg/m^3), ρ_l is the density of water (1.0 g/cm^3, or 1.0 kg/m^3), and η is the viscosity of water (1×10^{-3} Newton-sec/m^2 at 20 °C). Because all of the factors on the right side of the equation, except for d, are relatively constant under laboratory conditions, this equation can be rearranged and simplified to:

$$t = \frac{h}{9 \times 10^5 \, d^2}$$

where the constant 9×10^5 has units of sec^{-1}m^{-1}. Using this formula, it would take a sand particle 0.05 mm in diameter (the smallest sand particle) 44 seconds to fall 10 cm through a column of water:

$$t = \frac{0.10 \text{ m}}{9 \times 10^5 \, (0.05 \times 10^{-3}\text{m})^2} = 44 \text{ sec}$$

Using this information, it is relatively simple to determine the relative amounts of different particle size separates in a sample by suspending the dispersed sample in a column of water (Figure 4-8), then determining the amount of particles that remain suspended in the water after prescribed time periods. This determination is commonly made either by hydrometer, which measures the density of the liquid, or by removing a small volume of liquid from the water column, drying it in an oven to drive off all the water, and weighing the particles that remain. For example, after 44 seconds a measurement of the amount of particles that remain in the water column at a depth of 10 cm would tell you the amount of silt + clay in the sample. There would be no sand particles left in the upper 10 cm.

1. Disperse 50 g of dry soil in water and dilute to 1 L in a glass cylinder.
2. Add a dispersing agent. Stir and insert a hydrometer after 30 seconds.
3. Read the g of silt and clay in solution on the hydrometer stem after exactly 45 sec (sand-sized particles have already settled out).
4. Read the g of clay in solution after 478 min (silt-sized particles have settled out).
5. Determine the g of sand and silt by difference

50 g – t_{45sec} = g sand

t_{45sec} – t_{478min} = g silt

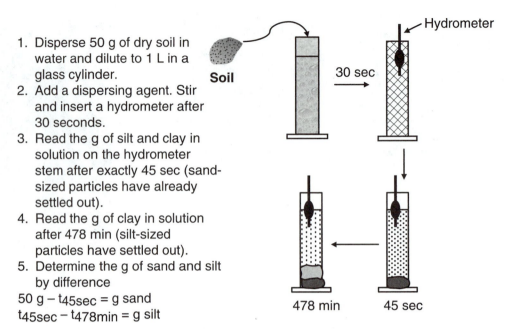

FIGURE 4-8   Procedure for determining particle size by sedimentation. In this case a hydrometer is used to read the particle density directly. Samples could be taken with time as described in the figure to determine particle density by weight.

GEORGE GABRIEL STOKES

Sir George Gabriel Stokes (1819–1903) was an Irish mathematician and physicist. He was born in County Sligo, Ireland and was the youngest son of the Reverend Gabriel Stokes, who was rector at Skreen. Stokes was recognized as one of the premier physicists of his time. In 1849 Stokes was named Lucasian Professor of Mathematics at Cambridge, a post once also held by Sir Isaac Newton. In addition, Stokes was president of the Royal Society from 1885–1890.

Stokes made contributions in many areas of mathematics and physics including fluid dynamics, optics, and mathematical physics. From the perspective of soil science he is best remembered for his 1851 law that related the terminal settling velocity of smooth rigid spheres in viscous fluid of known density to the diameter of the sphere when it was subjected to a known force field (to put it more plainly, he figured out how to predict how fast objects sink depending on their size). "Stoke's law" forms the basis for particle size analysis by the pipette, hydrometer, or centrifuge methods:

$$V = gD^2(\rho_1 - \rho_2)/18\eta$$

where:

V = velocity of fall (m sec^{-1})
g = gravitational force (9.81 Newton kg^{-1})
D = the "equivalent" diameter of the particle (m)
ρ_1 = the density of the particle (usually 2.65 kg m^{-3})
ρ_2 = the density of the medium (usually 1.0 kg m^{-3})
η = the viscosity of the medium (10^{-3} Newton-sec m^{-2})

Gravitational force and viscosity are essentially constants in standard laboratory conditions (settling through water at 20°C). Likewise, the difference between the density of soil particles (2.65 kg m–3) and water (1.0 kg m^{-3} at 20°C) is a constant. Consequently, Stoke's law simplifies to:

$$V = \frac{\text{height}}{\text{time}} = (D^2)\frac{(9 \times 10^5)}{\text{sec m}}$$

When determining texture by the pipette or hydrometer method, you typically completely suspend a soil sample in a column of water and determine how much of the solids remain after a given interval. So, based on Stoke's law, if you wait long enough for all the sand to settle below a certain depth (say 10 cm), then all that remains in solution is the silt and clay fraction. The total soil mass minus this amount gives you the mass of sand-sized particles. Likewise, if you wait a little longer (sometimes a lot longer) all the silt-sized particles will settle below a given depth, and all that will remain in the solution is clay-sized particles. So, the total mass of soil minus the sand-sized particles minus the clay-sized particles will give you the mass of silt-sized particles.

SOIL SEPARATES AND THEIR DISTRIBUTION

Soil texture is not the same everywhere. It is affected by the distribution of soil parent materials, from which the soil particles originate, and it is affected by the distribution of soil-forming processes that lead to the deposition, formation, movement, and destruction of soil particles. Differing textures with depth and from one area to another can influence land use because they affect:

> Many factors determine the soil texture that develops at a site.

- ❖ water relations, such as infiltration, percolation, water storage, and aeration
- ❖ tillage properties, such as consistency (stickiness, hardness), power requirements
- ❖ erosion processes, such as **crusting** potential, **erodibility**
- ❖ chemical relations, such as nutrient storage, buffering capacity, surface activity (adsorption potential)
- ❖ solute transport and leaching potential

The important role of soil texture in soil use and management must be viewed in light of how soil properties change both with depth within a soil profile and in space across landscapes. These differences may be attributable to geologic processes that deposited materials of different textures in layers in the soil (such as in the floodplains of rivers and streams), or because of pedologic processes that move soil particles within the profile and across the landscape.

Distribution and Depth

There are many changes in soil profile properties over time as a result of soil development. One of the most prominent is the movement of clay-sized particles downward within the profile. Over many thousands of years, this process leads to the development of Bt horizons in the subsoil of many mature soils (Figure 4-9a). The presence of such a high clay layer, sometimes less than 15 to 30 cm (6 to 12 in) from the soil surface, can profoundly influence the capabilities of a soil. For example, plowing a soil with a shallow Bt could incorporate high-clay, low-organic-matter soil into the plow layer. The Bt may also be denser than the overlying soil, with fewer large pores. This may impede root growth or slow percolation of water into the soil profile. This wet soil layer may pose problems for plant roots, buried pipes, or below-ground structures (basements, foundations, etc.).

> Soil texture typically becomes more silty or clayey with depth.

IMPROVING CLAY SOIL

A common problem in many places is soil dominated by clay-sized particles. When wet, these soils are extremely sticky and susceptible to compaction. When dry, they can become hard and cloddy. Either way, these soils can be difficult to manage. The most effective way to improve the condition of clay soils is to regularly add organic matter, such as incorporating plant residues or adding manure, compost, or other organic amendments. The organic matter and the organisms that thrive from it will promote greater structural stability and improve water and air movement.

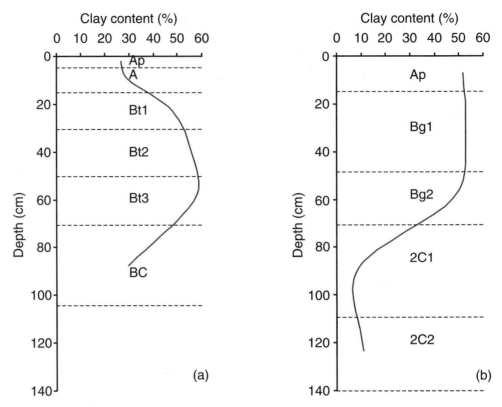

FIGURE 4-9 Changes in soil texture with depth of (a) a typical mature soil, and (b) a soil with a pronounced textural discontinuity.

It is best to add the organic matter after the growing season to avoid competition between plants and microorganisms for available nutrients. Adding sand does not have much of an effect because it does not readily encourage biologic activity or increase aggregation.

Many soils, though, do not demonstrate such a clay "bulge" in their profile. This may be attributable to a lack of weathering and soil development (for example, in dry climates or on steep slopes), or may be because of a change in parent material with depth. A common example of such a textural discontinuity would be in soils formed from river, lake, or ocean sediments (Figure 4-9b). As the water conditions changed, different materials may have been deposited. Such soils can have one, two, or more significant textural changes within the profile, with sand over clay, clay over sand, or any other possible combination.

| Textural discontinuities often develop because of sedimentation by wind or water. |

Distribution and Landscapes

Just as water redistributes soil particles within the profile, water moving over and through the soil can produce differences in soil texture along hillslopes and across landscapes. There are no general rules that can be applied to all landscapes, but patterns do exist and can be recognized and explained. In landscapes with mature soils that have experienced more recent erosion, the topsoil of the eroded shoulder positions may have the highest clay content because the original A horizon has been eroded and the top of the Bt horizon is being incorporated into the plow layer (Figure 4-10).

| Erosion and tillage can affect texture. |

The floodplains of rivers can also exhibit distinct patterns in soil surface texture as a result of differences in the depositional environment across

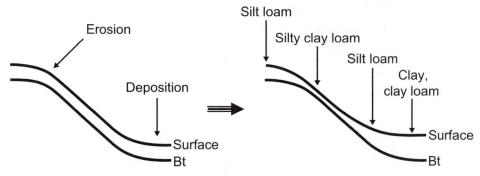

FIGURE 4-10 Changes in surface soil texture along a hill slope where erosion and subsequent tillage has incorporated subsoil material in the plow layer.

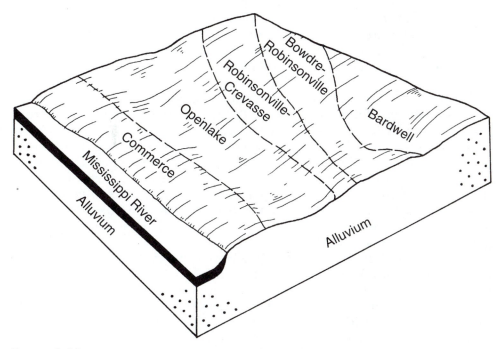

FIGURE 4-11 A typical soil sequence in a floodplain, with sandy materials deposited on the current (Commerce) and former (Robinsonville) levees. The Openlake, Bowdre, and Bardwell soils are finer in texture because their lower position allows water to stagnate and clays to deposit.

the floodplain (Figure 4-11). The coarsest textures tend to be directly adjacent to the river channel. As the velocity of flood waters is reduced upon overtopping the bank, the largest materials settle out first, just as is seen in a sedimentation column when determining texture in the lab. Slack water areas tend to have higher clay contents because the water stagnates and the finer sediments can be deposited.

FOCUS ON . . . SOIL TEXTURE

1. From the textural triangle (Figure 4-3), what is the minimum percentage of clay necessary for a soil to be in a clayey textural class (sandy clay or silty clay or clay)?

2. What is a loam?

3. What is texture-by-feel?

4. What is the minimum size particle soil sieving is good for?

5. What is the principle behind Stoke's law?

6. How does erosion and deposition affect soil texture?

SUMMARY

Soils are separated into three classes: sand, silt, and clay. The combination of these classes in varying proportion gives each soil its unique texture. A soil textural triangle is used to divide soil textures into twelve different soil textural classes: coarse-textured soils are sandy, loamy sand, or sandy loam; medium-textured soils are sandy clay loams, loams, silt loams, and silt; fine-textured soils are sandy clays, clay loams, silty clay loams, silty clays, and clays. The differences in the physical and chemical properties of these different textural classes greatly affect how they can be used.

Texture is determined using three main methods: by feel, by sieving, and by sedimentation of the soil particles. Texture-by-feel is a field method that evaluates the stickiness and grittiness of a soil as it is molded into ribbons and strings by hand. Sieving uses sieves of different aperture to separate soil particles and can be performed either dry or wet. Sedimentation methods are based on Stoke's law, which describes how rapidly particles will fall through a viscous solution like water; sand particles will fall fastest and clay particles will fall slowest. Either a hydrometer or dried sample of soil solution is used to determine how much soil remains in solution over a set period, and this is used to determine the texture.

Texture is not uniform in soils. Clay content tends to increase with soil depth and the Bt horizon is a diagnostic horizon in some soils, indicating the accumulation of clay. When hillslopes erode it tends to change the texture of the remaining material. In floodplains the migration of water tends to deposit clay particles farthest from the main river channel, and this can lead to bands of different-textured soils adjacent to a river.

In Chapter 5 you will explore how the different-sized particles that make up texture are organized to create soil structure.

END OF CHAPTER QUESTIONS

1. The largest clay particle is 0.002 mm in diameter, the largest silt particle is 0.05 mm (25 times larger than the largest clay particle), and the largest sand particle is 2.0 mm (1000 times larger than the largest clay particle). Name three familiar items with the same relative size differences as sand, silt, and clay particles.

2. Why is it difficult to change the texture of soil in the field?

3. Why do you think that it would be important to know about the texture of a soil?

4. Which type of soil would be better for growing plants, sandy loam or clay loam? Explain your answer.

5. Why is clay sticky?

FURTHER READING

The methodology of textural analysis is explained in:

Soil Survey Division Staff. 1993. *Soil survey manual.* USDA-NRCS Agricultural Handbook 18. Washington, DC: U.S. Government Printing Office.

The underlying physics and assumptions of Stokes's law are explained in:

Hillel, D. 1998. *Environmental soil physics.* San Diego, CA: Academic Press, pp. 66–69.

REFERENCE

Thien, S. J. 1979. A flow diagram for teaching texture-by-feel analysis. *Journal of Agronomic Education* 8: 54–55.

Chapter 5

SOIL SOLIDS: STRUCTURE, AGGREGATION, AND POROSITY

"We know more about the movement of celestial bodies than about the soil underfoot."

Leonardo da Vinci

OVERVIEW

You have begun to examine how the individual soil particles in the mineral fraction of the soil influence the bulk soil properties. However, it is rare to have a soil or soil horizon in which the individual particles are loose. Normally, the sand, silt, and clay particles are bound together into secondary particles or structures. The properties of these secondary particles similarly influence the way the soil solids interact with the water and air in the soil, and, therefore, how a soil is used and managed.

In this chapter you continue to examine the physical properties of soil. You will examine the architecture of the soil and answer several key questions about it:

✔ How do sand, silt, and clay particles bind together to form secondary soil particles (peds)?

✔ How do the secondary particles influence interactions with water, plants, and soil organisms?

✔ What natural and artificial processes influence the formation or destruction of stable soil aggregates?

✔ What is the role of pores in the soil?

✔ How does soil structure influence other soil characteristics that determine the use and management of soils, such as soil stability, permeability, and soil strength?

OBJECTIVES

After reading this chapter, you should be able to:

✔ Name and recognize different structural types.

✔ Describe the effects of structure and aggregation and explain the role of soil pores in water movement, root penetration, gas exchange, and soil stability.

✔ Calculate bulk density, porosity, and other related quantities using simple field and laboratory measurements.

KEY TERMS

bulk density
flocculate
macropores

mesopores
micropores

pore size distribution
porosity

SOIL STRUCTURE

Soil structure describes the combination and arrangement of primary soil particles (sand, silt, and clay) into secondary particles, or units, called aggregates or peds. The spaces between the peds are the pores that conduct and store water and allow air exchange between the soil and atmosphere. The characteristics of these pores are determined by the size and shape of the peds. Therefore, it is useful to be able to describe soil structure in standard terms.

> Peds reflect the organization of primary soil particles.

Describing Soil Structure

Soil aggregates are described in terms of their shape, size, and stability. The most influential of these attributes is shape because it strongly influences the pore structure in the soil.

Structure Type (Shape)

Soil scientists recognize six basic types of soil structure: granular, platey, angular blocky, subangular blocky, prismatic, and columnar (Figure 5-1).

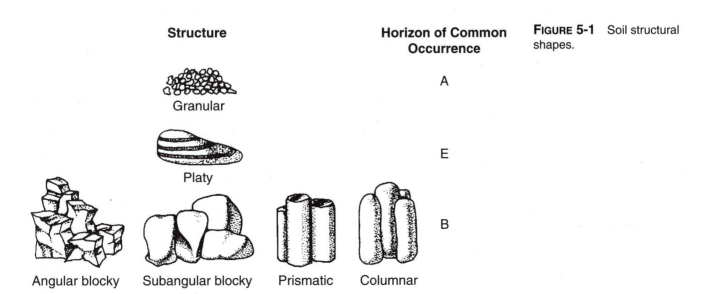

Structure	Horizon of Common Occurrence
Granular	A
Platy	E
Angular blocky Subangular blocky Prismatic Columnar	B

FIGURE 5-1 Soil structural shapes.

Granular. Granular structure refers to relatively small, rounded, and porous aggregates. Aggregates average about 2 to 5 mm in size. They form mainly in soil horizons that have abundant organic matter and relatively dense concentrations of roots spread through the soil. As a result, this structural type is most commonly associated with the A horizon.

Platey. Platey structure refers to aggregates that are proportionally wider than they are tall. The flat shape of these aggregates can form naturally in the soil. Platey structure can result from layering in parent material, such as depositional layers or rock stratification. Platey structure can also form in response to tillage, traffic, or other management that compresses and compacts the soil. Platey structure can develop anywhere in the soil profile, but it is most commonly associated with E horizons.

> Blocky structure occurs where clay concentrations tend to be greater.

Blocky. The two blocky structure types, angular blocky and subangular blocky, are aggregates that are block-like in shape, meaning that they are roughly the same size in all dimensions. The difference between the two is that angular blocky peds have relatively sharp or well-defined edges, while subangular blocky peds are more rounded with less-distinct edges. Generally larger than granular structure, these aggregates form where organic matter and roots are less abundant, but where clay content is higher. Typically, blocky structure is found in the B horizon, with angular blocky peds associated with higher clay contents.

> A typical sequence of structure in the soil profile is: granular, blocky, prismatic or columnar.

Prismatic and Columnar. Both prismatic and columnar structure types refer to aggregates that are taller than they are wide. Prismatic aggregates have relatively flat tops, while the tops of columnar aggregates are rounded. These relatively large aggregates are not commonly distributed in the soil, and tend to be found deeper in the soil profile where roots and organic matter are less abundant. Prismatic structure is most commonly associated with dense, impermeable fragipan horizons, although it is not restricted to fragipans. Columnar structure is mainly found in more arid climates where soluble salts can accumulate in the soil.

Structureless. Not all soil horizons have recognizable aggregate structure. Soils with extremely sandy texture do not have enough clay or other cementing agents to bind the individual particles together. In this case the soil is described as *single grain.* However, finer-textured soils can also be structureless. In younger soils, particularly in the subsoil, the soil may not have had enough time to develop stable aggregates. While the soil will break apart, there is no regularity to it, and that soil is typically referred to as being massive. Structureless soil is restricted to the C horizon—by definition, if structure was present it would be a B horizon.

Structure Size (Class) and Stability (Grade)

Structure can also be described in terms of its size and stability. Aggregate size is described in terms of five different classes: very fine, fine, medium, coarse, and very coarse. The size limits to these classes vary among the different structure types (Table 5-1). Granular structure is typically small, while prismatic and columnar structure is much larger.

> Strong structure refers to aggregates that are readily apparent in the profile and do not fall apart when removed.

The *grade* of structure refers to the stability of the aggregates. Because of the importance of the pores between the peds, the strength of the structure will also indicate the stability of the pores. Structure classified as strong refers to aggregates that are readily apparent while in the profile and do not fall apart when removed from the profile and handled. Structure of moderate grade is apparent in the profile, but some aggregates fall

TABLE 5-1 Relationships between structure shape and size.

Aggregate Size Class	Shape of Structure			
	Platey[a]	Prismatic and Columnar	Blocky	Granular
	mm	mm	mm	mm
Very fine	< 1	< 10	< 5	< 1
Fine	1–2	10–20	5–10	1–2
Medium	2–5	20–50	10–20	2–5
Coarse	5–10	50–100	20–50	5–10
Very coarse	> 10	> 100	> 50	> 10

[a]In describing plates, "thin" is used instead of "fine" and "thick" instead of "coarse."

apart when removed from the profile and handled. Structure is considered to have weak grade when peds are barely apparent in the profile and readily fall apart when removed from the profile and handled. These distinctions in the stability of structural units help indicate how soil aggregates will respond to disturbance, with weak structure disintegrating under even gentle pressure, but strong structure being able to withstand greater degrees of disturbance.

FOCUS ON . . . SOIL STRUCTURE

1. What does soil structure refer to?
2. How can you describe soil structure?
3. How many basic shapes of soil structure are there?
4. What is a structureless soil?
5. What does grade of structure refer to?

PROCESSES THAT INFLUENCE STRUCTURAL STABILITY

Formation of soil structure is encouraged by the presence of clay, organic matter, and iron and aluminum compounds, and by biological activity such as plant root growth, microbial activity, and organic matter decomposition. The small, flat shape and charge properties of clay particles cause the individual particles to **flocculate** together into small stacks or clumps. (The negatively charged clay particles are linked by positively charged ions between the individual layers of clay, or to positively charged edges of other clay particles.) Large groups of clay particles can form bridges between larger soil particles to create relatively stable soil aggregates. Similarly, iron and aluminum oxide minerals and organic compounds, alone or in conjunction with clay particles, bind sand and silt particles together.

Many factors, such as organic matter, clay content, and root growth, influence soil structure.

The growth of thread-like roots and other soil organisms (fungi, bacteria) can physically bind soil particles together as they extend around particles and through the soil. These organisms also exude gummy organic substances that can cause particles to stick together. Similarly, the decomposition of organic matter in the soil also produces sticky substances that promote aggregation.

Any process that causes individual soil particles to be pressed together, such as root pressure, animal burrowing, wetting and drying cycles, and freeze and thaw cycles, further promotes aggregation. The pressure exerted on adjacent soil particles as roots and animals move through the soil encourages closer contact between soil particles, enabling the clay particles and other compounds that promote aggregation to cohere.

> Physical processes such as wetting and drying also shape soil structure.

Other processes readily destroy soil structure. The activity of burrowing animals can also produce significant mixing within the soil profile. More commonly, human activity such as tillage, digging, or other soil disturbance can break apart stable aggregates. Traffic on the soil can lead to compaction that alters the soil structure. To help maintain soil structure in managed soils, amendments can be added, such as gypsum (calcium sulfate) or organic polymers, which actively stabilize soil aggregates, or organic materials (compost, manure), which encourage biological activity.

FOCUS ON . . . PROCESSES THAT AFFECT STRUCTURAL STABILITY

1. What kinds of things help develop soil structure?
2. How do living organisms help develop structure?
3. What kinds of activities destroy structure?

WHY IS STRUCTURE IMPORTANT?

Soil structure is important for many reasons. In general, more stable structure means more rigid soil, which can support more weight. The most important effect of structure, though, is its influence on **porosity**—the spaces between the soil solids. Quite simply, water and air will flow preferentially around peds rather than through them. The size, shape, and integrity of the aggregates influences the size, shape, and integrity of the soil pores. These pores are the channels through which water and air move through the soil. Thus, structure influences the infiltration of water into the soil, the percolation of water through the soil, and the exchange of air between the soil and the atmosphere. The pores between the peds are also the preferred pathways for root elongation. A soil with strong structure with ample pores between peds (such as granular or blocky types) will have greater permeability to both air and water. It will have higher rates of infiltration and percolation, and have greater aeration. It should also have a more extensive distribution of roots throughout the soil. Often, positive aspects of well-developed soil structure can overcome the negative aspects of clayey texture. Similarly, a soil with a sandy texture that is massive or with weak structure may be just as permeable as a well-structured clay soil.

> The size, shape, and integrity of the aggregates influence the size, shape, and integrity of the soil pores.

Another effect of structure is that a soil with strong structure erodes less than a soil with weaker structure. Because the individual soil particles are bound together they are less likely to be detached and transported away by rainfall and running water. Overall, a soil with stronger structure is easier to manage, easier to dig or till, and more beneficial for plant growth.

Focus on . . . Why Structure Is Important

1. What are two important reasons for having good soil structure?
2. How can good structure overcome the influence of clayey soil?
3. Why do well-structured soils resist erosion?

Soil Porosity and Pore Size Distribution

Porosity

Though it is often easier to describe the solids within the soil, the voids between the solids (which represent from 40 to 60 percent of the volume of most soils) are equally, if not more, important in determining the uses and limitations of many soils. It is through the soil pores that water and air move. Water is also stored for later plant use in the pores. Roots extend through the soil pores, and other soil organisms travel through the pores. Promoting the maintenance and development of pores enhances these processes. Conversely, destruction of pores will diminish them.

> Water and air move through the soil pores.

The total amount of pore space, or porosity, can be defined as the percentage of the total soil volume not occupied by solid particles (mineral or organic). Soil porosity conveys information about ease of root penetration, water-holding capacity, and soil strength. Porosity, though, does not convey any information about the sizes of the many pores within the soil.

Pore Size

The total pore space in the soil is actually made up of pores of many different sizes. As with the primary soil particles, you can divide these pores into different pore size classes (Table 5-2): macropores, mesopores, micropores, ultramicropores (or nanopores), and cryptopores. These distinctions are important because the different-sized pores perform different functions within the soil.

> Soil pores come in all shapes and sizes.

TABLE 5-2 Summary of properties of different pore sizes.

Pore Size Class	Diameter Range (μm)	Dominant Functions
Macropores	> 75	Air movement, water drainage, microbial habitat
Mesopores	30 to 75	Available water storage, microbial habitat
Micropores	5 to 30	Unavailable water storage, microbial habitat
Ultramicropores (Nanopores)	0.1 to 5.0	Unavailable water storage, persistent C storage
Cryptopores	< 0.1	Unavailable water storage, persistent C storage

Macropores

Macropores are the largest pores (> 75 μm). They accommodate most of the water drainage from the soil and they facilitate air movement and exchange with the atmosphere. These relatively large pores also are commonly the habitat for animals living in the soil. Because macropores drain quickly they are most often filled with air, except immediately after a precipitation event, or when water tables are high. In well-aggregated soils, macropores are formed by the spaces between individual peds. As a result, the amount and stability of macropores in a soil is most affected by soil structure.

> Macropores drain quickly and are often filled with air.

Mesopores

Mesopores are medium-sized pores (30–75 μm). These pores are large enough for fungi and root hairs to penetrate. Compared to macropores, water moves more slowly through the mesopores. These pores hold water for longer periods of time between precipitation events, slowly releasing water for plant uptake, evaporation, or drainage. As such, the mesopores function mainly for short-term storage of water that is mostly available to plants.

Micropores and Smaller

Micropores are the smallest pores (5–30 μm) that typically hold soil organisms. These pores hold water tightly, such that much of this water is unavailable to plants. The water in micropores is mainly seen as long-term water storage. The largest micropores are large enough for bacteria to grow and move. However, the smaller micropores (approximately < 5 μm) are small enough to exclude most if not all soil organisms. Micropores are formed by the spaces between individual sand, silt, and clay particles within peds. As a result, the amount and stability of micropores in a soil is most affected by soil texture. Ultramicropores and cryptopores have such small diameters that even microbes are excluded. But this does not mean that the pores can't hold water, which they do. One of the reasons that soil carbon and organic contaminants in soil (such as pesticides) are thought to persist in soils is because they have diffused into these very small pores that protect them from microbial degradation.

> Micropores are involved in long-term water storage.

Pore Size Distribution

Pores of these different size ranges occur in varying amounts in different soils. This **pore size distribution,** or relative amounts of pores of different sizes, has an important influence on the ability of a soil to transmit water, exchange air with the atmosphere, and accommodate roots and other soil organisms. Most soils have a wide range of pore sizes, although differences in soil texture, soil structure, soil depth, and management history can lead to differences in both the average pore size and the range in pore sizes.

Bulk Density

Porosity is the percentage of the total soil volume not occupied by solid particles, and you can also consider the inverse of this measure: or the amount of solids in a given volume of soil. This is known as the **bulk density.** It is important to note that because bulk density and porosity are inversely related it is an absolute rule that as porosity increases, bulk density decreases—or as bulk density increases, porosity decreases. This relationship with porosity is why bulk density is such a useful piece of information about soil. It is also easy to measure directly (unlike porosity), so it is commonly used in many soil investigations. Like porosity, bulk density is related to the ability of water, air, and roots to move into and through

> The ratio of soil solids to soil volume is called bulk density.

the soil. This in turn influences water infiltration and percolation, aeration, and rooting depth and distribution.

Factors Affecting Porosity and Bulk Density

There are many factors affecting the porosity, pore size distribution, and bulk density of a soil, including soil texture, organic matter content, soil depth, and management practices. Each of these is important because of its influence on the formation and maintenance of stable soil aggregates, and therefore the pore structure between and within these aggregates.

Soil Texture

In general, sandy soils tend to be less porous than silty or clayey soils (Table 5-3). Sandy soils have much lower porosity and much higher bulk density values compared to those of clayey soils. These differences arise mainly because of the influence of clay content on soil structure. In a sandy soil there is less aggregation and closer packing of particles. Additionally, in a clayey soil, there are many micropores within the aggregates in addition to the larger pores found between peds (Figure 5-2).

However, despite the lower porosity and higher bulk density associated with sandy soils, it is important to note that the average size of individual pores is greater in sandy soils. Conversely, the pore space in finer-textured soils is dominated by the preponderance of micropores within the peds. If

> Clay soils have smaller pores but greater porosity than sandy soils.

TABLE 5-3 The influence of soil texture on porosity and bulk density.

Texture Class	Porosity	Bulk Density
Sand	32–42%	1.55–1.80 g cm^{-3}
Loam	43–49%	1.35–1.50 g cm^{-3}
Clay	51–55%	1.20–1.30 g cm^{-3}

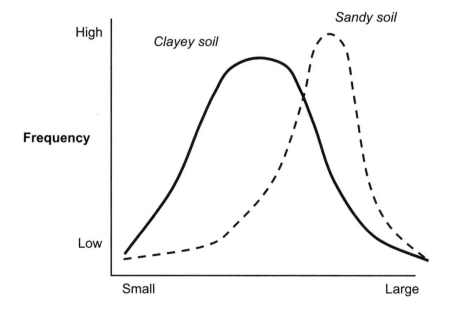

FIGURE 5-2 Pore size distribution of soils of differing soil texture. Sandy soils have relatively larger pores, but pore size is more narrowly distributed than in clayey soils.

Relative pore size

you consider the pore size distribution of soils of differing texture you see that in sandy soils, most pores are medium-sized, but in silty and clayey soils, pore size is much more widely distributed.

Organic Matter Content

Soil organic matter is important in the formation and stabilization of soil structure. Increasing amounts of organic matter in the soil promote more stable structure and lead to increased porosity and decreased bulk density. Additionally, the particle density of organic matter is approximately half that of the mineral material. The relationship between organic matter and structural development is, in part, why porosity and bulk density are influenced by management practices and soil depth.

Soil Depth

> Porosity decreases and bulk density increases with depth in soil.

In general, porosity decreases and bulk density increases with depth in the soil (Figure 5-3). This change can be attributed to several factors, including changes in (1) organic matter content, (2) the distribution of soil organisms, and (3) aggregation. First, organic matter content in the topsoil is generally greater, which promotes structural development and stability, and decreases particle density. The action of plant roots and burrowing soil fauna such as earthworms, mainly in the topsoil, can create macropores that increase porosity. As a result of the higher-organic-matter contents and denser root networks in the topsoil, soil structure is mainly granular, which has abundant interaggregate pore space. The coarser blocky structure associated with the subsoil naturally has less interaggregate pore space. Also, the movement of clay particles into the subsoil, where the translocated clays accumulate as linings of soil pores, can reduce porosity and increase bulk density (Figure 5-3a).

However, the trend of increasing bulk density with increasing depth within the soil profile is not always found. In intensively weathered soils, such as those found in the southeastern United States, the development of sandy surface horizons (A and E) over clayey subsoils can produce

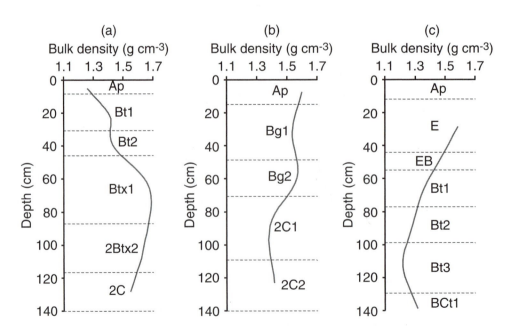

FIGURE 5-3   Changes in soil bulk density with depth of (a) a typical mature soil, (b) a soil with a pronounced textural discontinuity, and (c) an intensively weathered soil with a pronounced textural profile.

strongly contrasting textural profiles in which the effects of texture far out-weigh the other factors that influence bulk density (organic matter content, overburden pressures; Figure 5-3c). This creates soil profiles with higher bulk densities at the sandy surface layers, and lower bulk densities in the clayey subsoil layers. Other exceptions can also be seen. The presence of textural discontinuities in profiles with two parent materials of different textures (e.g., clayey over sandy) can produce soils with higher bulk densities in the upper (clayey) horizons (Figure 5-3b). This occurs because the wetter clays are more easily compacted under vehicle traffic than are the sandy layers in the subsoil. These depth relationships are important because soil layers with high bulk densities and low porosities can restrict water movement and inhibit root extension.

> Sudden changes in soil texture with depth in soil can create management problems.

FOCUS ON . . . SOIL POROSITY AND PORE SIZE DISTRIBUTION

1. What is porosity and why is it important?
2. What are the characteristics of the three sizes of pores in the soil?
3. What happens to bulk density as porosity increases?
4. Name three factors that will affect soil porosity.
5. What factors will affect soil porosity changes with depth in soil?

HOW ARE BULK DENSITY AND POROSITY DETERMINED?

You have seen the importance of porosity and bulk density in assessing soil use and management. Simple relationships between soil mass and volume provide quick and easy methods to quantify soil structural relationships.

Mass-Volume Relationships

If you collect an intact portion of soil, it has both mass and volume. You know, however, that within that volume of soil, there exists both solids and pores (voids; Figure 5-4).

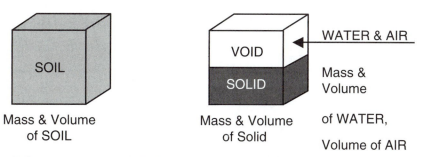

FIGURE 5-4 Components of bulk density. Bulk density measurements reflect both the mass and void space in a soil. The void space can be filled with water, so *all* bulk density measurements are made on oven-dry soil, so that the weight of water in void space can be eliminated.

Particle Density

The mass per unit volume of the soil particles is *particle density*, often abbreviated as *PD*. How much would a 1 cm^3 particle weigh?

$$\text{Particle density} = \frac{\text{Mass of SOLID}}{\text{Volume of SOLID}}$$

mean = 2.6–2.7 g cm^{-3}

Particle density increases with the presence of iron oxides and decreases with the presence of organic matter.

Bulk Density

Bulk density (often abbreviated as BD) is the mass of dry soil per unit of bulk volume, including the air space. How much would 1 cm^3 of soil weigh?

$$\text{Bulk density} = \frac{\text{Mass of SOLID}}{\text{Volume of SOIL}}$$

Bulk density is *always* less than particle density. The mean for most soils is typically 1.3–1.35 g cm^{-3}. Bulk density is affected by the structure of the soil, including the degree of compaction. Even in extremely compacted soil, though, BD < PD.

Porosity

Porosity is the volume percentage of the total soil bulk not occupied by solid particles. How much of the soil is void?

$$\% \text{ porosity} = 100 \left(1 - \frac{\text{Bulk density (BD)}}{\text{Particle density (PD)}} \right)$$

The range of porosity is typically 30 to 60 percent. Coarse-textured soils tend to be less porous than fine-textured soils. *Qualitatively,* how are porosity and bulk density related? Porosity increases as bulk density decreases.

Data Collection and Analysis

Making measurements of soil porosity and bulk density is relatively simple and straightforward because the calculations require only a few easily determined masses and volumes. The procedure begins by collecting a known volume of soil. This is easily done using a cylindrical can, such as an empty soup can. You can collect a sample of known volume by pounding the can into soil, then excavating it. Once back in the laboratory, the moist sample is weighed, then allowed to dry overnight in an oven. The moisture in the sample evaporates, leaving only the soil solids. The sample is weighed a second time. The water content is calculated as:

$$\text{Gravimetric water content} = \frac{(\text{Mass of moist soil} - \text{Mass of oven dry soil})}{\text{Mass of oven dry soil}}$$

The volume of the oven dry soil is calculated (assuming you have a nice cylinder) as:

$$\text{Volume of soil} = (\text{Height})(\pi r^2)$$

You now have all the information you need to calculate bulk density and total porosity: soil volume and soil dry weight.

FOCUS ON ... SOIL STRUCTURE AND DENSITY

1. Why might it be important to describe the size and grade of soil structure in addition to aggregate shape?

2. Curious about the density of the soil in your garden, you dig a small hole (about 10 cm deep and 20 cm wide), placing the excavated soil in a small container. You then line the hole with plastic and fill it with water. You subsequently weigh the soil when moist and after allowing it to dry. You have collected the following information: mass of moist soil, 4480 g; mass of dry soil, 3670 g; volume of water needed to fill hole, 3810 cm^3. What is the bulk density of your garden topsoil? What is the percent pore space (total porosity) of your garden topsoil?

3. Why does a sandy soil, in general, allow for greater movement of air and water than finer-textured soils?

4. How does agriculture (or most soil disturbance) affect the porosity and pore size of a soil? Why?

SOIL STRUCTURAL MANAGEMENT

Common management practices, such as tillage, traffic, and land use/land cover, often alter the pore structure of the soil, which affects porosity, pore size distribution, and bulk density.

Tillage

Plowing, disking, and other cultivation methods are employed to create more favorable seedbed conditions. The short-term effect is to increase the soil porosity and lower the bulk density in the plow layer by loosening and "fluffing up" the soil. However, these disturbances destroy soil structure and reduce aggregate stability. Over the long term, tillage will decrease porosity and increase bulk density (Table 5-4). It is important to note that not only is the total pore space reduced by long-term disturbance, but this reduction in porosity is mainly attributed to the loss of the vital macropores that occur between the peds. Part of the reduction is due to the accelerated decomposition of organic matter in cropped soil (Table 5-5).

> Soil disturbances usually destroy soil structure and reduce aggregate stability.

TABLE 5-4 Traffic compacts soil; it decreases porosity and increases density.

Surface Soil	Bulk Density	
	Cropped	Uncropped
	- - - - - - - - - - g cm^{-3} - - - - - - - - - -	
Hagerstown loam	1.25	1.07
Marshall silt loam	1.13	0.93
Sutherland clay	1.28	0.98

TABLE 5-5 Uncropped soils tend to have more organic matter and better structure than cropped soils.

Land Use	Bulk Density
	g cm^{-3}
Forest	1.0–1.2
Pasture	1.2–1.3
Cultivated	≥ 1.3

SUMMARY

It is rare to have a soil in which the individual particles are loose. Most soils are structured, meaning that the individual particles are arranged to provide a mixture of solid space and void space. It is through this void space that air and water move in soil. Soil scientists recognize six types of soil structure: granular, platey, angular blocky, subangular blocky, prismatic, and columnar. The structure is held together by a combination of inorganic cementing agents, biologically produced gums, and roots. If the soil structure tends to withstand manipulation it is referred to as *strong*. If it falls apart easily it is called *weak*. Tillage, organic matter decomposition, and animal burrowing are all processes that can destroy structure.

The void space in soil is called porosity. Three classes are recognized: macropores, mesopores, and micropores. But each class is not discrete and there is a whole range of pore sizes in soil. As texture becomes finer the pores tend to decrease in size but become more numerous and occupy a greater fraction of the soil. The opposite is true as soils become coarser. Bulk density refers to the mass of soil relative to the volume of soil. Bulk density usually increases with depth in soil, but not always. As porosity increases, bulk density decreases and vice versa. Porosity, particle density, and bulk density are three simple and common measurements made of soils. One of the effects of tillage is to increase bulk density, which is an indication that soil structure is being affected.

END OF CHAPTER QUESTIONS

1. Define soil structure.

2. What are the six major types or subtypes (shapes) of soil structure and the distinguishing features of each?

3. Why is particle density usually not measured but assumed to be 2.65 g/cm^3 (= 2.65 Mg/m^3)?

4. What is the difference between particle density and bulk density?

5. How does soil texture influence soil structure?

6. What four factors contribute to increased bulk densities deeper in the soil profile?

7. Does long-term tillage increase or decrease soil bulk density? Why? (Note: There are at least *three* reasons.)

8. Why is root growth inhibited by soil with a high bulk density?

9. Which type of soil would be better for growing plants, well structured or poorly structured? Explain your answer.

10. What is the difference between porosity (pore space) and pore size?

11. What is the general difference between macropores and micropores (e.g., size, location)?

12. What are the roles of macropores and micropores in the soil?

13. Which are more stable, smaller aggregates or larger aggregates? Which are better, smaller or larger aggregates?

14. What physical-chemical processes contribute to the formation of stable soil aggregates?

15. What is flocculation? Why is it a problem if the clay is dispersed (*not* flocculated)?

16. How do wet-dry and freeze-thaw cycles affect clay domains?

17. What biological processes contribute to the formation of stable soil aggregates? How does organic matter promote better soil structure?

18. How does tillage promote better soil physical properties in the short-term? How does it promote poor soil physical properties in the long-term?

19. Which is better for plant growth: A soil with a wide range of soil pore sizes, or one where all the pores are about the same size? Explain your answer.

FURTHER READING

The following articles describe how soil biology affects soil structure and the development of soil structure:

Tisdall, J. M. 1994. Possible role of soil microorganisms in aggregation in soils. *Plant and Soil* 159: 115–121.

Tisdall, J. M., and J. M. Oades. 1982. Organic matter and water stable aggregates in soils. *Journal of Soil Science* 33: 141–163.

REFERENCE

Soil Survey Division Staff. 1993. *Soil survey manual.* USDA-NRCS Agricultural Handbook 18. Washington, DC: U.S. Government Printing Office.

Chapter 6

SOIL WATER, TEMPERATURE, AND GAS RELATIONS

"Life is animated water."
Vladimir Vernadsky, 1929

OVERVIEW

Water makes life on Earth possible. When you think of water, you think of life. Water is no less important to the way a soil functions. Water is involved in weathering and dissolving soil minerals. Water regulates the climate of the soil environment. Water controls the amount of gas in a soil and the amount of oxygen that is available for plant roots and other organisms. Water controls the rate at which nutrients and other chemicals diffuse in soil. Because the amount and availability of water is crucial to the way a soil behaves, it is the focus of this chapter.

> Water is important because it affects: weathering, temperature, aeration, and diffusion.

OBJECTIVES

After reading this chapter, you should be able to:

✔ Identify key physical properties of water.

✔ Distinguish between water quantity and water availability (potential).

✔ Evaluate how water moves in soil.

✔ Determine how water affects soil temperature and gas transport.

KEY TERMS

air dry	gravitational potential	unsaturated flow
equivalent surface depth	hysteresis	vadose zone
field capacity	matric potential	volumetric water content
hygroscopic coefficient	osmotic potential	water potential
gravimetric water content	saturated	wilting point
	saturated flow	

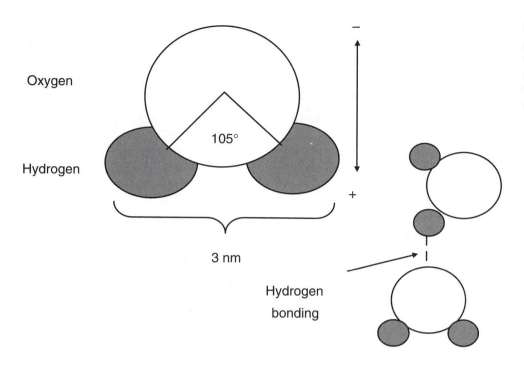

FIGURE 6-1 The water molecule. Unique properties of the water molecule include its dipole nature, which gives it positive and negative poles, and its ability to form hydrogen bonds.

THE WATER MOLECULE

Water has unique and useful chemical properties because it is a dipole. Water is a liquid at room temperature even though it has a low molecular weight and is composed of two elements, oxygen and hydrogen, which are normally gases. Water can be a solid (ice) and a liquid within a temperature range suitable for life. Water becomes more viscous and dense as the temperature declines. It reaches a maximum density at 4°C, and then proceeds to become less dense as it crystallizes. Thus, ice floats, rather than sinks, and lakes freeze from the top down. Soils in which ice freezes and thaws continuously compress and rebound, which changes the soil structure. Water has a high surface tension, which contributes to its movement by capillarity through soil pores. Water has an extremely high heat capacity, which allows it to resist changes in soil temperature.

Some of the unique properties of water are due to its molecular structure (Figure 6-1). Because the H-O-H bond is bent at an approximately 105° angle, the water molecule forms a molecular dipole; one end of the molecule has a slight if transitory negative charge while the other has an equally slight and transitory positive charge. This polarity makes water molecules mutually attractive, it allows water to adsorb readily to solid surfaces, hydrate ions and colloids, act as a nearly universal solvent, and generate intermolecular H bonds (Hillel, 1998).

> Water has unique and useful chemical properties because it is a dipole.

WATER CONTENT AND AVAILABILITY

Water Content

Soil scientists have several ways of describing water content depending on the scale at which they are working. The most common ways to describe water content in small soil samples are percent by weight (gravimetric

Water content is quantified by weight, volume, and equivalent surface area.

water content) and percent by volume (volumetric water content). At field scales, equivalent surface depth, percent available water depleted, and field capacity deficit are frequently used terms.

The gravimetric water content is the mass of water in soil as a percent of the total mass of oven-dry soil:

$$\text{Gravimetric water content } (\%_w \text{ or } \theta_w) = \frac{\text{Wet mass} - \text{Oven-dry mass}}{\text{Oven-dry mass}} \times 100$$

The volumetric water content of soil is the volume of water in soil as a percent of the total soil volume:

$$\text{Volumetric water content } (\%_v) = (\%_w)(\text{Bulk density})$$

Volumetric water content is a very useful term because it can be used to express equivalent surface depth (depth of water per depth of sample zone), the amount of water needed to replace water to a given depth in soil:

$$\text{Equivalent surface depth} = (\%_v)(\text{Sample depth})$$

In terms of irrigation, for example, an acre-inch of water is the equivalent depth term describing the amount of water required to fill one acre of soil with one inch of water.

FOCUS ON . . . WATER CONTENT

1. What are three ways of expressing water content in soil?

2. What determines how much water a soil can hold?

3. How would you express the water content of 10 g of oven-dry soil that contained 2.5 g of water?

4. If the bulk density of the soil in question 3 was 1.3 g cm^3, could you express the water content a different way?

5. If a hectare of the soil in questions 3 and 4 was going to be wet to a 2 cm depth, how much water would it take?

WATER AVAILABILITY

Descriptive terms of water availability are: saturation, field capacity, wilting point, and hygroscopic coefficient.

Although there is some water chemically bound to minerals in soils, the water of significance to us is found in the soil's pore space. The soil is **saturated** when the soil pores are completely full of water. Gravity will cause some of the water to drain from a saturated soil. When the soil is holding all the water it can against the force of gravity it is at **field capacity.** Loams and silt loams will have about 50 percent of their pore space full of water at field capacity. Water won't continue to drain at field capacity; some of the water will be taken up by plants (transpired) and some will evaporate. Without more rain or irrigation the water content of soil will steadily decrease. It becomes more difficult to remove water from soil as the water content decreases, because the remaining water is held with increasing force by the soil particles. The soil is at the **wilting point** when the water is so tightly held that plant roots can no longer extract it. Plant available water is the water held in soil pores after drainage by gravity and the amount

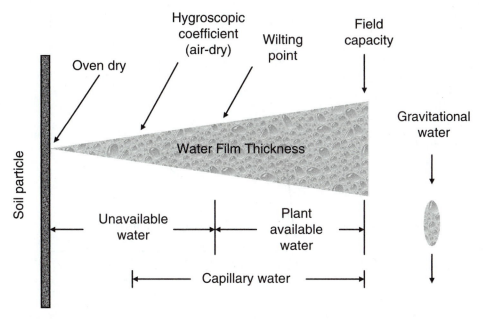

FIGURE 6-2   Benchmarks of water availability. *(Adapted from Thien and Graveel, 1997)*

remaining in soil at the wilting point. However, water at the wilting point can still be lost through evaporation. When no more water can evaporate the soil is **air dry,** or at its **hygroscopic coefficient.** Some additional water can still be removed from air-dry soils by heating them to 105°C for twenty-four hours. This leaves oven-dry soil that will not lose further water except through mineral decomposition (Figure 6-2).

The water content at these various benchmarks of water availability is used to make decisions about irrigation (Thein and Graveel, 1997). For example, irrigation is frequently started when 50 percent of the available water for plants is depleted. This can be calculated as follows:

Practical decisions about irrigation can be made by knowing water content.

$$\text{Available water depleted (\%)} =$$

$$\frac{\%_w \text{ at Field capacity} - \%_w \text{ at Existing conditions}}{\%_w \text{ at Field capacity} - \%_w \text{ at Wilting point}} \times 100$$

To figure out how much water to add, or how much is needed by rain, the field capacity deficit is calculated:

Field capacity deficit, depth =
($\%_v$ at Field capacity − $\%_v$ at Existing conditions)(depth of soil)

FOCUS ON . . . WATER AVAILABILITY

1. What term is used to describe water in soil that is freely draining?

2. Which soil has more available water, one at field capacity or one at the hygroscopic coefficient?

3. In what way is texture related to water availability and content?

4. How is *field capacity deficit* a useful term?

5. Is all capillary water available for plant growth?

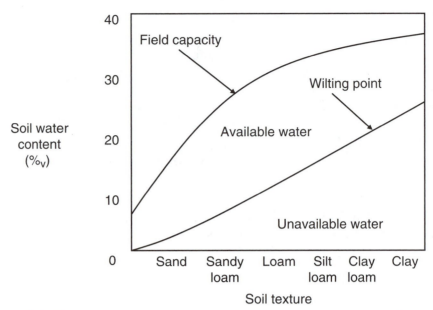

FIGURE 6-3 The effect of soil texture on available water and water content. Although the amount of water in clay soils at field capacity is greater than in coarser textured soils, less is available. In a sandy soil almost all the water may be available, but the water-holding capacity is low. Loam and silt loam soils typically have the most available water for plants.

WATER POTENTIAL

Soil water potential is a mathematical description of soil water availability.

There's a problem with describing water in soil in terms of its content or benchmarks for availability, and this is demonstrated in Figure 6-3. Soil texture affects porosity and different soil particle fractions differ in the strength with which they attract water. Consequently, water availability differs greatly from soil to soil, even though the soils may have the same volumetric or gravimetric water content.

Pure water at the soil surface is completely available. However, the water will percolate through soil because of gravity. As soon as the water enters soil it comes into contact with particles, which have an attraction for water, and soil solutes, which also have an attraction for water. These attractions act like chains on water and prevent it from moving as freely as before. The attraction of soil particles for water is called the **matric potential** (Figure 6-4). The smaller the particle, the greater its surface area, and the greater its matric potential. This is one reason why clayey soils have a greater matric potential than sandy soils. The attraction of soil solutes (salts and dissolved organic compounds) for water is called the **osmotic potential.** The greater the concentration of solutes in the soil solution, the greater the osmotic potential, and the less available the water. The combined effect of these attractions, along with the **gravitational potential,** are the three main components of what is called **water potential** (ψ).

Water is available at high water potential and unavailable at low water potential.

Soil water potential is a mathematical description of water availability with units in terms of pressure. The higher the water potential, the more available the water. Water potential ranges from 0 MPa (0 atmospheres) at saturation to approximately -3 MPa (-30 atmospheres) in air-dry soil. The key point is that water in two soils, regardless of their differences in tex-

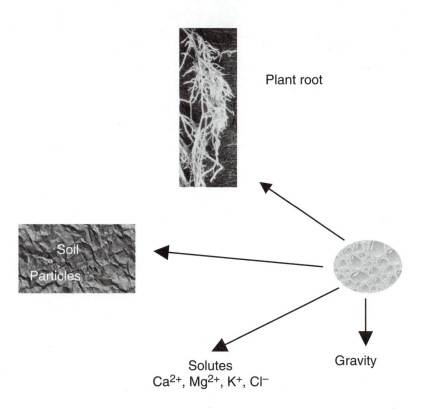

Plant root

Soil Particles

Solutes
Ca^{2+}, Mg^{2+}, K^+, Cl^-

Gravity

FIGURE 6-4 Gravity, soil solids, and soil solutes all restrict water availability and compete for water with plant roots.

ture, elevation, or salinity, will have the same availability if those soils have the same water potential.

To avoid using negative values, water potential is frequently described in terms of tension, the overall strength of attraction of soil for its water. Soils with high tension have water that is unavailable and soils with low tension have water that is very available (Figure 6-5).

The relationship of soil water potential (or tension) to water content is described by constructing water retention curves for different soils. An example is shown in Figure 6-6.

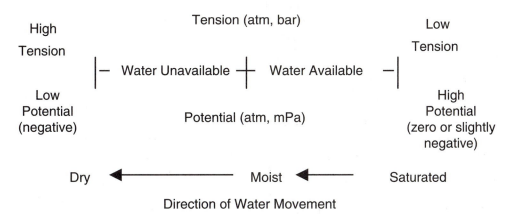

FIGURE 6-5 The relationship of water potential and tension to water movement. Water always moves from areas of high potential and high water availability to areas of low potential and low water availability.

FIGURE 6-6 Typical soil water retention curve for coarse- and fine-textured soils. For the same water potential, the water content of the two soils will be very different, although the availability of water will be exactly the same.

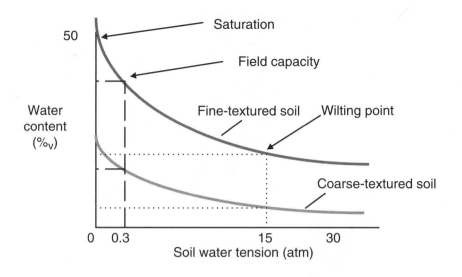

FOCUS ON . . . WATER POTENTIAL

1. What is water potential?
2. What forces in soil limit the availability of water?
3. Explain why one soil can have a higher water content but a lower water availability than another soil.
4. Which soil has higher water availability, one with a water potential of −1 atm or one with a water potential of −5 atm?
5. What does a moisture retention curve show, and how is it useful?

WATER FLOW

Liquid water moves through soil by saturated and unsaturated flow.

Water moves in soil as a vapor and as a liquid, moves from zones of high relative humidity to low relative humidity, and moves from locations of high water potential to low water potential. When water flows as a liquid it may be either through **saturated flow,** in which pores are completely full of water, or **unsaturated flow,** in which water moves by capillary action.

Saturated Flow

Saturated flow is the type of flow that you observe after a sudden rain or when a field or furrow is irrigated. The water potential of the overlying water is much higher than that in the soil, so the water will flow into the soil through large and small pores. Water flow through large pores is quite rapid, 6–15 cm/hour or more. Water flow through small pores is much slower (Figure 6-7). If you think of the flow in terms of infiltration, then it is understandable why sandy soils with large pores have much higher infiltration rates than loamy or clayey soils. Saturated flow is dominated by gravity, so water movement is primarily vertical.

Darcy's law is used to predict water flow through saturated soils:

$$q = K_{sat} \, A\Delta H/L$$

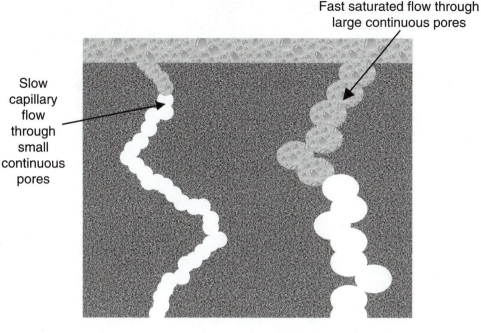

Slow capillary flow through small continuous pores

Fast saturated flow through large continuous pores

FIGURE 6-7 Saturated flow is more rapid through large pores than small pores.

where

q = volume (cm^3) per unit time

K_{sat} = saturated hydraulic conductivity (cm/sec or cm/hr)

A = cross-sectional area (cm^2)

ΔH = hydraulic head or potential causing flow (cm)

L = length of the flow path (cm)

In practice, soil scientists are much less interested in the value of q than they are in the value of K_{sat} for the simple reason that if rain or irrigation exceeds a soil's saturated hydraulic conductivity, then runoff or ponding will occur. Consequently, Darcy's law is rearranged to:

$$K_{sat} = (q/A)*(L/\Delta H)$$

which means the K_{sat} is directly proportional to the flow of water into soil when the flow length and hydraulic head are kept constant. One method of measuring K_{sat} in the field is by use of the double-ring infiltrometer (Figure 6-8). Saturated hydraulic conductivities vary greatly from soil to soil and help to determine the types of uses to which those soils can be put.

FIGURE 6-8
Measuring saturated hydraulic conductivity with a double-ring infiltrometer.*(Photograph courtesy of R. Barnhisel)*

MEASURING K_{SAT} WITH A DOUBLE-RING INFILTROMETER

The double-ring infiltrometer is a simple device used to measure saturated hydraulic conductivity in soils. It consists of two concentric rings inserted into the soil and filled with water (Figure 6-8). The inner ring is filled with water to a known depth and the rate at which the water level declines is measured continuously until it is constant. The outer ring is kept full of water to minimize the amount of lateral flow of water that can occur from the inner ring (see Figure 6-9). When water flow is constant it is assumed that all the pores capable of conducting water are saturated and that water flow into the soil is as rapid as it can be under those conditions. Any additional water added to the soil would exceed the soil's capacity for infiltration and the additional water would pond or run off.

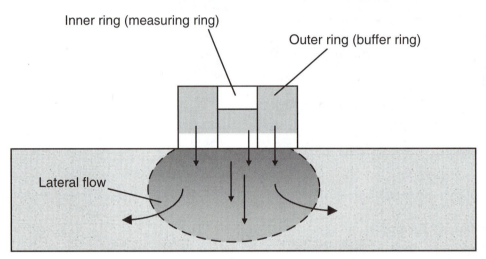

FIGURE 6-9   Diagram of water flow during hydraulic conductivity measurements. Water infiltration in the outer ring surrounds the inner ring with a saturated zone so that water movement is vertical rather than lateral.

Capillary Flow

Capillary flow refers to the movement of water through the small, interconnected pores in soil. Unsaturated flow occurs when the soil pores are not completely full of water. The unsaturated zone between the water table and the lower end of the rooting zone is called the **vadose zone.** In contrast to saturated flow, capillary flow can occur downward as water percolates through a soil, upward as water rises through soil above a water table, and horizontally. In each case the rate at which water flows is approximately equal.

When water rises above a permanent or fluctuating water table by capillary rise, the rise is determined by the diameter and continuity of the soil pores. The height to which water rises is determined by the capillary equation:

$$h = 0.15/r$$

where

h = the height of the water column in cm

0.15 is a constant

r = the radius of the capillary (or pore) in cm

> The capillary equation describes how high water can rise in soil pores.

Obviously, as the radius of a capillary becomes smaller, the height to which water rises increases.

The attraction of water for soil surfaces helps to hold water in soil capillaries against gravity, contrary to what it may seem; however, water's attraction to soil particles (adhesion) and the attraction of water molecules for one another (cohesion) does not drag water through capillaries. Rather, the water is pushed through the capillaries because the pressure within each capillary is less than the pressure exerted on the pool of water supplying the capillary with water, be that a water table or a wetting front.

> Water is pushed, not pulled, through capillaries.

There is an inverse relationship between pore size and capillary flow. Capillary flow will be most rapid through sandy soils with large pores and slowest through clayey soils with many fine pores. In part this is because clayey soils have a much higher surface area than sandy soils, which results in greater friction for water as it moves.

FOCUS ON . . . WATER FLOW IN SOIL

1. What are the differences between saturated and unsaturated flow?
2. What kind of water movement does Darcy's law describe?
3. What are the consequences when the rainfall rate exceeds a soil's K_{sat}?
4. What does the capillary equation describe?
5. Which soil, sandy or silty, will have more rapid capillary water movement?

SOIL TEXTURE EFFECTS ON WATER FLOW

The differences in pore sizes in soil lead to an interesting property called **hysteresis**. Hysteresis describes the phenomenon in which soils at the same water potential will have different volumetric water contents depending on whether they are wetting or drying. Figure 6-10 illustrates what happens. As a soil wets, capillary movement may encounter sudden large pores. Because of the mechanics of capillary movement, it takes more energy to fill these pores, and they are frequently bypassed. In contrast, water in draining soils will resist movement until the capillary pull of the narrowest part of the pore is overcome. Put in terms of potential energy, it takes more energy to fill open pores as well as to drain open pores connected to small necks.

> Water content in wetting and drying soils is not the same at the same water potential.

Filling   Draining

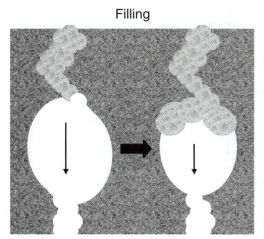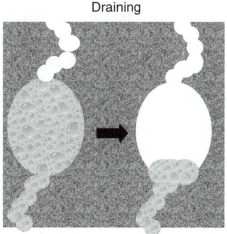

FIGURE 6-10 Differences in the way soil pores fill and drain cause a phenomenon called hysteresis.

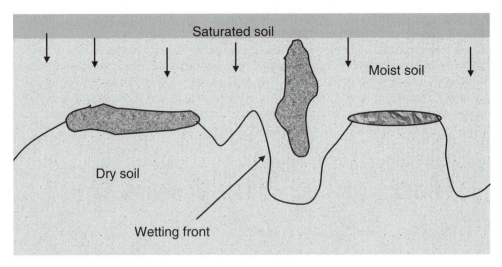

FIGURE 6-11   Differences in soil texture can influence the way water moves in soil. The textural differences can enhance or retard water transport depending on where they occur in soil. Coarse material in contact with overlying surface water for example, can rapidly channel water into the soil. Very fine textured material, as seen on the right-hand side of the figure, can restrict water movement and lead to dry zones in soil.

Porosity and texture have other effects on water flow. Although rain may uniformly wet a field, it will not necessarily uniformly wet the soil as it infiltrates. Soil layers of contrasting texture can influence how rapidly the water moves and where it moves. For example, a clay layer or compacted layer will impede water movement because of the low permeability of these materials. Water will pond above them and produce a temporary water table, while the soil beneath the layers remains dry. Likewise, lenses of coarse-textured soil or residue in soil can divert water because the infiltrating water lacks the energy to enter the large pores. These lenses can produce soil zones beneath them that are much drier, at least initially, than the surrounding soil (Figure 6-11). However, if these coarse-textured lenses extend to the soil surface and come into contact with free-standing water, they can serve as excellent channels to move water into the soil with rapidity. Recall that saturated and unsaturated flow is much greater in coarse-textured material once the soil water has entered.

> Clay and sand lenses in soil can divert water flow in wetting fronts.

SAMPLING SOIL WATER

Two types of water a soil scientist can sample are macropore water and micropore water. Macropore water is the water flowing through soils by gravity during saturated flow through macropores. It is a good reflection of the type of potential nutrients or contaminants that are flowing through soil after a heavy rainfall or irrigation. Micropore water samples the soil solution in the unsaturated soil. It is a good reflection of the type of solutes moving by diffusion.

To sample macropore water a device called a pan lysimeter is often used. A pan lysimeter is basically a large pan composed of inert material that is buried in a chamber excavated from within the soil. Figure 6-12a–b shows an example of a pan lysimeter being installed. A pipe reaches from the pan to the soil surface to allow sampling of any water that percolates into the pan.

A suction lysimeter (Figure 6-12c–d) operates by a different principle. The suction lysimeter typically has a porous ceramic cup at its tip. The tip is buried in soil so that it has good soil contact. Then the inside of the suction lysimeter is evacuated to create a vacuum. The vacuum pulls soil water through the ceramic tip into a cavity, which can be sampled. Although many nutrients can be sampled in the soil solution, the ceramic tip is sufficiently fine that it filters out larger particulates such as bacteria.

Fluctuating water tables are sampled by devices called piezometers. These are the simplest of all soil

(a)

(b)

(c)

FIGURE 6-12 Two methods of sampling soil water: Pan lysimeter (a, b); Piezometer; Suction lysimeter (c,d). The pan lysimeter samples freely draining water in the soil profile (water that moves through big pores). The pan lysimeter consists of a large open pan covered with a barrier to keep out roots and soil (a). Pan lysimeters are buried horizontally into the soil at various depths and attached to an access pipe that rises to the soil surface (b). The piezometer (not shown) samples water from fluctuating water tables. It consists of an open-ended pipe rising from the water table to the soil surface. The suction lysimeter samples the soil solution (water influenced by the soil matrix). A suction lysimeter consists of a porous ceramic cup attached to a closed cylinder (c). Tubes attached to the closed cylinder allow a vacuum to be imposed and the sample to be removed. The suction lysimeter is buried in the soil at various depths (d) and the hole back-filled with soil.

(d)

water sampling devices because they consist of a single tube, sealed at the bottom, with a series of openings at specified depths. As the water table rises, the water flows into these openings, where it can be sampled.

In permanently saturated soils, such as swamps and wetlands, a diffusion porometer can be used. The porometer is simply a series of open cells surrounded by a permeable barrier that traps water but allows the free movement of gases and solutes into and out of the cells. The porometers are buried in the sediment, allowed to come to equilibrium with the soil solution, and then excavated. The contents of each individual cell can then be sampled to investigate changes in soil water chemistry by depth.

FOCUS ON . . . TEXTURE AND WATER FLOW

1. What is hysteresis?
2. Why is it difficult for water to enter large pores by capillary flow?
3. How do clay lenses or pockets in soil impede water flow?
4. Why is soil often drier beneath buried residue than elsewhere after rain?

WATER EFFECTS ON THE SOIL CLIMATE

One of the most significant effects of soil water is in moderating soil temperature. Heat capacity reflects the change in temperature of a substance per unit change in temperature or input of energy. Water has an extremely high heat capacity, 1 cal cm^{-3} °C^{-1}, which means that the moist soil can absorb considerably more energy in the form of solar radiation than dry soil without the temperature rising appreciably. This is a disadvantage in cooler regions because it delays planting. However, it is a distinct advantage in summer when soils have the potential to become too hot to sustain optimum plant growth.

WATER AND SOIL HEAT CAPACITY

The volumetric heat capacity of soil is calculated by a fairly simple equation:

$$C = \Sigma f_{si}C_{si} + f_wC_w + f_aC_a$$

where

C = the total volumetric heat capacity

f_{si}, f_w, and f_a = the fraction of solids, liquid, and air, respectively

C_{si}, C_w, and C_a = the heat capacity of solids (on average 0.48 cal cm^{-3} °C^{-1}), water (1 cal cm^{-3} °C^{-1}), and air (0.003 cal cm^{-3} °C^{-1}), respectively

How much will the temperature of two soils increase, one at 25 percent and the other at 10 percent volumetric water content, if the soils receive 15 calories of energy?

Let's assume you had 100 g of soil with a bulk density of 1.2. The total porosity is 1 − bulk density/particle density = 1 − 1.2 g cm^{-3}/2.65 g cm^{-3} = 54.7 percent. So the total volume of soil is 83.3 cm^3 (100 g/1.2 g cm^{-3}) of which 54.7 percent or 45.56 cm^3 is pore space. The volumetric water content is already given as 25 percent and 10 percent, so the volume of water in each case is 20.8 cm^3 and 8.3 cm^3, respectively. The rest of the pore space is air. A table will clarify matters:

Soil	Vol. water content	cm^3 solid	cm^3 water	cm^3 air
A	25%	37.7	20.8	24.8
B	10%	37.7	8.3	37.3

Next we need to figure out the heat capacity for each soil.

Soil A: C = (37.7 cm^3)(0.48 cal cm^{-3} °C^{-1})
　　　　+ (20.8 cm^3)(1.0 cal cm^{-3} °C^{-1})
　　　　+ (24.8 cm^3)(0.003 cal cm^{-3} °C^{-1})
　　　　= 38.5 cal °C^{-1}

Soil B: C = (37.7 cm^3)(0.48 cal cm^{-3} °C^{-1})
　　　　+ (8.3 cm^3)(1.0 cal cm^{-3} °C^{-1})
　　　　+ (37.3 cm^3)(0.003 cal cm^{-3} °C^{-1})
　　　　= 26.5 cal °C^{-1}

Dividing the incoming energy by the volumetric heat capacity gives the estimated temperature increase:

Soil A: 15 cal/38.5 cal °C^{-1} = 0.39 °C

Soil B: 15 cal/26.5 cal °C^{-1} = 0.57 °C

Soil A, with the greater water content, has the smallest increase in temperature.

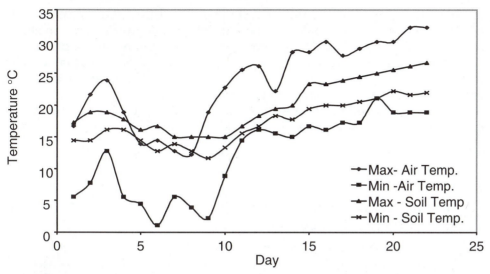

FIGURE 6-13 Seasonal air and soil temperatures at Lexington, KY in 2003. Notice how the maximum and minimum air temperature straddle the soil temperatures at 10 cm depth. You should also notice that although the temperature of the soil goes up with time, the day to day variability of the soil temperature is not nearly as great as it is for air temperature. This variability would be even less if we moved deeper into the soil. In deep cave systems such as Mammoth Cave in Kentucky the air temperature is constant at 12.8°C (55°F).

The soil itself undergoes distinct temperature changes that vary during the day (diurnally) and seasonally. These temperature fluctuations are affected by soil depth. As one goes deeper into the soil profile the magnitude of variation in soil temperature on a daily or seasonal basis decreases. Furthermore, the deeper one goes in soil the longer the lag between the temperature maxima and minima at the soil surface and that in the soil (Figure 6-13). At approximately 3 m depth in soil the temperature scarcely varies at all (Hillel, 1998). This results in the idealized condition of the soil being cooler in summer and warmer in winter than the surrounding environment, and explains why burying homes in soil is a popular approach to insulation.

> At 3 m depth the soil temperature is relatively constant all year.

FOCUS ON . . . WATER AND THE SOIL CLIMATE

1. Why are wetter soils generally cooler than drier soils?
2. What is heat capacity?
3. Which soil is most likely to be cooler on the same day, a soil with 25 percent gravimetric water content or the same soil with just 20 percent gravimetric water content?

GAS AND GAS TRANSPORT IN SOIL

The predominant gases in soil are the same as those found in air: N_2, O_2, and CO_2. The only real differences are that O_2 is somewhat depressed and CO_2 somewhat enriched because of root and soil respiration (Hillel, 1998). Carbon dioxide, for example, will typically have concentrations ten times higher than air (> 3600 ppm/v). There are a variety of other gases as well:

> N_2, O_2, CO_2, and water vapor are the main gases in the soil atmosphere.

methane (CH_4), hydrogen sulfide (H_2S), nitric oxide (NO), and nitrous oxide (N_2O), which reflect the biological activity of aerobic and anaerobic zones in soil. Another significant difference between the greater atmosphere and soil atmosphere is that soil atmospheres are nearly always close to 100 percent relative humidity despite the appearance of being air dry and long periods of drought (Coleman and Crossley, 1996).

Gases move through soil by mass flow, following concentration gradients, which is most simply described by Fick's law:

| | Fick's Law describes gas transport through soil. |

$$q = D\frac{dc}{dz}$$

where

q = the diffusion rate (g cm^{-2} sec^{-1})

D = the diffusion constant in soil (cm^{-2} sec^{-1})

dc = the concentration gradient (g cm^{-3})

dz = the depth (cm)

| Gas diffusion is typically 10,000 times slower through water than air. |

At this point it is important to note the critical influence that water has on gas diffusion in soil. As Table 6-1 shows, the diffusion of N_2, CO_2, and particularly O_2 is 10,000 times slower through water than through air. Thus, depending on the water-filled porosity of soil, plant roots and soil organisms can rapidly create O_2-free environments because their respiration simply consumes O_2 faster than it can diffuse to them.

Another impediment to gas transport through soil is the lack of porosity and the size of soil pores. As the bulk density increases due to compaction, the total porosity decreases and a greater fraction of that porosity is water-filled rather than air-filled, depending on water content. The same is true as pore size decreases—a greater fraction of the total porosity will be occupied by water rather than air. Consequently, because air is excluded from the pore space by water, and because its diffusion through water is slowed, aeration is poor in compacted or clayey soils. A rule of thumb for most soils is that soil respiration begins to decrease due to poor aeration when the water-filled pore space exceeds 60 percent (Paul and Clark, 1996).

FOCUS ON . . . GAS AND GAS TRANSPORT

1. What are unique features of soil atmospheres?
2. How does water control gas transport in soil?
3. What does Fick's law describe?
4. What are the consequences for soil aeration if bulk density increases?
5. At about what water content do soils show signs of diminished aeration?

TABLE 6-1 Diffusion constants of soil gases in air and water. *(Adapted from Paul and Clark, 1996)*

	Diffusion Constant (cm^2 sec^{-1})	
	Air	**Water**
N_2	0.205	0.164 × 10^{-4}
O_2	0.205	0.180 × 10^{-4}
CO_2	0.161	0.177 × 10^{-4}

SUMMARY

Water has unique chemical properties that are important for its activity in soil. It can be described qualitatively by terms such as *field capacity* and *wilting point,* but it can also be described quantitatively by terms such as *gravimetric* and *volumetric water content.* Because soils differ in terms of texture and porosity, a more exact description of water is based on its availability. *Water potential* is the term that describes the potential energy of water, and which also describes the direction in which water flows because water always flows from high to low potential.

Water moves through soil by saturated and unsaturated flow. Saturated flow is dominated by gravity and mainly occurs in large pores. Unsaturated flow occurs primarily by capillary movement. The pore diameter is critical in determining how high and how fast water will move in unsaturated flow. Soil texture can sometimes control the distribution of water in soil because lenses of coarse-textured soil can remain unsaturated. Compacted or clay layers in soil can also redirect water movement.

Among water's other important roles in soil, it also helps to control soil temperature because it has a high heat capacity. Moist soils can absorb more energy than dry soils without the same temperature increase. Soil water also controls gas diffusion through soil by occupying pore space, because gas diffusion is much slower through water than through air.

END OF CHAPTER QUESTIONS

1. What is the gravimetric water content of a soil that weighs 14 g when wet, 12.5 g when oven dry, and has a bulk density of 1.26 g cm^{-3}?

2. What is the volumetric content of a soil that has a bulk density of 0.95 g cm^{-3} and a gravimetric water content of 20 percent?

3. What is the equivalent surface depth of water in the first 15 cm of a soil at 33 percent volumetric water content?

4. What are four terms that can be used to describe water availability in soil?

5. If a soil reaches its wilting point at 10 percent gravimetric water and has 33 percent water at field capacity, what is its percent available water depletion?

6. Calculate the field capacity deficit of the first meter of a soil that has 40 percent volumetric water content at field capacity but only has 20 percent volumetric water content in its present condition.

7. Draw a moisture potential curve that illustrates two soils with the same water potential.

8. What will adding fertilizer (salts) do to the availability of water in a soil?

9. If a soil is at 15 atm tension, is it likely to have a lot of water or be deficient?

10. What is the flux of water (q) into soil if the hydraulic head is 10 cm, the path length is 20 cm, the cross-sectional surface area is 100 cm^2, and the K_{sat} is 5 cm h^{-1}?

11. What is the anticipated capillary rise in a tube with a diameter of 1 cm under standard conditions?

12. What is the heat capacity of 50 g of soil with a bulk density of 1.0 g cm^{-3} and a volumetric water content of 50 percent?

13. By how much will the temperature rise if the soil in question 12 receives 50 cal of energy?

14. What is the air-filled porosity of a soil that has a bulk density of 1.3 and a volumetric water content of 45 percent?

15. Is the soil in question 14 likely to be well aerated?

16. What does Fick's law describe and how is it influenced by soil water content?

FURTHER READING

Water Potential Relations in Soil Microbiology, SSSA Special Publication Number 9 (1985), is an invaluable guide to the basic principles of water potential. The first chapter has everything you need to get started.

REFERENCES

Coleman, D. C., and D. A. Crossley. 1996. *Fundamentals of soil ecology.* San Diego, CA: Academic Press.

Hillel, D. 1998. *Environmental soil physics.* San Diego, CA: Academic Press.

Paul, E. A., and F. E. Clark. 1996. *Soil microbiology and biochemistry,* 2nd ed. San Diego, CA: Academic Press.

Thien, S. J., and J. G. Graveel. 1997. *Laboratory manual for soil science,* 7th ed. Dubuque, IA: Wm. C. Brown Publishers.

SECTION 3

SOIL CHEMICAL PROPERTIES

Chapter 7

SOIL MINERALS AND WEATHERING

"Probably more harm has been done to the science by the almost universal attempts to look upon the soil merely as a producer of crops rather than as a natural body worth in and for itself of all the study that can be devoted to it, than most men realize."

C. F. Marbut, 1920

OVERVIEW

You have finished reading about many important physical properties of soil and the physical interactions among soil, water, and air. However, it is the chemical properties of soil that greatly determine the role of soil in plant growth, waste treatment, and other important areas. The minerals in the soil and their associated chemical properties control the ability of soil to adsorb and retain ions and water. Soil chemistry influences soil fertility and plant nutrition by controlling nutrient-holding capacity and leaching potential. Soil chemistry influences the capacity of soil to remediate waste effluent. Soil chemistry influences soil physical properties such as soil shrinkage, swelling, and cohesiveness. Soil chemistry influences the fate of chemicals added to soil, and how much and by what methods amendments should be applied.

To introduce the chemistry of the soil environment, in this chapter you will examine the types of minerals found in soils and how those soil minerals develop through weathering. You will examine the chemical properties of these minerals and how these mineral properties, as well as soil management, influence the chemical properties of the soil solution.

> Soil chemistry is important because it affects fertility, leaching, waste remediation, and shrink-swell potential.

OBJECTIVES

After reading this chapter, you should be able to:

✔ List five ways that soil chemistry influences soil properties.

✔ Understand the difference between primary and secondary minerals.

✔ Identify key chemical properties of soil minerals.

✔ Know four important outcomes of weathering in soil.

✔ Identify several types of colloids.

✔ Describe two ways by which soil colloids develop charge.

KEY TERMS

base cations
colloid
dissolution
hydration
illite
isomorphous substitution

kaolinite
octahedral
primary mineral
secondary mineral
smectite

specific surface area
tetrahedral
vermiculite
weathering

ROCKS AND MINERALS

The mineralogic composition of parent material determines the mineralogic composition of the resulting soil.

Soils form from rocks, and rocks are composed of minerals. The chemical properties of a soil can therefore be traced back to the chemical properties of the minerals within the original rocks. This can be seen if the original minerals still persist in soil and if weathering has transformed the **primary minerals** into **secondary minerals**. The nature of the primary minerals influences the type of secondary minerals that may develop. In many cases, soil chemical properties such as nutrient availability and pH can be predicted from the parent material mineral properties. Physical properties such as texture and degree of development may also be predicted from parent material mineralogy. You will begin this study by briefly examining rock and mineral types and weathering processes before you turn your attention to the chemical properties of the soils that result from this weathering.

Mineral Composition

Approximately 75 percent of the weight and approximately 95 percent of the volume of the Earth's crust (upper 10 miles) consists of oxygen (O) and silicon (Si; Table 7-1). Consequently, the types of minerals that dominate the rocks and soils of the Earth's surface are silicates and aluminosilicates (Table 7-2).

As was discussed in Chapter 4, there are differences in the mineral composition among the different-sized soil particle separates. Sand and silt particles predominantly contain primary minerals. Sand particles are

TABLE 7-1 The most common elements in the Earth's crust. *(Gabler et al., 1991)*

Element	%, by Weight	%, by Volume
O	46.6	93.8
Si	27.7	0.9
Al	8.1	0.5
Fe	5.0	0.4
Ca	3.6	1.0
Na	2.8	1.3
K	2.6	1.8
Mg	2.1	0.3

TABLE 7-2 Some common primary and secondary minerals and their relative ease of weathering.
(Adapted from Brady and Weil, 2002)

Primary		Secondary		
		Goethite	FeO(OH)	*More*
		Hematite	Fe_2O_3	*resistant*
		Gibbsite	$Al(OH)_3$	
Quartz	SiO_2			
		Aluminosilicates clays	variable	
Mica, muscovite	$KAl_2(AlSi_3O_{10})(OH)_2$	(e.g., kaolinite, smectite)		
Microcline	$KAlSi_3O_8$			
Feldspar, orthoclase	$K(AlSi_3O_8)$			
Mica, biotite	$K(Fe,Mg)_3AlSi_3O_{10}(OH)_2$			
Feldspar, albite	$NaAlSi_3O_8$			
Hornblende	$Ca_2(Al_2Mg_2Fe_3Si_6O_{22})(OH)_2$			
Augite	$Ca_2(Al,Fe)_4(Mg_2,Fe)_4Si_6O_{24}$			
Feldspar, anorthite	$Ca(Al_2Si_2O_8)$			
Olivine	$(Mg,Fe)_2SiO_4$			
		Dolomite	$CaMg(CO_3)_2$	
		Calcite	$CaCO_3$	*Less*
		Gypsum	$CaSO_4 (2H_2O)$	*resistant*

mostly quartz grains, which are highly resistant to weathering. Silt particles are also typically grains of quartz, with other resistant primary silicate minerals (such as orthoclase feldspar) also being common. Clay particles, however, are dominated by secondary minerals. These develop from the weathering of the less resistant primary minerals (such as anorthite feldspar or olivine).

Weathering

Weathering is a complex process by which rock is transformed into regolith and regolith is transformed into soil. Physical weathering transforms larger particles into smaller particles. Chemical weathering transforms primary minerals into secondary minerals.

Physical Weathering

Physical weathering is characterized by the mechanical disintegration of rocks. Physical weathering alone does not change the chemical nature of the minerals in the weathered materials. Physical weathering occurs when pressure or stress is applied to the rock or other mineral material and then released. The source of this stress may be from an external agent, such as ice. During freezing and thawing cycles, for example, the expansion of ice that forms within small cracks in a rock may exert a pressure of 150 T/ft^2 (1465 Mg/m^2) as it expands. Expanding roots may also exert physical stress on cracks in rocks. The stress may come from within the rock as it expands and contracts with heating and cooling.

Abrasion by water, ice, or sediment-carrying wind can contribute to the physical disintegration of rocks. Rocks that have been buried under large masses of material (including other rock layers) will expand following the release of the overburden stress. Chemical weathering may also change the volume of a mineral and put stress on the mineral structure, leading to physical weathering. In general, physical weathering decreases the size of mineral particles and increases their surface area. The increased surface area makes the minerals more susceptible to chemical weathering reactions, which mainly occur at mineral surfaces.

Chemical Weathering

Chemical weathering is the decomposition of minerals in response to H_2O, O_2, CO_2, and other chemicals. During the weathering process the elemental composition of the rocks or soil changes (Table 7-3). Initially, the basic elements (K, Na, Mg, Ca) are readily lost, along with some loss of Al and Fe. Over time, though, as weathering proceeds to greater extents, Si is also lost. In intensively weathered soils, Al and Fe have become concentrated in the soil while other elements have been lost.

Common weathering processes include hydration, dissolution, and hydrolysis, all of which are driven by the presence of water. **Hydration** occurs when one or more water molecules are attached to the molecular structure of a mineral. A simple hydration reaction is the addition of water to hematite:

$$Fe_2O_{3(s)} + 3H_2O_{(l)} \rightarrow 2Fe(OH)_{3(s)}$$

This reaction changes the hematite from a red color to a yellowish-brown color.

A more important result of some hydration reactions is that adding water changes the volume of the mineral. As mentioned before, this puts stress on the mineral structure, leading to physical weathering or to further chemical weathering. Among common minerals in the soil, gypsum is normally found as a hydrated mineral:

$$CaSO_{4(s)} + 2H_2O_{(l)} \rightarrow CaSO_4 \cdot 2H_2O_{(s)}$$

With further addition of water, the gypsum will undergo dissolution.

$$CaSO_4 \cdot 2H_2O_{(s)} + 2H_2O_{(l)} \rightarrow Ca^{2+}_{(aq)} + SO_4^{2-}_{(aq)} + 4H_2O_{(l)}$$

Mineral **dissolution** occurs when the solid materials dissolve in water. This is commonly observed when salt dissolves in water:

$$NaCl_{(s)} + H_2O_{(l)} \rightarrow Na^+_{(aq)} + Cl^-_{(aq)} + H_2O_{(l)}$$

TABLE 7-3 Chemical composition of average igneous rocks, a slightly weathered soil, and an intensively weathered soil. *(Bohn et al., 1985)*

	Average of Igneous Rocks	Slightly Weathered Soil %, by Weight	Intensively Weathered Soil
SiO_2	60	77	26
Al_2O_3	16	13	49
Fe_2O_3	7	4	20
TiO_2	1	0.6	3
MnO	0.1	0.2	0.4
CaO	5	2	0.3
MgO	4	1	0.7
K_2O	3	2	0.1
Na_2O	4	1	0.3
P_2O_5	0.3	0.2	0.4
SO_3	0.1	0.1	0.3
TOTAL	100.5	101.1	100.5

While not shown explicitly, in this and similar reactions hydration may be an important precursor to dissolution. The addition of the water molecules weakens the bonds between the elements within the mineral.

Hydrolysis involves splitting water into H^+ and OH^-:

$$H_2O_{(l)} \rightarrow H^+_{(aq)} + OH^-_{(aq)}$$

This reaction sometimes follows a hydration reaction. For example, when aluminum is released from a mineral it readily undergoes hydrolysis:

$$Al^{3+} + 6H_2O_{(l)} \rightarrow Al(H_2O)_6{}^{3+}_{(aq)}$$

$$Al(H_2O)_6{}^{3+}_{(aq)} \rightarrow Al(H_2O)_5(OH)^{2+}_{(aq)} + H^+_{(aq)}$$

This reaction will be given more attention later when you examine soil acidity.

In terms of common soil minerals, calcite weathering provides another example of hydrolysis:

$$CaCO_{3(s)} + H_2O_{(l)} \rightarrow Ca^{2+}_{(aq)} + HCO_3{}^-_{(aq)} + OH^-_{(aq)}$$

This reaction, though, is a simplified view of a more complex series of reactions that likely occur during weathering, including hydrolysis, dissolution, and hydration. Particularly when hydrolysis is involved, there is an exchange of hydrogen for other cations in the mineral structure (for example, H^+ for Ca^{2+} in calcite). Consequently, the presence of H^+ tends to accelerate weathering.

In the more complex soil system, weathering is commonly through the attack of H^+ on the Al-O bonds in clay minerals:

$$2\ NaAlSi_3O_8 + 9\ H_2O + 2\ H^+ = Al_2Si_2O_5\ (OH)_4 + 2\ Na^+ + 4\ Si(OH)_4$$

{albite}   {kaolinite}   {soluble silica}

Here, the weathering of a primary mineral, albite, is driven by the presence of H^+ in an aqueous system. Further hydrolysis of water alters the albite to a secondary mineral, kaolinite, and releases the sodium and some soluble silica.

There are at least four important products of the combined weathering reactions in the soil. The results of weathering reactions include: (1) formation of clay-sized secondary mineral particles [for example, $Al_2Si_2O_5(OH)_4$], (2) the release of ions into the soil solution (for example, Na^+), (3) the formation of other soluble compounds that may be leached from the soil [for example, $Si(OH)_4$], and (4) the concentration of residual resistant minerals.

Factors Affecting Weathering Rates

The factor that most influences weathering rates is climate, including temperature and precipitation. With increasing temperatures there is an increase in chemical reaction rates, so at higher temperatures weathering reactions are more rapid. Similarly, greater precipitation and greater moisture content increase weathering rates. Water is an important reactant in almost all chemical weathering reactions. In arid environments there is little weathering beyond physical weathering.

The physical and chemical characteristics of the initial minerals are also important. Some minerals are more resistant to weathering than others (Table 7-3). This resistance is derived from the strength of bonding in the mineral structure and the particle size of the mineral grains in the material. For sedimentary rocks, the degree of cementation is also important.

There is also a biological contribution to weathering. Carbon dioxide dissolves in water, becomes hydrated, then hydrolyzes the water:

$$CO_{2(g)} + H_2O_{(l)} \rightarrow H^+_{(aq)} + HCO_3^-{}_{(aq)}$$

Root and soil organism respiration provides a ready source of CO_2 in soil environments for this reaction to occur, and lowers the pH of the soil solution. Similarly, the presence of organic acids from organic matter decomposition increases acidity and promotes weathering. Plant nutrient uptake may also change the chemical equilibrium in soil and promote further mineral decomposition reactions that release additional nutrient ions.

FOCUS ON . . . WEATHERING

1. What is weathering? What are the two general types of weathering? What is the primary difference between the two?

2. What is the difference between primary minerals and secondary minerals?

3. Do all chemical weathering reactions involve water in some way? In regard to the water, what is the difference between hydration and hydrolysis? Between these and dissolution?

4. Why would chemical weathering proceed much more slowly without the living organisms on Earth?

5. What new substances may be formed from those released during chemical weathering?

CLAY MINERALS AND SOIL COLLOIDS

Clay formation is an important outcome of weathering reactions. You have already seen the term *clay* used to refer to a soil particle size separate and a soil textural class. You will also often use the term *clay* to refer to a class of soil minerals, namely the layer silicate clays. Chemical weathering processes alter primary minerals into these clay-sized secondary minerals. These particles impart many important properties to the soil.

Clays and Colloids

Many fundamental soil processes are surface phenomena.

The defining property of a clay particle, as discussed in Chapter 4, is size. By definition, clay particles are small, with effective diameters less than 0.002 mm. The smallest of the clay particles are termed **colloids**. These particles are less than 0.001 mm in diameter. In addition to their small size, clay particles differ in shape from the larger sand and silt particles. Clay particles have a platelike shape (Figure 7-1), which is due to the chemical structure of the minerals. Individual clay particles will often become oriented in the soil, forming stacks, or domains (Figure 7-2).

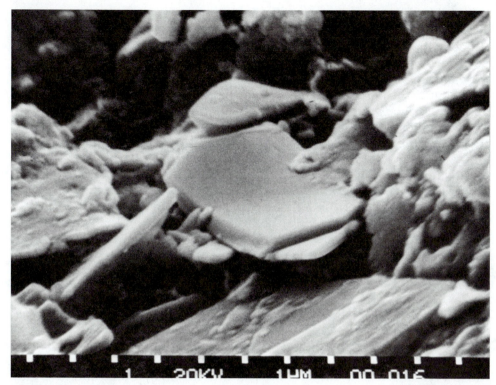

FIGURE 7-1 Electron micrograph of clays showing plate-like structure. Authigenic Chlorite flake from the Watahomigi Formation in Andrus Canyon, Grand Canyon National Park, AZ; 20,900X magnification. *(Photograph courtesy of the USGS)*

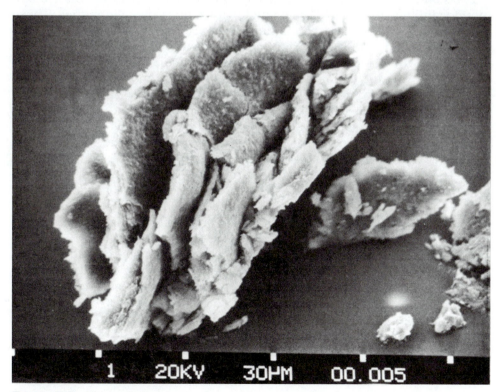

FIGURE 7-2 Electron micrograph of clay showing clay domains. Illite from the Manakacha Formation in SB Canyon, Grand Canyon National Park, AZ; 680X magnification. *(Photograph courtesy of the USGS)*

The small size and flat shape of the clay particles imparts an extremely large **specific surface area** (surface area per gram of soil) to these particles. Because the clay particles are often found in domains, the surface area can be divided between internal surface area (between individual particles) and the external surface area (on the exposed edges of the individual particles; as in Figure 7-2). The vast majority of the surface area of clay particles is internal surface area.

Layer silicate clay minerals also have important chemical properties. Due to an imbalance of charge in the chemical structure of the minerals, many clay particles develop surface charges. These charges are predominantly negative, but under some circumstances a positive charge can develop. These charges are important because they result in a soil's ability to adsorb positive ions (cations such as H^+, K^+, Ca^{2+}, Mg^{2+}). Polar water molecules are also attracted to the negative charges on the clay and colloid surfaces. As a result, the clay and colloid particles are important components of the soil that lead to cation and anion exchange, promote soil structure formation and stability, and influence water retention and movement.

> Many clay particles develop surface charge because of changes in chemical structure.

Types of Colloids

There are actually several types of mineral and organic soil colloids. The most common in most soils are the layer silicate clays. The properties of these crystalline clay minerals are quite variable. Consequently, clay type is similar to clay content in its influence on soil physical and chemical properties. All are characterized by the presence of an ordered crystalline structure, most commonly forming thin mineral layers that stack upon one another. You will get more details on clay mineral types in later sections.

> Colloids can be mineral or organic and crystalline or amorphous.

A second general type of mineral colloid is referred to as amorphous (noncrystalline) minerals (for example, imogolite and allophane). These minerals lack the ordered crystalline structure found in the layer silicate clays. They often have high surface charges, which may be negative or positive, giving them the ability to adsorb nutrients and water. These minerals, though, are not common in most soils. They are most often associated with soils formed from volcanic parent materials.

Other types of mineral colloids are the oxides of iron and aluminum (for example, hematite, goethite, and gibbsite). These minerals are most common in highly weathered soils, such as those found in the southeastern United States and tropical regions of South America, Africa, and southeast Asia. Unlike other soil colloids, these minerals have low surface charges and therefore low ability to adsorb nutrients and other cations. The iron oxides in particular are important because they are the source of red, orange, and yellow colors in most soils, with these colors becoming more pronounced as the amount of iron oxides increases.

> Iron and aluminum oxides are important soil colloids.

A fourth important soil colloid type is the organic soil colloids, or humus. These colloids are found in varying amounts in almost all soils—the product of plant and animal decay in or on soil. Humus is noncrystalline, but has a high surface charge that is predominantly negative. The source of this charge is the hydroxyl (—OH), carboxyl (—COOH), and similar functional groups that are components of these large, complex organic molecules. As you will see in later discussion, the charge on these organic colloids highly depends on the soil pH.

> Hydroxyl and carboxyl groups are important sources of charge on humus.

FOCUS ON . . . COLLOIDS

1. What is a colloid?

2. What properties distinguish colloids from other soil particles?

3. In reference to soil silicate clay particles, what is the difference between internal and external surface area?

4. What are the four major types of colloids present in soils? What are some important properties of each?

5. In what areas/environments is each type of colloid most commonly found?

6. What are common cations that are adsorbed to colloids? In what areas/environments is each most commonly found?

LAYER SILICATE CLAYS

Now it's time to turn your attention to the dominant type of colloid found in most soils. The characteristic physical and chemical properties of the layer silicate clays are directly related to the chemical structure of these minerals. You will examine the basic structure of these minerals, which will then explain the source of their charge and the cause of the stickiness, plasticity, and swelling capacity commonly associated with clay.

Layer Silicate Clay Structure

Layer silicate clays are ordered crystalline minerals, consisting mostly of layered sheets of Si, Al, Mg, and/or Fe atoms surrounded and held together with oxygen atoms and hydroxyl groups. The differences among the various clay mineral types are dictated by the composition and arrangement of these sheets.

> Layer silicate clays consist of sheets of Si, Al, Mg, Fe, and other elements held together by oxygen atoms and hydroxyl groups.

Tetrahedral and Octahedral Sheets

Two types of crystalline sheets may be found in the crystalline structure of a layer silicate clay. The silicon **tetrahedral** sheet has a silicon atom bonded to four oxygen atoms as its primary unit (Figure 7-3). The four oxygen atoms, seeking maximum separation from each of the other oxygen atoms, can be seen as the points of a four (tetra-) sided (-hedron) solid (Figure 7-4). The tetrahedron is sometimes described as a three-sided pyramid. A silicon tetrahedral sheet is formed when a Si atom shares each of its oxygen atoms that make up the base of the pyramid (basal oxygen atoms) with three adjacent Si atoms (Figure 7-5).

FIGURE 7-3 The primary unit of the silicon tetrahedron sheet. Silicon, in the center of the molecule, is bound to four oxygens.

FIGURE 7-4 The silicon and oxygen atoms form a four-sided unit called a tetrahedron.

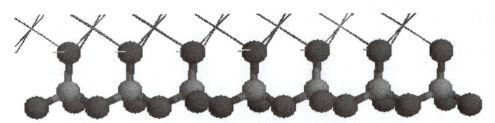

FIGURE 7-5 Silicon tetrahedral sheets are made when silicon tetrahedrons link together by sharing oxygen atoms.

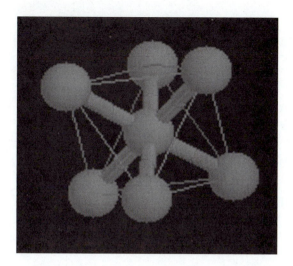

FIGURE 7-6 An aluminum-magnesium octahedral sheet has an aluminum or magnesium atom bound to six oxygen atoms.

FIGURE 7-7 Six oxygen atoms form the points of the octahedron surrounding the aluminum or magnesium.

The Al-Mg octahedral sheet has an Al or Mg atom bonded to six oxygen atoms as its primary unit (Figure 7-6). In this case, the six oxygen atoms can be seen as the points of an eight (octa-) sided (-hedron) solid (Figure 7-7). This is sometimes described as two four-sided pyramids joined at their bases. An **octahedral** sheet is formed when an Al or Mg atom shares four or six of its oxygen atoms with three or four adjacent Al or Mg atoms (Figure 7-8).

Layers

Soil clay minerals are not made up of individual tetrahedral or octahedral sheets. Layer silicate clay crystals consist of two or three of these sheets joined together (through the sharing of oxygen atoms) in one of several specific arrangements. When two or more sheets are joined together the resulting unit is termed a *layer*. The most simple configuration is for one octahedral sheet to be joined to one tetrahedral sheet. This occurs through the sharing of the apical oxygen atoms in the tetrahedral sheet.

Two or more sheets of layer silicate crystals joined together are called a layer.

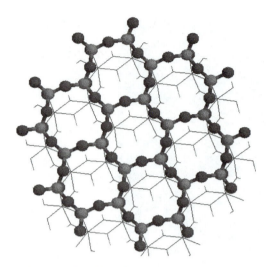

FIGURE 7-8   An octahedron can be described as two tetrahedrons joined at the base.

Another common configuration is for one octahedral sheet to be joined to two tetrahedral sheets. The octahedral sheet is sandwiched between the two tetrahedral sheets, again joined through the sharing of the apical oxygen atoms of the tetrahedral sheets. For both of these layer combinations it is important to note that the structural arrangement of the atoms is very precise. However, the atoms that occupy these arrangements (not including the oxygen atoms) can change in response to various factors during the crystallization, alteration, weathering, and recrystallization of these minerals.

FOCUS ON . . . LAYER SILICATE CLAYS

1. What is a tetrahedron? What atoms form the silicon tetrahedra in a layer silicate clay crystal?

2. What is an octahedron? What atoms form the aluminum-magnesium octahedra in a layer silicate clay crystal?

3. Tetrahedra or octahedra are joined together in sheets (tetrahedral or octahedral). When two or more sheets are stacked together, what is this called?

SOURCE OF CHARGE

Under ideal circumstances, both the individual sheets and the combined layers of the layer silicate clay minerals have no net charge, because the negative charges of the oxygen atoms are exactly balanced by the Si, Al, Mg, and hydrogen atoms in the chemical structure of the mineral. In reality, as these minerals form, the makeup of the crystals may diverge from the ideal. The crystals are also influenced by the surrounding conditions (mainly pH), which may change the structural makeup. In both of these cases the number of oxygen atoms remains unchanged. Adding or replacing the other atoms alters the number of positive charges in the crystal. With the number of negative charges from the oxygen atoms unchanged, the result is a change in the net charge on the crystal. These sources of charge on clay minerals and organic colloids lead to the net negative charge that attracts cations and water to these colloids.

Constant Charge

In layer silicate clay minerals, the primary source of charge arises when the Si, Al, or Mg atoms in the crystal are replaced by atoms that are approximately the same size, but of different charge. When this happens, the balance of positive and negative charges changes, but the arrangement of atoms in the crystal does not. For this reason, the process is called **isomorphous substitution** (iso = same; morpho = shape). For example, replacing a silicon atom (Si^{4+}) with an aluminum atom (Al^{3+}) in the tetrahedral sheet (Figure 7-9) produces a net charge of -1. Conversely, replacing a magnesium atom (Mg^{2+}) with an aluminum atom (Al^{3+}) in the octahedral sheet (Figure 7-9) produces a net charge of $+1$. In both of these cases, these substitutions produce permanent charges within the clay mineral structure. However, not all substitutions affect the net charge. Replacing an aluminum atom (Al^{3+}) with an iron atom (Fe^{3+}) in the octahedral sheet (Figure 7-9) produces no net charge.

> Isomorphous substitution occurs when atoms of similar size but different charge replace elements in silicate clay layers.

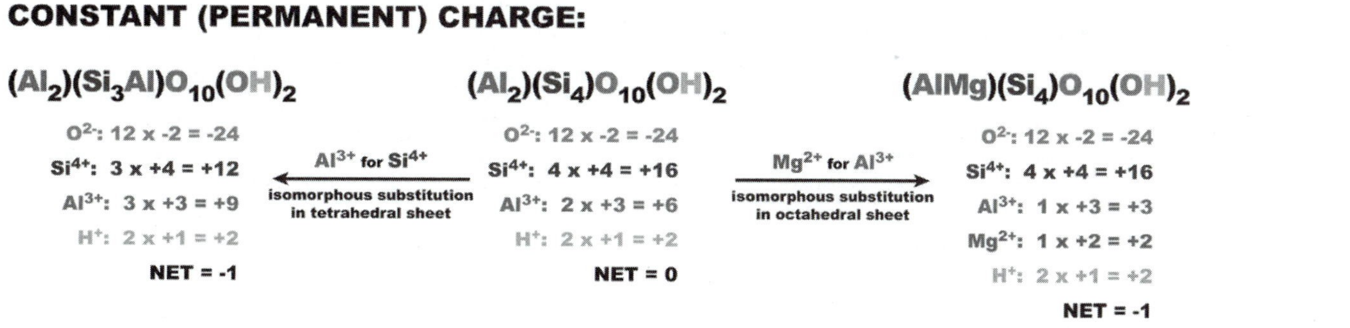

CONSTANT (PERMANENT) CHARGE:

$(Al_2)(Si_3Al)O_{10}(OH)_2$

O^{2-}: 12 x -2 = -24	
Si^{4+}: 3 x +4 = +12	
Al^{3+}: 3 x +3 = +9	
H^+: 2 x +1 = +2	
NET = -1	

← Al^{3+} for Si^{4+}
isomorphous substitution in tetrahedral sheet

$(Al_2)(Si_4)O_{10}(OH)_2$

O^{2-}: 12 x -2 = -24	
Si^{4+}: 4 x +4 = +16	
Al^{3+}: 2 x +3 = +6	
H^+: 2 x +1 = +2	
NET = 0	

→ Mg^{2+} for Al^{3+}
isomorphous substitution in octahedral sheet

$(AlMg)(Si_4)O_{10}(OH)_2$

O^{2-}: 12 x -2 = -24	
Si^{4+}: 4 x +4 = +16	
Al^{3+}: 1 x +3 = +3	
Mg^{2+}: 1 x +2 = +2	
H^+: 2 x +1 = +2	
NET = -1	

FIGURE 7-9 Development of permanent charge in silicate clays. Constant charge develops when atoms of lower charge substitute for atoms of higher charge.

Variable Charge

Hydroxyl (−OH) groups are found throughout the crystal structure of layer silicate clay minerals. Similarly, humus molecules contain many hydroxyl groups. As was seen with the water molecule (Chapter 6), the strong electronegativity of the oxygen atom leads to unequal sharing of the electron in the covalent bond between hydrogen and oxygen in the hydroxyl group. The electron of the hydrogen atom is drawn relatively closer to the oxygen atom (polar covalent bonds). This creates a negative charge concentration near the oxygen atom in the hydroxyl group. Because of this polarity, the hydroxyl groups of the colloid can interact with hydrogen ions or hydroxyl ions in the soil solution.

At low pH, when the concentration of H^+ ions in the soil solution is high, the electronegative oxygen atoms in the colloid can attract an electropositive H^+ ion in the soil solution (Figure 7-10). By adding a +1 hydrogen ion to the composition of the colloid, the net charge becomes more positive (less negative). At high pH, when the concentration of OH^- ions in the soil solution is high, the electronegative oxygen atoms of the OH^- ion can attract an electropositive H^+ ion in the colloid (Figure 7-10). By removing a +1 H^+ ion from the composition of the colloid the net charge becomes more negative (less positive).

The various soil colloids have differing amounts of total charge, as well as differing amounts of constant versus variable charge. Organic colloids have the greatest charge per gram of colloid, with most of this charge being pH-dependent. Iron and aluminum oxides also have predominantly variable charge, but their total charge is quite low. The layer silicate clays, in general, derive most of their charge as permanent charge through isomorphous substitution. However, each of the clay mineral types varies in their total charge and in their charge distribution between constant and variable forms. For this reason it is important to appreciate the different types of clay minerals. They are just as important as the total amount of clay in determining the chemical properties of a soil.

> Variable charge depends on the pH of the soil environment—positive when acid and negative when basic.

> Most of the permanent charge of clay minerals comes from isomorphous substitution.

VARIABLE (pH-DEPENDENT) CHARGE:

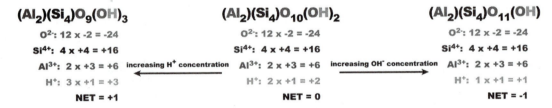

$(Al_2)(Si_4)O_9(OH)_3$

O^{2-}: 12 x -2 = -24
Si^{4+}: 4 x +4 = +16
Al^{3+}: 2 x +3 = +6
H^+: 3 x +1 = +3
NET = +1

← increasing H^+ concentration

$(Al_2)(Si_4)O_{10}(OH)_2$

O^{2-}: 12 x -2 = -24
Si^{4+}: 4 x +4 = +16
Al^{3+}: 2 x +3 = +6
H^+: 2 x +1 = +2
NET = 0

increasing OH^- concentration →

$(Al_2)(Si_4)O_{11}(OH)$

O^{2-}: 12 x -2 = -24
Si^{4+}: 4 x +4 = +16
Al^{3+}: 2 x +3 = +6
H^+: 1 x +1 = +1
NET = -1

| INCREASING H+ | | | | | | | | | | | | DECREASING H+ |

1   2   3   4   5   6   7   8   9   10   11   12   13   14
pH

FIGURE 7-10 Development of pH-dependent charge in silicate clays. As the pH goes up or down the hydroxyls that are part of the silicate clay layers protonate and deprotonate.

FOCUS ON . . . SOURCE OF CHARGE

1. What are the two major sources of charge on soil colloids? What is the fundamental difference between them?

2. What is isomorphous substitution? How is it related to the charge of soil clay particles?

3. If there is a substitution of Mg^{2+} for Al^{3+} in an octahedral sheet of a layer silicate clay, does the net negative charge of the clay increase or decrease?

4. What is more common among layer silicate clays, a net negative or net positive charge?

5. What is the source of pH-dependent, or variable, charge on soil colloids?

6. Chemically, what happens to the hydroxyl (−OH) group under basic (pH > 7) conditions? How does this change the net charge on the soil colloid?

TYPES OF SILICATE CLAYS

There are two general types of silicate clays. Minerals with one tetrahedral sheet and one octahedral sheet are termed 1:1 clays. Minerals with two tetrahedral sheets for each octahedral sheet are termed 2:1 clays. All are considered secondary minerals, formed by weathering of primary minerals. In fact, there are general geologic and climatic conditions that promote formation of the different clay minerals. Accordingly, each of the different silicate clay minerals has different elemental compositions. We have noted that some of the important properties of clay particle include relatively high surface area, the presence of negative surface charges, stickiness, and the ability to swell upon wetting. Each of the common clay minerals differs in these key properties. We consider here the four most common types of soil clay minerals.

> There are several different types of silicate clay minerals: illite, vermiculite, smectite, and kaolinite.

Illite

Illite (Figure 7-11) is a 2:1 type clay with moderate surface area (100–120 m^2/g) and moderately high negative surface charge. The charge arises through isomorphous substitution of Al^{3+} for Si^{4+} in the tetrahedral sheets. This concentration of charge on the outer surfaces of the individual clay layers promotes the strong adsorption of cations, mainly potassium (K^+). The presence of the K^+ creates strong bonds between individual clay layers, causing illite to be only slightly sticky with low to moderate swelling potential.

The illite clay minerals are normally associated with less intense weathering environments where the parent materials were relatively high in **base cations** (Ca^{2+}, Mg^{2+}, K^+, and Na^+) and climatic conditions have not favored the leaching of these base cations.

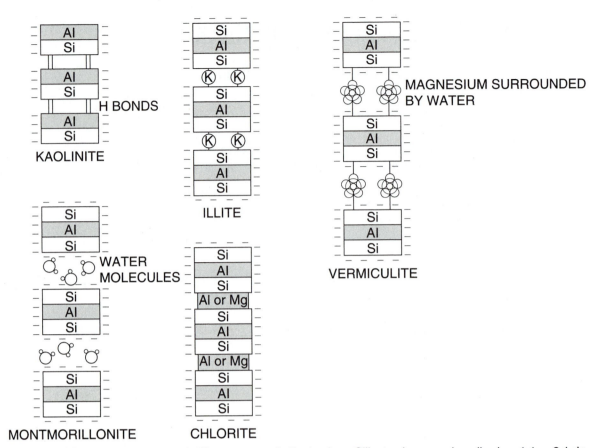

FIGURE 7-11   Micelle structure of different types of silicate clays. Silicate clays are described as 1:1 or 2:1 depending on whether they have one silicon tetrahedral layer to octahedral layer or two silicon tetrahedral layers sandwiching one octahedral layer. *(Adapted from Plaster, 1997)*

Vermiculite

Vermiculite (Figure 7-11) is also a 2:1 type clay, but with a high surface area (700–800 m^2/g) and a high negative surface charge (100–200 meq/100 g soil or 100–200 cmol$_+$/kg soil). Similar to illite, this charge originates mainly from isomorphous substitution in the tetrahedral sheets, although some substitution in the octahedral sheet does occur. Though not as prominently as with illite, this charge configuration promotes stronger adsorption of interlayer cations. There exist moderately strong bonds between the individual crystal layers, yielding a sticky consistency and a moderate swelling potential for vermiculite clays. Vermiculite minerals are also associated with less intense weathering environments, but have begun with or lost more of the potassium, which helps hold the clay layers together in illite.

Smectite (Montmorillonite)

Smectite (Figure 7-11) is the third prominent 2:1 type clay found in many soils. Similar to vermiculite, it has a very high surface area (especially when wet; 600–800 m^2/g) and a high negative surface charge (80–150 meq/100 g soil; 80–150 cmol$_+$/kg soil). Unlike illite and vermiculite, this charge originates from isomorphous substitution in the inner octahedral sheet. Consequently, the interlayer cations that can help bond clay layers together are not held as tightly as with illite and vermiculite.

Smectite clays have weak bonds between layers, which gives them a very sticky consistency and a high swelling potential. This propensity to swell upon wetting makes management of soils with high smectite clay contents a significant challenge. Construction activities are particularly hampered by the swelling and shrinking cycles that occur, which can damage foundations, crack walls, shift poles, and warp roadbeds.

Smectite minerals are more associated with moderately weathered environments where there has been greater removal of base cations over time, particularly K and Mg.

Kaolinite

Kaolinite (Figure 7-11) is a 1:1 type clay with low surface area (5–30 m^2 per gram of soil) and low negative surface charge (1–15 milliequivalents per 100 grams of soil; 1–15 cmol$_+$/kg soil). The low charge is a result of limited isomorphous substitution in the crystal structure. When stacks of individual kaolinite crystal layers occur, the oxygen atoms of the exposed tetrahedral sheet are adjacent to the hydrogen atoms of the hydroxyl groups of an octahedral sheet. As a result, relatively strong hydrogen bonds develop between the layers, causing kaolinite to have a very low swelling potential (almost none) and to be only slightly sticky.

Soil kaolinite is mainly found in leached environments with acid conditions associated with hot, wet climates. Consequently, it is the most common clay mineral in the southeastern United States.

The nonexpansive nature of kaolinite clay contributes to its many and varied uses as an extractable resource. It has long been used for making bricks, tiles, and other earthenware, from simple pottery to expensive china and ceramics. Kaolinite can also be used to produce smooth, shiny coatings on glossy paper and soothing coatings on your stomach to combat diarrhea, and is sometimes used in paints as a pigment and in nondairy products (shakes and ice cream) as a thickening agent.

Nonsilicate Clays

The various Fe and Al oxide minerals, including hematite, goethite, and gibbsite, are the most common nonsilicate clay colloids found in most soils. There is no isomorphous substitution in these minerals, so they possess very low negative surface charge (0–15 meq/100 g soil; 0–15 cmol$_+$/kg soil), and this charge is entirely pH-dependent. At low pH, the oxygen atoms may readily adsorb hydrogen ions from solution and a net positive charge may develop. These minerals lack the defined crystal structure associated with the layer silicate clays so they are nonsticky and have no swelling potential. These minerals are the result of highly intense weathering environments where most if not all of the base cations and much of the silicon has been lost. The Fe oxide minerals are noteworthy because they are a primary source of color in most soils. This explains the typical strong red colors associated with the highly acid, highly leached soils common in the humid tropical regions, as well as the general lack of nutrient-holding capacity and phosphorus availability.

> Nonsilicate minerals don't have isomorphous substitution but they do have pH-dependent charge.

Clay Minerals and Their Distribution

Soil mineralogy is not the same everywhere. It is affected by the distribution of soil parent materials, from which the soil minerals originate, and it is affected by the distribution of soil-forming processes that lead to the deposition, alteration, decomposition, formation, and movement of soil particles. Differing mineralogy with depth and from one area to another can influence land use because it affects:

- ❖ water relations, particularly moisture availability
- ❖ physical properties, such as consistency (stickiness, plasticity) and expansiveness
- ❖ chemical relations, such as nutrient storage, buffering capacity, surface activity (adsorption potential)
- ❖ solute transport and leaching potential

Distribution and Landscapes

The assortment of minerals in a soil changes with time in response to the intensity of chemical weathering. Although conditioned by the mineralogy of the soil parent materials, some general trends exist (Table 7-2; Table 7-4). The most soluble minerals, gypsum and carbonates (calcite, dolomite), are normally only found in young sediments or in arid soil environments where moisture is lacking. Among the clay minerals, you have already seen the general weathering progression from illite and vermiculite, to smectite, to kaolinite, to Fe and Al oxides.

As weathering intensity increases, the layer silicate clays are removed and the relative abundance of Fe and Al oxides increases. Quartz, a rather resistant primary mineral, is relatively abundant in mildly weathered soils. Its relative abundance increases as less-resistant minerals are weathered and removed from the soil. However, over time, even quartz begins to be weathered away. These general trends in weathering sequences and products let you make inferences about soil-forming environments and weathering intensity based on the assortment (and amounts) of various minerals in a soil.

Water moving over and through the soil can produce differences in the local weathering environment and yield differences in soil mineralogy along hill slopes and across landscapes. There are no general rules that can be applied to all landscapes, but patterns do exist and can be recognized and explained. In upper landscape positions the leaching intensity is greater, so weathering products (bases, soluble silica) are easily lost. These products may then accumulate in lower landscape positions, where leaching is less intense. As a result, the upper slope positions tend to have more of the clay minerals associated with intensive weathering (Al oxides, kaolinite) and the lower horizons may have more of the clay minerals associated with less intensive weathering (smectite, vermiculite).

Distribution and Depth

Weathering reduces the amount of silicate clay minerals and increases the amount of iron and aluminum oxides.

The intensity of weathering reactions and the results of subsequent soil-forming processes may also alter the distribution of soil minerals with depth. In intermediate to strongly weathered soils in humid climates, the formation of secondary clay minerals and subsequent downward translocation of clay not only alters soil texture (as discussed in Chapter 4), but will also concentrate primary minerals in the upper horizons (A and E) and enrich the subsoil horizons in secondary clay minerals (Table 7-5).

Variation in the specific types of clay minerals that occur at different depths within the soil profile is controlled by the leaching intensity. In arid areas with little leaching, or in high-precipitation areas with excessive

TABLE 7-4 Summary of mineralogic data of soils from various different soil environments. *(Derived from data in Allen and Hajek, 1989)*

Soil Environment	Weathering Intensity	Dominant Mineral[a]	Subordinate Minerals
Young sediments, resistant parent materials	Mild	Smectite	Illite, carbonates
Hot and dry (arid)	Mild	Carbonates	Smectite, gypsum
Subhumid grassland	Intermediate	Smectite	Carbonates
Warm and humid forest	Intermediate	Smectite	Kaolinite, vermiculite
Hot and humid forest	Strong	Kaolinite	Fe and Al oxides
Tropical	Strong	Fe and Al oxides	Kaolinite

[a]In all cases the occurrence of mixed mineralogy is not uncommon. These represent the generally recognized dominant minerals in many such soils.

TABLE 7-5 Partial profile description of a soil from the Frondorf series from Christian County, Kentucky, which formed in parent materials of loess over sandstone residuum. *(Adapted from Froedge, 1980)*

Horizon	Depth	Sand	Silt	Clay	Primary Minerals	Secondary Clay Minerals
	cm	------------------%------------------			-----------------------%-----------------------	
A	0–5	15.9	75.1	9.0	88	12
E	5–18	20.1	73.0	6.9	86	14
Bt1	18–33	14.5	60.1	25.4	63	37
Bt2	33–53	18.3	54.2	27.5	65	35
2Bt3	53–84	24.6	52.6	25.8	70	30
2R	>84					

leaching, clay type is more or less uniform throughout the profile. In a more moderate leaching environment the weathering products from the upper horizons are translocated to the subsoil. As a result, upper horizons may have more of the clay minerals associated with intensive weathering (Al oxides, kaolinite) and the lower horizons where base cations and silicates accumulate may have more of the clay minerals associated with less-intensive weathering (smectite, vermiculite).

FOCUS ON . . . CLAY MINERALS

1. What is the difference between a 1:1 and a 2:1 layer silicate?

2. How are the sheets of a layer silicate held together? How are adjacent layers of a clay crystal (or micelle) held together?

3. What causes clay layers to expand when wetted? Why are micas nonexpanding?

4. Rank the following clay minerals in terms of their net charge: vermiculite, smectite, mica.

5. Which clay minerals are most common in soils with low weathering intensity?

6. Which clay minerals are most common in highly weathered soils?

Summary

Soil chemistry begins with the minerals of soil and how they form through weathering, which influences soil fertility and plant nutrition by controlling nutrient-holding capacity and soil physical properties such as soil shrinkage, swelling, and cohesiveness. Primary minerals in sedimentary, metamorphic, and igneous parent materials weather by physical and chemical processes to yield secondary minerals. Secondary minerals have smaller size and greater surface area, which increases their chemical reactivity.

The secondary minerals produced by weathering are composed of layer silicate materials or amorphous Fe and Al oxides. The layer silicates are composed of oxygen, Si, Al, and other elements arranged into tetrahedral or octahedral sheets. These sheets are organized into layers, and the layers are organized into stacks or domains.

Layer silicates can have charge because of isomorphous substitution in which atoms of similar size, but lesser charge, substitute for elements within the layer silicates. This is the most important source of permanent charge in clay minerals. There is also variable pH-dependent charge in organic and mineral colloids that is primarily due to hydroxyl groups.

Weathering forms secondary minerals and soil colloids, but as weathering proceeds the secondary minerals begin to be translocated from the soil surface into the soil profile. As weathering proceeds, even Si in the clay minerals is leached and most of the colloids in soil are composed of Fe and Al oxides, which have low cation and anion exchange capacity and therefore reduced nutrient-holding capacity. Illite, vermiculite, smectite, and kaolinite are four significant types of 1:1 and 2:1 clay minerals in soil.

End of Chapter Questions

1. Why is physical weathering more prevalent in dry, cool regions, while chemical weathering is more prevalent in wet, hot regions?

2. How does temperature contribute to physical weathering? How does abrasion contribute to physical weathering? How do plants and animals contribute to physical weathering?

3. Which type of soil would be better for growing plants: an immature soil high in primary minerals, or a mature soil high in secondary minerals? Explain your answer.

4. Why do you think that Al^{3+} is held more tightly by soil colloids than Na^+?

5. Why does the relative concentration of cations in the soil solution influence the proportion of cations adsorbed to colloids?

6. Which is better for building roads or foundations, kaolinite or smectite? Explain your answer.

7. How is the relative amount of interlayer expansion related to the net charge of the clay layers?

8. How do a soil's local climate and parent materials influence the weathering stage of its clay minerals and, therefore, the geographic distribution of clay minerals?

9. Most soils exhibit a net negative charge. Under what conditions might a soil exhibit a net positive charge?

10. Consider a clay mineral with the following structural formula: $(Al_{1.5}Mg_{0.5})(Si_{3.7}Al_{0.3})O_{10}(OH)_2$. Which are the tetrahedral atoms? Which are the octahedral atoms? What is the net charge on this clay mineral?

11. Consider a soil with 4 percent organic matter and 20 percent clay (5 percent smectite and 15 percent kaolinite). Using the representative charges shown in Table 8-1, calculate an estimate of the cation exchange capacity (at pH 7) of this soil.

12. Chemically, what happens to the hydroxyl ($-OH$) group under acidic (pH < 7) conditions? How does this change the net charge on the soil colloid?

13. If there is a substitution of an Al^{3+} for a Si^{4+} in a tetrahedral sheet of a layer silicate clay, does the net negative charge of the clay increase or decrease?

14. If there is a substitution of an Al^{3+} for a Mg^{2+} in an octahedral sheet of a layer silicate clay, does the net negative charge of the clay increase or decrease?

FURTHER READING

Buseck, P. R. 1983. Electron microscopy of minerals. *American Scientist* 71: 175–185. (For a close-up examination of soil minerals.)

Dixon, J. B., and S. B. Weed. 1989. *Minerals in the soil environment*. Madison, WI: Soil Science Society of America. (A standard reference text in soil mineralogy.)

REFERENCES

Allen, B. L., and B. F. Hajek. 1989. Mineral occurrence in soil environments. In J. B. Dixon and S. B. Weed (eds.), *Minerals in soil environments,* pp. 199–278. Madison, WI: Soil Science Society of America.

Bohn, H. L., B. L. McNeal, and G. A. O'Conner. 1985. *Soil chemistry,* 2nd ed. New York: John Wiley & Sons.

Brady, N. C., and R. R. Weil. 2002. *The nature and properties of soil.* Upper Saddle River, NJ: Prentice Hall.

Froedge, R. D. 1980. *Soil survey of Christian County, Kentucky.* Washington, DC: Dept. of Agriculture, Soil Conservation Service.

Gabler, R. E., R. J. Sager, and D. L. Wise. 1991. *Essentials of physical geography,* 4th ed. Philadelphia: Saunders College Publishing.

Plaster, E. J. 1997. *Soil science & management.* Clifton Park, NY: Thomson Delmar Learning .

Chapter 8

SURFACE CHEMISTRY OF SOILS

*". . . for only rarely have we stood back and celebrated our soils
as something beautiful, and perhaps even mysterious. For
what other natural body, worldwide in its distribution, has
so many interesting secrets to reveal to the patient observer?"*

Les Molloy, *Soils in the New Zealand Landscape: The Living Mantle,* 1988

OVERVIEW

In Chapter 7 you looked at how soil weathering influences the type of particles that form. You saw how the chemical characteristics of each particle influence that particle's behavior in soil. You also examined how changes in the chemical characteristics of these particles could create either permanent or variable charge. In this chapter you will examine what happens because soils have charged surfaces. You will examine how this property affects the capacity of soils to accumulate and release important plant nutrients. You will also examine how charged soil surfaces play a role in regulating the acidity or alkalinity of the soil environment.

OBJECTIVES

After reading this chapter, you should be able to:

✔ Indicate two important features of ion exchange reactions.

✔ List a simplified order of ion strength of adsorption.

✔ Know the difference between cation and anion exchange capacity.

✔ Describe why CEC is an important soil property.

✔ Identify four soil factors that control CEC.

✔ Describe the effect of weathering on CEC.

✔ Estimate CEC based on soil constituents.

✔ Specify the difference between acidity and alkalinity.

✔ Know the difference between active, exchangeable, and residual acidity.

✔ Describe how buffering capacity works.

✔ Identify sources of acidity in soils.

KEY TERMS

acidity
active acidity
alkalinity
anion exchange
   capacity (AEC)

base cation saturation
   (BCS)
buffer capacity
cation exchange
   capacity (CEC)

equilibrium
exchange capacity
exchangeable acidity
pH
residual acidity

ION EXCHANGE

> Ion exchange is the ability to hold and release cations, anions, and water, and make them available to plants.

Ion exchange is the exchange of ions in soil solution with ions adsorbed to the surface of soil clay minerals, organic matter, or other soil colloids. The presence of surface charges and unsatisfied bonds on soil colloidal materials gives rise to a soil's ion retention capability. Because most soil colloids have a net negative charge they can hold positively charged ions, which gives soil the ability to retain nutrients and water, and then make these available to plants. The surface charge of soil also allows the retention of organic and inorganic contaminants applied to soil (either intentionally or accidentally). Without the ability to exchange and retain ions with the surrounding soil solution, the role of soil would be greatly diminished.

There are ions attached to the colloid and in the soil solution (the water layer around the soil particle and the water in the pores). Ion exchange takes place when one of the ions in the soil solution replaces one of the ions on the soil colloid. This exchange only takes place when the ions in the soil solution are not in equilibrium with the ions on the soil colloid, which is almost always the case because leaching, fertilizer addition, plant removal, and many other soil processes all keep this system from remaining static. Two important things to remember about ion exchange reaction are: (1) these reactions are rapid and reversible; (2) these reactions are "charge for charge" not "ion for ion." A few simple examples will illustrate these reactions.

> Ion exchange reactions are rapid and reversible and "charge for charge" not "ion for ion."

Exchange Reactions

> Cation exchange reactions are the most common type of exchange reaction in soil.

In most soils, the net surface charge is negative, so cation exchange is the most common type of exchange reaction. A negatively charged soil colloid will retain enough cation charge to balance the net negative charge on the colloid. A simple cation exchange reaction is illustrated by:

$$\text{colloid} \| \text{Ca} + 2\text{NH}_4^+ \leftrightarrow \text{colloid} \left\| \begin{matrix} \text{NH}_4 \\ \text{NH}_4 \end{matrix} \right. + \text{Ca}^{2+}$$

Here, the soil colloid retains a single Ca^{2+} cation. In the presence of excess ammonium ions (NH_4^+), the calcium ion may be removed from the exchange complex and replaced by two ammonium ions. Note that this reaction is charge-for-charge—it took two ammonium ions at +1 each to replace a calcium ion with a charge of +2. Also, note that this reaction is reversible. Just as easily as the calcium was replaced by the two ammonium ions, the two ammonium ions may be replaced by a calcium (or any other combination of ions with a total charge of $+2$).

Anion exchange reactions are less common in soil than cation exchange reactions. In such instances, a soil colloid with positive exchange sites in the ion exchange complex may retain negatively charged ions. These anions may be replaced by other anions in the soil solution. A simple anion exchange reaction is illustrated here:

$$\text{colloid}\begin{Vmatrix} NO_3 \\ NO_3 \end{Vmatrix} + Cl^- \leftrightarrow \text{colloid}\begin{Vmatrix} Cl \\ NO_3 \end{Vmatrix} + NO_3^-$$

The chloride (Cl^-) anion in the soil solution replaced one of the two nitrate (NO_3^-) anions held in the exchange complex of the colloid. Again, it was a charge-for-charge reaction, and may easily be reversed.

These examples focus on single exchange reactions on simplified soil colloid exchange complexes. With most soils, the exchange complex is not constituted by one or even two elements, but is a complex mixture of various ions that change rapidly in response to changing conditions in the soil solution. A more complicated yet still relatively simplified example is seen with a colloid with four different cations (here the subscripted numbers by each element refer to the actual number of ions of that element that are held on this model cation):

> The cations on the exchange complex are in equilibrium with the cations in soil solution.

$$\text{colloid}\begin{Vmatrix} Ca_{40} \\ Al_{20} \\ H_{20} \\ Mg_{20} \end{Vmatrix} + 7KCl \leftrightarrow \text{colloid}\begin{Vmatrix} K_7 \\ Ca_{38} \\ Al_{20} \\ H_{19} \\ Mg_{19} \end{Vmatrix} + 2CaCl_2 + HCl + MgCl_2$$

This example also helps to illustrate that exchange reactions are influenced by the types of cations present on the adsorption sites and in soil solution because of differences in affinity, or strength of adsorption. Aluminum is held tightly within the exchange complex because of a high positive charge and a relatively small size. A simplified order of strength of adsorption is $Al^{3+} > Ca^{2+} > Mg^{2+} > K^+ = NH_4^+ > Na^+$.

Another important consideration is the concentration of cations in the surrounding soil solution. As the concentration of a given cation increases in the soil solution, it will more readily replace cations on the exchange sites. Conversely, if the cations that are released from the exchange complex readily react with anions in the soil solution, further exchange reactions will occur.

> The strength of association of cations on the exchange complex are different.

FOCUS ON . . . CATION EXCHANGE

1. What is cation exchange?
2. Is cation exchange a rapid or slow reaction?
3. Is cation exchange a reversible or irreversible reaction?
4. Do cation exchange reactions increase or decrease the leaching of cations from the soil? Why?

Exchange Capacity

Exchange capacity describes the ability of soil to hold ions.

Exchange capacity is the sum total of exchangeable ions that a soil can adsorb. It is also used to describe the ability of a soil to hold ions. The vast majority of the exchange complex is occupied by cations because of the net negative charge on most colloids, so one often focuses on **cation exchange capacity (CEC)**. There may also be at least a small **anion exchange capacity (AEC)**. The exchange capacity is normally expressed in units of $cmol_c$ kg^{-1} (centimoles of charge per kilogram of soil).

Cation Exchange Capacity (CEC)

The CEC is the dominant form of exchange capacity because most soil colloids have more unsatisfied negative charges than positive charges. Anion exchange may become significant in soils high in Fe oxides, such as highly weathered soils of the tropics. The CEC is an important soil property because it affects:

❖ Soil fertility and plant nutrition because of its influence on nutrient-holding capacity and leaching potential

❖ The capacity of the soil to remediate waste effluent

❖ Soil physical properties such as soil shrinkage, swelling, and cohesiveness

The CEC of a soil is important in determining the amount and timing of soil amendments.

The CEC is also related to the fate of chemicals added to the soil, such as how much and by what methods amendments should be applied. For example, application of lime and fertilizers is governed by a soil's CEC. With a low-CEC soil, low rates of lime and potassium fertilizer are applied, but on a more frequent basis. At high rates, the low cation adsorption of the low-CEC soils would promote leaching loss of the lime or fertilizer. Because of the high leaching potential of low-CEC soils, ammonium-based fertilizers are also not recommended. Similarly, CEC can influence application protocols for organic wastes and pesticides.

Based on what you have seen about colloids and colloid charge, it should be apparent that the CEC is controlled by four soil factors:

❖ Soil texture, or more specifically clay content

❖ Clay type

❖ Organic matter content

❖ Soil pH

Texture, clay type, organic matter, and pH all help to control a soil's CEC.

The colloids in the soil are the source of charge. The greater the content of colloids in the soil, either mineral or organic, the greater the CEC. However, you have seen that not all colloids are equal in their amount of charge. Organic colloids have more charge per gram of material than do mineral colloids. Furthermore, some mineral colloids (smectite, vermiculite) have more charge than others (kaolinite, iron oxides). Finally, most of the colloids gain or lose charge with changes in **pH.** The net negative charge increases as pH increases. So, a soil may have a high clay content but have a low CEC if that clay is mainly kaolinite. Or a soil with just a little organic matter may have a high CEC if the pH is high. To fully appreciate the charge characteristics of a soil it is important to understand all four of the properties influencing CEC.

Remember that clay type changes as a soil weathers. Thus, it is interesting to note that because of these relationships between clay type and

TABLE 8-1 Cation exchange capacity for typical soils and colloids. *(Adapted from Brady and Weil, 2002; Thomas and Hargrove, 1984)*

Soil or Colloid	CEC (cmol$_+$ kg^{-1})	Typical Soil Order
Humus	150–300	
Vermiculite	120–150	Alfisols, Aridisols, Entisols, Inceptisols, Mollisols, Ultisols
Smectite (montmorillonite)	80–120	Alfisols, Entisols, Mollisols, Vertisols
Typical clay soil	30	
Mica	20–40	Alfisols, Aridisols, Entisols, Inceptisols, Mollisols,
Chlorite	20–40	
Allophane	30	Andisols
Typical loam soil	15	
Kaolinite	1–10	Alfisols, Ultisols
Halloysite	5	Alfisols, Inceptisols, Oxisols
Gibbsite (Al) and goethite (Fe)	4	Oxisols
Typical sandy soil	3	

CEC, there are relationships between degree of soil development and CEC. In general, as weathering intensity increases, the dominant soil clay minerals change from vermiculite to smectite to kaolinite to Fe and Al oxides. There is a corresponding decrease in CEC between each of these clay minerals (Table 8-1). As weathering intensity increases there is a general trend for decreasing CEC. Older, more weathered soils in hot and wet environments tend to have a lower CEC and a lower nutrient-holding capacity than soils in drier, more temperate climates. This creates particular problems for soil in areas such as the southeastern United States and tropical regions around the world.

> As a general rule, the greater the weathering, the lower the CEC.

Determining CEC

The CEC of a soil can be measured experimentally or it can be estimated from other known properties of a soil. Laboratory determination of CEC is based on the knowledge that all of the negative changes on the soil colloids must be exactly balanced by positive charges in the exchange complex. Because CEC is affected by pH, the measurement of CEC is usually done at pH 7. First, you measure a known mass of soil. This soil contains a mixture of different cations within the exchange complex. The soil is mixed with a solution containing a high concentration of a single known cation (usually a cation, such as ammonium, not normally found on the exchange complex in significant quantities). Filtration is used to separate the solution from the soil. At this point you can measure the amount of each of the various cations in the liquid filtrate. The sum of these charges of these cations is the CEC of the soil.

If more advanced instrumentation is not available, the CEC can also be determined with a few additional steps. First, you rinse excess cations from soil (cations in solution). Now all of ammonium (or other replacement cation) in the soil is associated with the exchange complex. Next, the soil is mixed with a solution containing a high concentration of a second known cation (for example, barium). Again, filtration is used to separate the solution from the soil, and you measure the amount of ammonium in this filtrate. The sum of the charges on the ammonium is the CEC of the soil.

> CEC is typically measured by exchanging all of the cations on a soil with a single cation at high concentration.

As an example, consider the following data from a soil sample collected from the A horizon of a soil from a humid environment:

$$Ca^{2+}: \qquad 10.8 \ cmol_c \ kg^{-1}$$
$$Mg^{2+}: \qquad 0.12 \ cmol_c \ kg^{-1}$$
$$K^+: \qquad 0.94 \ cmol_c \ kg^{-1}$$
$$Na^+: \qquad 0.07 \ cmol_c \ kg^{-1}$$
$$H^+: \qquad 3.87 \ cmol_c \ kg^{-1}$$
$$Al^{3+}: \qquad 0.00 \ cmol_c \ kg^{-1}$$

Note that the values are given in concentration of charge ($cmol_c \ kg^{-1}$), indicating that the charge on the individual cations has already been multiplied by the concentration of the cation in the soil. The cation exchange capacity is calculated as just the sum of all cations:

$$CEC = Ca^{2+} + Mg^{2+} + K^+ + Na^+ + H^+ + Al^{3+}$$

So, for this example you find:

$$CEC = 10.8 + 0.12 + 0.94 + 0.07 + 3.87 + 0.00 = 15.80 \ cmol_c \ kg^{-1}$$

You can calculate the percent saturation with a given cation. For example the percent Ca saturation is:

$$\% \ \text{Calcium saturation} = \left(\frac{Ca^{2+}}{CEC}\right) \times 100 = \left(\frac{10.80}{15.80}\right) \times 100 = 68.4\%$$

or the percent **base cation saturation (BCS)**:

$$BCS = \left(\frac{Ca^{2+} + Mg^{2+} + K^+ + Na^+}{CEC}\right) \times 100 =$$
$$\left(\frac{10.80 + 0.12 + 0.94 + 0.07}{15.80}\right) \times 100 = 75.5\%$$

The base cation saturation is important because it conveys information about: (1) the ability of a soil to provide these important plant nutrients, and (2) the relative acidity of the soil solution. You will read more about this later when soil acidity is discussed.

Estimating CEC

CEC can be estimated if the amount of organic matter, clay, and dominant clay type is known.

You can also calculate an estimate of CEC because of the known relationships between clay content, clay type, organic matter content, and CEC (Table 8-1). Consider the following information about a soil:

organic matter content = 3.0%

clay content = 25%

clay type = smectite

The contribution of charge from both clay and organic matter can be calculated. Organic matter (humus) has a typical CEC of 200 $cmol_c \ kg^{-1}$ (Table 8-1). So, for a soil sample with 3.0% organic matter (or 0.03 kg organic matter per kg of soil):

$$200\frac{cmol_c}{kg \ OM} \times \frac{0.03 \ kg \ OM}{1 \ kg \ soil} = 6\frac{cmol_c}{kg \ soil}$$

Smectite clay has a typical CEC of 100 $cmol_c \ kg^{-1}$ (Table 8-1). So, for a soil sample with 25% smectite clay:

$$100\frac{cmol_c}{kg \ clay} \times \frac{0.25 \ kg \ clay}{1 \ kg \ soil} = 25\frac{cmol_c}{kg \ soil}$$

and the total CEC is:

$$6\,\frac{cmol_c}{kg\ soil} + 25\,\frac{cmol_c}{kg\ soil} = 31\,\frac{cmol_c}{kg\ soil}$$

Exchangeable Cations and Their Distribution

Exchangeable cation concentrations are not the same everywhere. They are affected by the distribution of soil parent materials, from which the exchangeable cations originate through weathering. Exchangeable cations are also affected by the distribution of soil-forming processes that lead to the movement of elements, mainly in migrating soil water. Differing exchangeable cation concentrations with depth and from one area to another can influence land use because it affects:

> Soil-forming processes affect the amount and distribution of exchangeable ions in soil.

❖ Nutrient availability

❖ Base cation saturation, which influences soil acidity

❖ Clay activity, such as expansiveness

Soil CEC also varies with depth and from one area to another, mainly in response to differences in clay mineralogy (as discussed in Chapter 7) and organic matter content.

Distribution and Depth

Cation concentrations vary with depth within soil profiles, mainly in response to weathering and leaching, which translocates (moves) the more soluble base cations from upper soil horizons to the subsoil. Nutrient cycling by plants may be able to bring some bases back to the soil surface as they are mineralized from decaying leaf litter.

Distribution and Landscapes

Water moving over and through soil can produce differences in cation concentrations among soils along hillslopes and across landscapes. The more mobile base cations are easily leached from the soils in the upper landscape positions and accumulate in lower landscape positions.

At larger scales, climatic variations can lead to distinct differences in the types of cations found in soils of different areas. In arid environments, where weathering and leaching are minimal, there are commonly greater concentrations of the base cations (Ca, Mg, Na, K). As precipitation increases, leaching intensity increases, and the soluble base cations are lost from the exchange complex and mainly replaced by hydrogen. The concentration of aluminum on the exchange complex may also increase with greater weathering.

> Soil leaching removes base cations and replaces them with H and Al.

FOCUS ON . . . CATION EXCHANGE CAPACITY

1. How is a soil's CEC related to soil texture? Organic matter content?

2. What is a typical range of CEC of soil organic matter? Of Fe . . . Al oxides?

3. Why is the type of clay important when considering the contribution of clay minerals to the total CEC of the soil?

4. As pH increases, does a soil's CEC increase or decrease? Explain this process.

5. What is the percentage of base saturation?

6. Which cations are included in the base saturation? Why is it an important soil property?

7. Which cations are most commonly found on the exchange complex of soils in humid regions, and which are most common in soils of more arid regions?

SOIL ACIDITY

Acidity refers to the concentration of hydrogen ions (H^+) in the soil. Conversely, **alkalinity** is the concentration of hydroxyl ions (OH^-) in the soil. The measurement of acidity is easy and rapid, and it is commonly used to diagnose plant growth problems. It can also be used to make inferences about the fertility status of a soil.

Acidity and pH

Acidity is expressed quantitatively using the pH scale. Recall that pH is calculated as:

$$pH = -\log [H^+]$$

where $[H^+]$ is the concentration of H^+ measured in moles of H^+ per liter of water (moles/L, or M). The term pH refers to the "power of Hydrogen" and the pH scale ranges from 0 to 14. This can be illustrated with an example:

$$[H^+] = 0.00001 \text{ moles/L} = 1 \times 10^{-5} \text{ moles/L}$$

$$pH = -\log (0.00001) = -\log (1 \times 10^{-5}) = 5$$

| pH measures the hydrogen ion concentration in soil. |

Thus, you see that except for very acidic soils, concentrations of H^+ are relatively small (Table 8-2).

There are several important things to remember about pH and the pH scale:

❖ Neutral is defined as pH 7—acidic is less than pH 7, and basic is greater than pH 7.

❖ As H^+ concentration increases, pH decreases—or as H^+ concentration decreases, pH increases.

❖ Decreases in H^+ concentration imply increases in OH^- concentration (or increases in alkalinity).

❖ A decline in one pH unit indicates a ten-fold decrease in H^+ concentration.

TABLE 8-2 Concentrations of H^+ in solution at various pH values.

pH	Concentration of H^+
	moles/liter
3	0.001
4	0.0001
5	0.00001
6	0.000001
7	0.0000001
8	0.00000001

Influence of pH on Soil Properties and Soil Use

You have seen the role of pH in mineral weathering and its influence on cation exchange capacity. Soil pH, though, affects many other soil functions and processes, such that it is sometimes referred to as a "master variable."

Nutrient Availability

In general, pH influences the solubility of many elements. Therefore, changes in pH influence the availability of many plant nutrients (see Chapter 14). Macronutrients, such as N, P, K, S, Ca, and Mg, all have reduced availability at low pH values (Clark, 1984). Phosphorus in particular also has reduced availability at low pH values as well as high pH values. Many micronutrients are also reduced in availability at low (Mo, Cu, B) or high (Fe, Mn, Cu, B, Zn) pH values. In these cases, reduced availability in the soil may result in nutrient deficiencies in plants growing in the soil. Conversely, some elements have increased availability at low (Al, Fe, Mn, Zn) or high (Ca, Mg, Mo) pH values. Aluminum, Fe, and Mn in particular may become toxic to plants and microorganisms when concentrations become high (Foy, 1984). To avoid both deficiency and toxicity conditions, it is best to maintain a pH somewhere in the range of 5.5 to 7.0.

> pH influences the availability and toxicity of soil elements.

Microbial Activity

In neutral to alkaline soils, bacteria flourish. At pH values below about 5.5 fungi are more adaptable and are able to prosper at the expense of bacteria. In extremely acid soils only specialized bacteria are able to grow.

Fungi do not grow well relative to bacteria when the pH is alkaline. In contrast, one group of bacteria, the actinomycetes, seem to grow better than other microbial groups in alkaline soils. However, when the soil is extremely alkaline only specialized bacteria called alkalophiles are able to grow.

Plant Adaptation

Most plants are adapted to thrive in specific ranges of pH values. In most cases, this is related to a plant's response to the availability or toxicity of different elements as pH changes. Examples of this are found at both low pH and high pH values. Many important crops, including alfalfa, sweet clover, sugar beets, asparagus, and lettuce have higher requirements for calcium, which is much more available at high pH. Other plants, mainly those that evolved in more acidic soil environments, require high iron availability. Examples include azalea, rhododendron, blueberry, and many of the species of pine. At low pH aluminum toxicity is a particular problem because high concentrations of exchangeable aluminum in acid soils restrict root elongation (Foy, 1984). Some plants, such as loblolly pine, have developed a certain level of resistance to high Al concentrations and are able to survive at low soil pH values.

> Plants have adapted to soils of varying pH.

FOCUS ON . . . ACIDITY AND pH

1. What is pH?

2. What is the relationship between pH and acidity?

3. What are some of the reasons why soil pH is considered a "master variable"?

4. Which plant nutrients are more available in acidic (low pH) soils? Which are more available in alkaline (high pH) soils?

5. What is the optimum pH range for nutrient availability?

TYPES OF SOIL ACIDITY

There are three important types of soil acidity: active acidity, exchangeable acidity, and residual acidity.

In many respects you can view soil acidity as a special case of cation exchange capacity in which you focus your attention on exchangeable forms of H and Al, the two elements that control pH and soil acidity. In this respect you can subdivide the total soil acidity into three different types based on the interactions of H^+ and Al^{3+} with the soil.

Active Acidity

Active acidity is just the H^+ activity in the soil solution. This is the acidity that is measured with litmus paper or a pH meter. It is also the pH that determines the solubility and availability of the many elements in the soil. However, this is a very small pool of acidity. It takes a very small amount of lime to neutralize this acidity, but as active acidity is neutralized, it is replenished by acidity from the exchange complex.

Exchangeable Acidity

Exchangeable acidity, which is technically referred to as salt-replaceable acidity, is composed of the H^+ and Al^{3+} on colloidal exchange sites in the soil. This is a larger pool of acidity than active acidity (100 times or more greater). Through cation exchange reactions, these ions are released into the soil solution to become active acidity. At any given pH, this pool is greater in soils with higher CEC, so a measure of active acidity alone does not tell you the size of this pool. This is important to remember when it is time to neutralize this acidity.

Residual Acidity

Residual acidity is the greatest source of acidity in soil.

By far the largest pool of acidity is the **residual acidity** (1000 to 100,000 times greater than active acidity). This pool is made up of H^+ and various forms of Al^{3+} bound in nonexchangeable forms. Nonexchangeable H is mainly in the form of hydroxyl ($-OH$) groups on clay minerals, organic matter, or Fe and Al oxides. Residual Al acidity comes from structural Al in clay minerals. When weathering and other soil reactions release this H and Al into the soil solution they become part of the active acidity, lowering the pH of the soil solution.

FOCUS ON . . . TYPES OF ACIDITY

1. What are the three different types of acidity?

2. What is active acidity? How does it compare in size to the other forms of acidity?

3. Why is active acidity important?

4. What is exchangeable acidity? How does it compare in size to the other forms of acidity?

5. How is exchangeable acidity related to a soil's CEC?

6. What is residual acidity?

7. How does residual acidity compare in size to the other forms of acidity?

THE ROLE OF ALUMINUM

Aluminum has been repeatedly referred to as an acidic element. Though Al contains no acid (no H^+), it is acidic because when aluminum is added to the soil, or when it is released from the exchange complex or from the mineral structure (during weathering of soil minerals), it can generate H^+ ions in the soil solution through hydrolysis (splitting of a water molecule into its H^+ and OH^- components):

$$Al^{3+}_{(aq)} + H_2O_{(l)} \rightarrow Al(OH)^{2+}_{(aq)} + H^+_{(aq)}$$

What makes aluminum so strong of an acid is that it does this not once, but three times:

$$Al(OH)^{2+}_{(aq)} + H_2O_{(l)} \rightarrow Al(OH)_2^{1+}_{(aq)} + H^+_{(aq)}$$

$$Al(OH)_2^{1+}_{(aq)} + H_2O_{(l)} \rightarrow Al(OH)_3^{0}_{(aq)} + H^+_{(aq)}$$

So the overall reaction is:

$$Al^{3+}_{(aq)} + 3H_2O_{(l)} \rightarrow Al(OH)_3^{0}_{(aq)} + 3H^+_{(aq)}$$

indicating that for every one aluminum released from the exchangeable or residual pool, three H^+ are generated. Ferric iron (Fe^{3+}) and copper ion (Cu^{2+}) are two other soil cations that can undergo hydrolysis and acidify soil (Thomas and Hargrove, 1984).

> Aluminum is an important acidifying element in soil because it can release three hydrogen ions into the soil solution during hydrolysis.

Acidity and Base Cation Saturation

When CEC was discussed you saw that base cation saturation is defined as the percentage of the total CEC that is made up of base cations (Ca^{2+}, Mg^{2+}, K^+, and Na^+). (Note that the base cations are not truly basic. They are merely the nonacidic cations.) You also saw how the presence of acidic cations (H^+ and Al^{3+}) on the exchange complex increases exchangeable and total acidity. Because the cations on the exchange complex are in **equilibrium** with the cations in soil solution, a higher exchangeable acidity (lower base cation saturation) indicates a higher active acidity and lower soil solution pH. Conversely, a higher base cation saturation (lower exchangeable acidity) indicates a lower active acidity and higher solution pH.

This relationship between soil pH and base cation saturation (BCS) is easily illustrated by examining soil data from various soils (Table 8-3). For these soils you see that there is a clear, direct relationship between pH and BCS: as BCS decreases, pH decreases. It is also interesting to note that the exchangeable cation composition of the more neutral soils is dominated by Ca, and there is no measurable Al on the exchange complex. However, with decreasing soil pH the Ca saturation decreases, and both the H and Al saturation increase. Note also that there is no specific relationship between CEC and BCS—one cannot be used to predict the other.

> As base saturation decreases, pH decreases.

TABLE 8-3 Exchangeable action composition, cation exchange capacity (CEC), base cation saturation (BCS), and pH of A horizons from various soils.

Soil Series	Ca	Mg	K	Na	H	Al	CEC	BCS	pH
				-------cmol^c kg^{-1}-------				----%----	
Maury	13.5	0.17	0.83	0.04	0.38	—	14.92	97.5	6.4
Lowell	8.25	0.72	0.12	0.03	4.48	—	13.60	67.0	5.9
Zanesville	3.00	0.90	0.52	0.03	3.51	—	7.96	55.9	5.8
Cecil	2.6	1.3	0.3	0.2	9.0	—	13.4	32.8	5.1
Riney	0.50	0.12	0.21	0.06	3.49	0.60	4.98	17.9	4.7
Buladean	0.3	0.1	TR	0.2	11.1	2.5	14.2	4.2	4.6
Goldsboro	0.3	0.2	0.3	0.2	26.7	5.9	33.6	3.0	4.0

TR = Trace

— = Not measured

FOCUS ON . . . ACIDIC AND NONACIDIC CATIONS

1. What two adsorbed cations lead to increased soil acidity (lower pH)?

2. How do aluminum cations, which are not directly acidic, contribute to soil acidity in both strongly and moderately acid soils? Be specific.

3. What cations dominate the exchange complex of acidic soils?

4. What cations dominate the exchange complex of alkaline soils?

5. What is the general relationship between soil pH and base cation saturation?

BUFFER CAPACITY

Buffer capacity is the capacity to resist changes in pH. Buffer capacity increases as the CEC increases.

Another important concept related to soil acidity is **buffer capacity.** Buffer capacity is defined as the ability of solid-phase soil materials to resist changes in ion concentration in the solution phase. In terms of soil acidity, this means that buffer capacity is a soil's resistance to change in pH. This can be illustrated with a simple example. Consider two acid soils (represented by the following colloids), to which is added a specific amount of lime, in this case $Ca(OH)_2$ (remember that *lime* is a generic term for a compound that increases the pH or alkalinity of soil):

$$\text{colloid}\left\|\begin{matrix}K_{10}\\H_{90}\end{matrix}\right. + 10Ca^{2+} + 20OH^- \leftrightarrow \text{colloid}\left\|\begin{matrix}K_8\\H_{72}\\Ca_{10}\end{matrix}\right. + 2K^+ + 18H_2O + 2OH^-$$

$$\text{colloid}\left\|\begin{matrix}K_2\\H_{18}\end{matrix}\right. + 10Ca^{2+} + 20OH^- \leftrightarrow \text{colloid}\|Ca_{10} + 2K^+ + 18H_2O + 2OH^-$$

Both soils have a low base cation saturation (10 percent), but the first soil has a much larger CEC. In both cases, the 20 moles of Ca replace 20 moles of charge on the exchange complex (2 moles of K and 18 moles of H), and in both the liberated H is quickly neutralized. However, the BCS of the two soils now differ significantly. With the high-CEC soil the BCS is still relatively low (28 percent), while the low-CEC soil has lost all of its exchangeable acidity and has a BCS of 100 percent.

In this example it is harder to raise the pH of the soil with the higher CEC. You can say that the soil is more buffered against changes in H^+ concentration in the soil solution because removal of H^+ ions from the soil solution results in their replacement by H^+ ions from the exchange complex.

Buffer capacity is determined solely by the CEC of a soil: The higher CEC the greater will be its buffer capacity. This is because with a higher CEC, at the same initial pH, there is more exchangeable (potential) acidity that must be neutralized to change the pH of the soil. Remember, though, that while this example illustrates a resistance to an increase in pH of an acid soil, a soil with a higher buffer capacity and a high pH will also resist a downward change in pH for the same reasons.

Remember that the factors influencing soil CEC (organic matter content, clay content, and clay type) will affect the buffer capacity of a soil:

❖ Clay soils have a high buffer capacity.

❖ Organic soils have a high buffer capacity.

❖ Kaolinitic soils have a low buffer capacity.

Buffering capacity is an important soil property because it must be taken into account along with the pH when estimating the amount of lime needed to raise the pH. This will be discussed further in Chapter 14.

Sources of Acidity

For most soils in most soil-forming environments, the natural trend is to gradually become more acidic with time. Weathering releases Al from many soil minerals, which directly increases soil acidity. Rainfall is naturally slightly acidic, so in humid climates there is a constant source of H^+ added to the soil. Acid rain, which contains strong acids such as dissolved H_2SO_4 and HNO_3, add even greater concentrations of H^+ (Likens et al., 1996). Microbial processes such as respiration, nitrification, and sulfur oxidation in aerobic soils also contribute to acidity. In addition to this, leaching preferentially removes base cations and further promotes the natural acidification of soils. There are, though, several other factors that may contribute to acid soil conditions.

Parent materials that are low in basic cations (for example, granite or sandstone) will naturally produce soils that are low in basic cations. Plant uptake of Ca, Mg, and K removes basic cations from the soil and replaces them with H. Nutrient cycling in natural ecosystems (through the deposition and decay of plant residues) will keep some of these basic cations in the soil. However, organic acids generated from decomposing plant residues will act to increase acidity. In managed systems, removal of plant materials through harvest will remove the basic cations stored in the plant tissues and concentrate H^+ in the exchange sites.

Other soil management practices increase soil acidity. Most notable is N fertilization using ammonium-based fertilizers (such as urea, ammonium

Most soils acidify with time because of weathering and leaching of base cations.

Soil management can affect the amount of acidification.

nitrate, and anhydrous ammonia). The nitrification process, which converts the ammonium (NH_4^+) to nitrate (NO_3^-), releases H^+:

$$NH_4^+{}_{(aq)} + 2O_2 \rightarrow NO_3^-{}_{(aq)} + H_2O + 2H^+{}_{(aq)}$$

As a result, overfertilization with N fertilizer can not only be an expensive waste of capital and lead to leaching losses of N to surface and groundwater supplies, it can contribute to acidification of soil, which will likely require subsequent applications of lime to raise the pH.

Phosphorus fertilization can also increase soil acidity during the hydrolysis of dicalcium phosphorus by this reaction, which essentially produces phosphoric acid:

$$Ca(H_2PO_4)_2 + H_2O \rightarrow CaHPO_4 + H_3PO_4$$

When phosphorus fertilizers are banded during application (applied as a strip), the localized acidification can be dramatic, which is one reason seed placement where fertilizer is banded is a bad idea (Thomas and Hargrove, 1984).

No-till agriculture also tends to produce very acidic surface layers because organic matter decomposition and N fertilization occur at the soil surface, with little subsequent mixing to redistribute the accumulated acids (Blevins et al., 1983).

These processes, along with other sources of acidity, such as the oxidation of mine spoils containing iron pyrite (FeS_2; see the equation below), or the exposure of acid sulfate deposits, lead to changes in soil pH that ultimately may affect nutrient availability (deficiency or toxicity) and plant growth.

$$2FeS_2 + 7H_2O + 7\tfrac{1}{2} O_2 \rightarrow 4\ SO_4^{2-} + 8\ H^+ + 2\ Fe(OH)_3$$

> It is essential to monitor and manage soil acidity in most plant production systems.

It is essential to monitor and manage soil acidity in most plant production systems, from intensive row crop agriculture to silviculture to home lawn care. The many aspects of liming and management of soil acidity are discussed in Chapter 14.

FOCUS ON . . . SOIL ACIDITY DYNAMICS

1. What is buffering capacity? Why is it important?

2. If all of the active acidity (only the H^+ in the soil solution) is neutralized by lime, why does the pH of the soil not change? (Explain your answer in terms of the other forms of soil acidity.)

3. What soil property most influences buffer capacity?

4. What are natural sources of acidic cations in the soil? What are natural sources of basic cations in the soil?

5. Why do soils of arid climates tend to have more basic cations than soils of more humid climates?

6. How do different soil management practices (fertilizer use, tillage) affect soil pH?

SUMMARY

Ion exchange is the exchange of ions in soil solution with ions adsorbed to the surface of soil clay minerals, organic matter, or other soil colloids. Because most soil colloids have a net negative charge they are able to hold positive-charged ions. This gives soil the ability to retain nutrients and water, and then make these available to plants. Because the net surface charge of most soil colloids is negative, cation exchange is the most common type of exchange reaction.

Cation exchange reactions are rapid and reversible. These reactions are charge-for-charge not ion-for-ion. Exchange reactions are influenced by the types of cations present on the adsorption sites and in the soil solution because of differences in affinity, or strength of adsorption.

Exchange capacity is the sum total of exchangeable ions that a soil can adsorb. Cation exchange capacity is controlled by soil texture, or more specifically clay content, clay type, organic matter content, and soil pH. As weathering occurs the CEC tends to decrease. The CEC can be determined by completely exchanging all of the cations from the surface of colloids with one or more known cations. The CEC can also be estimated from the constituents of soils. Exchangeable cations are not uniformly distributed throughout the soil environment. Their concentration can be influenced by water and landscape position.

Acidity refers to the concentration of hydrogen ions (H^+) in the soil. It is a special case of cation exchange. Soil scientists are typically concerned with three types of acidity: active, exchangeable, and residual. Soil acidity develops because of leaching, weathering, microbial processes, and inputs to soil such as fertilization and acid rain. The availability of nutrients is affected by soil pH, and some elements such as Al can become phytotoxic if the pH falls too much. Liming can reduce soil acidity, but is strongly affected by the CEC of soil—liming will have a much greater effect on the pH of soils with low CEC than those with high CEC. Buffer capacity is determined solely by the CEC of a soil: The higher the CEC the greater will be its buffer capacity.

END OF CHAPTER QUESTIONS

1. Why is it beneficial to have exchangeable cations released into the soil solution? Could this also have a negative effect?

2. Why is it necessary to know *both* the CEC and the base saturation to adequately assess the availability of a given nutrient in a soil?

3. In general, as soil weathering intensity increases, does the soil CEC tend to increase or decrease? Why?

4. Considering the relative strength of adsorption of Al^{3+}, Ca^{2+}, Mg^{2+}, K^+, NH_4^+, and Na^+, why do you think that Al^{3+} is held more tightly by soil colloids than Ca^{2+} or Mg^{2+}? Why is Mg^{2+} held more tightly than K^+ or Na^+?

5. What do you think would be a typical pH for a natural (unmanaged) soil in your area? Would this value be higher or lower if the soil were managed for agricultural production? Why?

6. As the soil weathering intensity increases, how does the relative importance of active versus exchangeable versus residual acidity change? Explain your answer.

7. Why do the upper soil horizons tend to have fewer basic cations than horizons lower in the soil profile? How might nutrient cycling or certain management practices reverse this trend?

8. What is acid rain? Why is soil acidification by acid rain more problematic in unmanaged forest land than in managed agricultural land?

9. How does the land application of organic waste materials affect soil pH?

10. When planning an ornamental garden, revegetating a disturbed soil, or managing multiple crops in an agricultural setting, how does pH influence plant selection and subsequent soil management?

FURTHER READING

For those not afraid of a little chemistry, Agronomy Monograph No. 12, "Soil Acidity and Liming," (1984, American Society of Agronomy, Madison, WI) is a thorough introduction to the chemistry of acidity, the physiological effect of acidity on plants, and the agronomics of raising soil pH in various parts of the United States and the tropics.

REFERENCES

Blevins, R. L., G. W. Thomas, M. S. Smith, W. W. Frye, and P. Cornelius. 1983. Changes in soil properties after 10 years continuous non-tilled and conventionally tilled corn. *Soil Tillage Research* 3: 135–146.

Brady, N. C., and R. R. Weil. 2002. *The nature and property of soils*, 13th ed. Upper Saddle River, NJ: Prentice Hall.

Clark, R. B. 1984. Physiological effects of calcium, magnesium, and molybdenum deficiencies in plants. In F. Adams (ed.), *Soil acidity and liming*, 2nd ed., pp. 99–170. Madison, WI: American Society of Agronomy.

Foy, C. D. 1984. Physiological effects of hydrogen, aluminum, and manganese toxicities in acid soil. In F. Adams (ed.), *Soil acidity and liming*, 2nd ed., pp. 57–97. Madison, WI: American Society of Agronomy.

Likens, G. E., C. T. Driscoll, and D. C. Buso. 1996. Long term effects of acid rain: Response and recovery of a forest ecosystem. *Science* 272: 244–246.

Thomas, G. W., and W. L. Hargrove. 1984. *The chemistry of soil acidity*. In F. Adams (ed.), *Soil acidity and liming*, 2nd ed., pp. 3–56. Madison, WI: American Society of Agronomy.

SECTION 4

SOIL BIOLOGICAL AND BIOCHEMICAL PROPERTIES

Chapter 9

SOIL ORGANISMS

"Essentially, all life depends upon the soil. . . . There can be no life without soil and no soil without life; they have evolved together."
Charles E. Kellogg, *USDA Yearbook of Agriculture,* 1938

OVERVIEW

Soil without organisms or biological activity is little more than glorified support media. The soil is not inert, because clays and soil minerals have chemical activity, but a soil's full capacity to support plant life and other environmental processes (such as decomposition and gas exchange with the atmosphere) are not realized without its biology. In this chapter you will begin examining the biological properties of soil by examining the organisms that are present in it.

OBJECTIVES

After reading this chapter, you should be able to:

✔ Identify the organisms in soil.

✔ Determine how these organisms grow and live.

✔ Describe how these organisms interact with their environment.

✔ Identify the ecological roles these organisms have in soil.

✔ List the activities these organisms perform.

KEY TERMS

acellular	encyst	microorganisms
aerobes	endomycorrhizae	mycorrhiza (pl.
ammensalism	eukaryotes	mycorrhizae)
archaea	fenestration	phototroph
autotroph	heterotroph	prokaryotes
bacteriophage	humification	proteobacteria
bioturbation	lithotroph	redox potential
chemotroph	macrofauna	rhizosphere
drilosphere	mesofauna	symbiosis
ectomycorrhizae	microfauna	synergism

WHAT ORGANISMS ARE PRESENT IN SOIL?

One way of classifying soil organisms is by size and complexity.

The soil is teeming with life. Some of it is visible but most of it is invisible. The easiest way to characterize these organisms is by size and complexity (Table 9-1). Soil organisms vary from large, multicellular insects you can catch in a jar to **acellular** viruses that require an electron microscope to see. One classification scheme (and there are many) is to divide soil organisms into the following groups, which become progressively smaller and less complex as you proceed: macrofauna, mesofauna, microfauna, microorganisms.

Macrofauna

Macrofauna are soil organisms that vary in width from 2 to 20 mm, and in length from 10 to > 80 mm. They can be transient, temporary, periodic, or permanent soil residents (Wallwork, 1970). Macrofauna are less directly affected by physical and chemical conditions in soil than are smaller, soil-dwelling organisms.

Insects and Arachnids

Crytozoans are macrofauna that dwell beneath rocks and litter.

Many of the insects found in the soil environment are cryptozoans, animals that dwell beneath rocks, bark, and debris and are easy to see when these shelters are disturbed. Typical cryptozoans include isopods (pillbugs, sowbugs, Figure 9-1) and diplopods (millipedes), which feed on dead and decaying plant material and leaf litter. Chilopods (centipedes) are generally predators, feeding on smaller soil insects.

The three most important insect groups in soils are the Isoptera, Hymenoptera, and Coleoptera. Isoptera (termites) are social insects that have a major impact on wood decay and soil restructuring. They are one of the

TABLE 9-1 Characterizing soil organisms by size.

Organism	Example	Typical Dimensions (μm)
Troglodyte/cave dweller	*Homo sapiens*	2×10^6
Earthworm	*Lumbricus*	1×10^5
Insects	Hymenoptera	$1 \times 10^4 - 1 \times 10^5$
Nematodes	*Pratylenchus*	$20 - 3000$
Protozoa	*Euglena*	15×50 (20 to 2000)
Algae	*Chlorella*	5×13
Fungi	*Mucor*	8.0 (diameter do hypha) 1×10^6 to 1×10^9 (fruiting bodies or hyphal spread)
Bacteria		
Actinomycete	*Streptomyces*	$0.5 - 2.0$ (up to 10,000)
Proteobacteria	*Pseudomonas*	0.5×1.5 (up to 1000)
Archaea	*Methanobacillus*	0.5×1.5
Mycoplasmas		$0.1 - 0.3$
Virus	TMV (tobacco mosaic virus)	$0.02 - 0.30$

FIGURE 9-1 Pillbugs, examples of isopods.

FIGURE 9-2 Soil excavated from an ant colony.

three major groups of earth-moving invertebrates along with ants and earthworms. Lower termites have symbiotic associations with protozoa in their gut, which enables them to digest cellulose. Higher termites lack the protozoa, but form symbiotic associations with bacteria and fungi, which also allows them to consume wood and other cellulose-rich material for food.

Hymenoptera (ants and wasps) are social insects, although the ground-dwelling wasps are usually solitary. Ants, like termites, burrow and excavate nests and in so doing restructure the soil and contribute to nutrient redistribution (Figure 9-2). Collectively, ants are omnivorous, but individual species may have very selective feeding habits ranging from strictly carnivorous to cultivating fungi for food (leaf cutter or Attine ants).

Coleoptera (beetles) are the largest insect order (Figure 9-3). They range from strict predators to plant pests. Some groups, such as scarab beetles, play important roles by burying animal wastes and carcasses in soil.

Arachnids (spiders and scorpions) are major predators in leaf litter layers. The smallest arachnids (mites) are the only members that can truly be said to reside in soil.

Termites, ants and wasps, and beetles are the three most important insect groups in soil.

FIGURE 9-3 Several beetles extracted from soil.

Earthworms and Potworms

Earthworms (annelids) and potworms (enchytraeids) are oligochaetes—multisegmented invertebrate animals. Potworms are much like miniaturized versions of earthworms, only 10 to 20 mm long. They feed on decaying plant material, humus, and fecal pellets deposited by larger organisms. Because of their size, they do not contribute much to restructuring soils. Earthworms, however, play a major role in restructuring and redistributing soil, a fact noted by Charles Darwin (1881). The most common earthworms you see in the United States, the Lumbricidae, are actually European invaders; most native earthworm species have been displaced and are only found in undisturbed environments.

Earthworms can be grouped into three categories from an ecological perspective: epigeic, endogeic, and anecic, which also indicates how much they influence soil (Bouché, 1977). Epigeic earthworms live and feed in litter layers. Endogeic earthworms are rarely at the soil surface and form deep and continually extending burrows in soil. Anecic earthworms such as nightcrawlers (*Lumbricus terrestris*) travel between the soil and the soil surface (Figure 9-4). The organic debris that they carry to their burrows forms middens in cultivated fields.

> Lumbricus terrestris, the common nightcrawler, is actually a European import.

> Epigeic, endogeic, and anecic are terms describing where earthworms reside in soil.

Mesofauna

Mesofauna are soil animals that range in length from 0.2 to 10 mm. For the most part they are permanent soil residents. The major groups are rotifers, tartigrades, collembolans, mites and other microarthropods, and

FIGURE 9-4 Earthworm and earthworm cast on the soil surface. *(Photograph courtesy of the USDA-NRCS)*

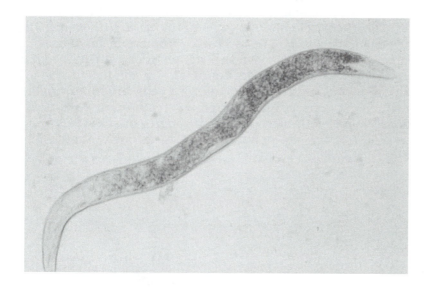

FIGURE 9-5 Nematodes (roundworms) are the most numerous multicellular animals in soil. *(Photograph courtesy of the Soil Science Society of America)*

nematodes. Rotifers and tartigrades ("water bears") are very small saprophytic soil animals < 2 mm in length and usually much smaller. Collembolans (springtails) graze on soil bacteria and fungi. Mites are numerous and diverse members of the spider family. They are the most abundant microarthropods in soils. The most significant members of the mesofauna, however, are the nematodes, which have important roles in soil nutrient cycles, plant disease, and human health (Figure 9-5).

Nematodes

Nematodes (roundworms) are among the most numerous multicellular animals in soil and inhabit water-filled pore spaces and water films. Most nematodes are between 160 and 1300 μm (1.3 mm) long. They are very simple animals with a smooth, tapered body cavity surrounding a digestive tract. At one end of the nematode is a distinctive mouth, which nematologists use to identify and place nematodes into one of several groups: bacterial feeders, fungal feeders, plant feeders, predators, and omnivores. Plant and fungal feeders, for example, are equipped with a specialized structure called a *stylus* that functions like a miniature hypodermic needle and lets nematodes stick and suck out the contents of their food. Predatory nematodes are equipped with a mouthful of teeth.

Although most nematodes in soil are harmless saprophytes, some are serious human and plant pathogens.

Most nematodes in soil are beneficial saprophytes. A few, such as soybean cyst nematode, are serious plant pathogens. Nematodes from human and animal wastes applied to soil are referred to as *helminths*, and are a major health concern.

Microfauna

The **microfauna** are the smallest soil animals, microscopic in size (20 to 100 μm long), and require magnification to observe. Nematodes and the very smallest microarthropods are sometimes placed in this group. However, microfauna, for the most part, consist of the ciliated, flagellated, naked, and testacean protozoa.

Protozoa

The most important microfauna are the protozoa.

There are four major classes of protozoa.

Protozoa are characterized on the basis of their mobility and whether they have an external exoskeleton (Figure 9-6). Flagellated protozoa move through water films in soil by means of one or more whiplike appendages called flagella. *Giardia lamblia* is an example of a pathogenic flagellated protozoa. Cilliated protozoa are covered by numerous small hairs, or cilia. The cilia beat in unison to move the organism or transport food into its mouth. *Paramecium* is a classic example of a ciliated protozoa. Naked amoeba move by means of an elastic, flowing cell wall, which enables them to enter otherwise inaccessible soil pores. Testate amoebas (testaceans) have a hard exoskeleton with openings through which the cell contents exude. The protozoa are saprophytes and feed extensively on microorgan-

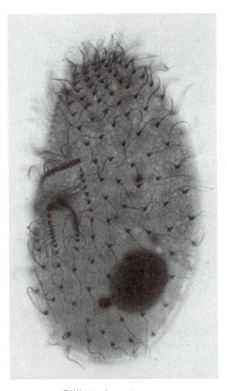

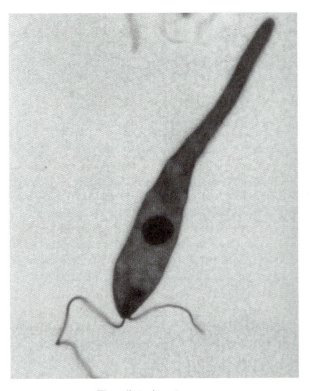

Cilliated protozoan   Flagellated protozoan

FIGURE 9-6   Examples of protozoa. *(Adapted from the 2001 Soil Planning Guide, NRCS/SSSA)*

isms. They are difficult to count because their populations constantly change in response to food availability and soil environmental conditions. Protozoa routinely become dormant (**encyst**) when soil conditions are unfavorable for growth.

Microorganisms

Four groups make up the microbial population in soil: algae, fungi, bacteria, and viruses. **Microorganisms** are generally invisible to the eye, but this belies their critical role in soil processes. While the larger organisms you have read about play a major role in introducing organic matter to soil and influencing soil structure, it is the microorganisms that play the principal role in significant biochemical reactions and processes that occur in soil.

> The microbial population in soil is composed of algae, fungi, bacteria, and viruses.

Algae

Algae are the smallest of the **eukaryote** plants. The algae in soil—green algae (chlorophyta), yellow-green algae (xanthophyta), and diatoms—can take atmospheric and dissolved CO_2 and convert it into sugars through photosynthesis. They begin the food chain in some soil systems. Algae are classified based on the type of photosynthetic pigments they have. Green algae, for example, contain chlorophyll. Yellow-green algae, in addition to chlorophyll, contain carotenoid pigments, which gives them their distinctive pigmentation. Diatoms are a unique algal group because they are surrounded by a hard, silica-rich shell called a frustule that often has an incredibly intricate and delicate structure. Algae are most noticeable in soil right after it rains, and algal blooms can often color the soil green.

Fungi

When you think of fungi in soil it is usually the most obvious manifestations of their growth to which you refer: mushrooms, toadstools, mildew, smuts, and mold (Figure 9-7). However, these fruiting bodies are not representative of the microscopic filaments and individual vegetative cells of fungi in the soil environment. These vegetative cells are 5 to 10 μm in diameter. In the case of vegetative filaments, or mycelia, they can be either septate or aseptate (coenocytic). These terms refer to whether a cross wall is present (septate) or absent (coenoytic) in the hypha. The three major fungal groups in the soil are the ascomycetes, basidiomycetes, and zygomycetes.

> The three major fungal groups in soil are ascomycetes, basidiomycetes, and zygomycetes.

Bacteria

Bacteria are **prokaryote** microorganisms belonging to two groups, the **proteobacteria** and the **archaea**, which have distinctive physiological characteristics but similar shapes and sizes. Bacteria are usually found as rods (bacilli), spheres (cocci), vibrios (short spirals), and spirals of various length (spirilla and spirochaetes; Figure 9-8). Actinomycetes are members of the proteobacteria that can have filamentous growth, like fungi.

> The shape of bacteria is an important basis for their classification.

FIGURE 9-8 Common shapes of soil bacteria.

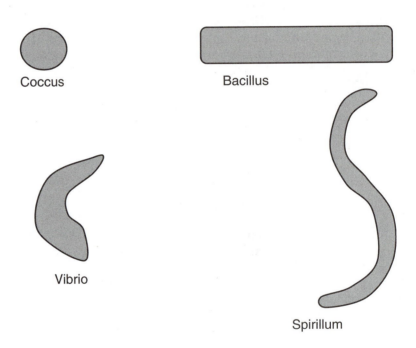

Coccus

Bacillus

Vibrio

Spirillum

Bacteria are very small, usually 2 to 5 μm in diameter. In contrast to fungi, which have great structural diversity, the bacteria have an incredibly rich array of biochemical transformations that they perform. Bacteria also have an incredibly rich variety of physiological capabilities that let them inhabit environments unsuitable for any other type of life. Bacteria can grow in extremes of temperature, pH, pressure, carbon availability, and airlessness; they can photosynthesize; they can metabolize inorganic and organic materials to reproduce and grow.

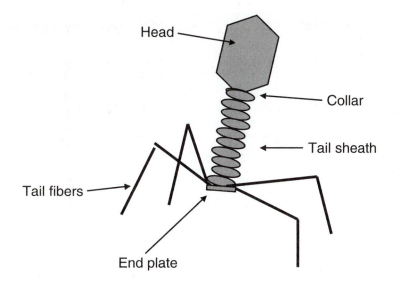

FIGURE 9-9 Bacteriophage—
a virus infecting soil bacteria.

Viruses

A virus is simply a nucleic acid core surrounded by a protein coat. Viruses are the smallest ($< 0.1\ \mu$m in diameter) and most numerous living organisms in soil (Figure 9-9). They are obligate parasites and rely on their hosts—all other larger organisms—to reproduce. The only organisms that viruses don't attack are other viruses because they are such simple organisms that they have lost their ability to reproduce without the assistance of a host. Viruses, also called phages, come in several shapes. Tailed viruses, which typically infect bacteria, have an icosahedral head (a structure with twenty faces), a tail, and a baseplate with appendages that attach to the surface of bacteria. Other viruses are cubic or cylindrical.

> Viruses are obligate parasites; they need a suitable host in order to reproduce.

FOCUS ON . . . TYPES OF SOIL ORGANISMS

1. Which of the macrofauna are mainly responsible for perturbing soil?
2. How does the habitat of earthworms affect their influence on soil structure?
3. What part of the nematode do you look at to characterize it, and why?
4. How do the four types of protozoa differ in terms of mobility?
5. How are fungi different from bacteria?

PHYSIOLOGICAL CHARACTERISTICS

You can characterize soil organisms based on size. You can also characterize soil organisms on the basis of basic cell characteristics such as whether they have a cell nucleus (eukaryotes vs. prokaryotes), or can hardly be considered cells at all (viruses). Proteobacteria, typical soil bacteria, can be distinguished from archaea based on the biochemistry of their cell wall and DNA composition. A classic approach to characterizing differences in soil bacteria is to determine whether they have a thick cell wall

---

THE GRAM STAIN—A CLASSIC APPROACH TO CHARACTERIZING BACTERIA

The Gram stain is a simple laboratory staining procedure that is used to categorize bacteria into two basic groups. The procedure itself is simple. Bacteria are deposited onto the surface of a glass slide, heat treated to kill and permanently fix them to the slide surface, and then stained with a series of stains. First, crystal violet is added. Second, Gram's iodine solution is added to precipitate the crystal violet. Third, the stained cells are briefly decolorized with ethanol. Fourth, the cells are counterstained with safranin. Gram-positive bacteria will appear purple under a microscope after counter staining and Gram-negative bacteria will appear red. The usefulness of the staining procedure entirely depends on bacteria having different thicknesses of cell walls. Gram-positive bacteria have a thick cell wall compared to Gram-negative bacteria. As a general rule, Gram-positive bacteria grow more slowly than Gram-negative bacteria but are more tolerant of adverse soil conditions and low available nutrients. Gram-negative bacteria are able to grow much quicker and respond rapidly to available food, but tend to die quicker if environmental conditions deteriorate.

(Gram-positive bacteria) or a thin cell wall (Gram-negative bacteria) based on a differential staining procedure—the Gram stain.

What Do Soil Organisms Need for Growth?

All organisms in soil require the same basic factors to grow:

1. A favorable environment in terms of temperature, water availability, and pH
2. Basic elements such as H, O, N, P, and S to build the framework of organic compounds such as proteins, lipids, and high-energy molecules for growth; Ca, Mg, K, and Fe to facilitate metabolism and electron transport; micronutrients as metal cofactors for enzyme activity
3. Oxidized and reduced molecules suitable for oxidation and reduction reactions in the cell
4. Growth factors such as vitamins if the organisms are unable to synthesize them independently
5. A source of carbon (C)
6. A source of energy

To get at the heart of what soil organisms do in the soil environment, it is best to characterize them based on their carbon and energy requirements. This allows us to look at general physiological groups carrying out similar sorts of metabolism regardless of organism size or type.

Carbon and Energy Sources

You can characterize all soil organisms based on where they get their C and their energy.

You can characterize all organisms based on where they get their energy and their carbon. If they get their energy from light through photosynthesis they are **phototrophs**. If they get their energy from oxidizing organic and inorganic molecules they are **chemotrophs**. For example, in these terms all plants and some bacteria are phototrophs, whereas all animals, fungi, and some bacteria are chemotrophs. Organisms that exclusively oxidize inorganic compounds are also called **lithotrophs.**

In terms of carbon, organisms that get their carbon from atmospheric or dissolved CO_2 are called **autotrophs,** and organisms that get their carbon from organic sources are called *organotrophs* or **heterotrophs.** Figure 9-10 outlines this scheme for you.

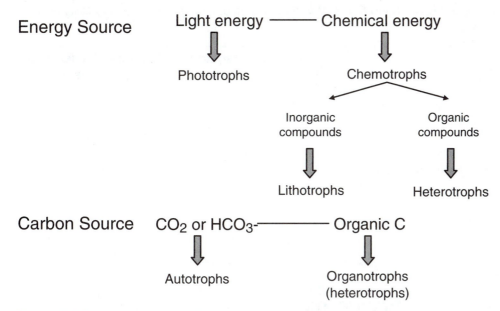

FIGURE 9-10 Classifying soil organisms by carbon and energy requirements.

For heterotrophic organisms the source of organic carbon for food can also be used as a defining characteristic. Organisms can be carnivores, herbivores, or omnivores, for example, depending on whether they consume animals, plants, or both. Another useful characterization is to distinguish between predators, which feed on living organisms, and saprophytes, which consume dead and decayed material.

Metabolism

The organisms in soil can be divided into two groups based on metabolism: those that respire and those that ferment. Respiration, the process that humans carry out, requires oxidizing electron-rich (reduced) compounds. Oxidation removes the electrons from the reduced compounds and as they travel through the cell membrane they are used to make ATP. The electrons are ultimately used to convert O_2 into H_2O. In anaerobic respiration other compounds besides O_2 are used to accept these electrons. In fermentation, reduced compounds are used to make a series of intermediates with high-energy phosphate bonds in a process called substrate-level phosphorylation. Respiration generates a lot more energy for growth than fermentation.

Growth takes energy, so the larger an organism is, the greater the energy it will take to move, grow, and reproduce. Bacteria, which are very small soil organisms, have the fastest growth rates in soil, as little as 0.5 hour per generation (Table 9-2). Earthworms, which are among the largest soil organisms, require almost a month to reproduce. If bacteria were allowed to grow uncontrollably, their mass would soon outweigh the entire mass of soil. This doesn't happen because bacterial growth is limited by the availability of carbon, nutrients, and other growth factors in soil.

Some soil organisms ferment, but most respire to produce their energy.

Numerous factors in soil keep soil organisms from reproducing as fast as they could.

Table 9-2 Generation times of various soil organisms. *(Adapted from Coleman and Crossley, 1996)*

Organism	Minimum Generation Time (hours)	Number of Generations per Season
Earthworm	720	3
Potworm	170	?
Mites	720	2–3
Springtails (collembola)	720	2–3
Nematodes (microbivorous)	120	2–4
Protozoa	2–4	10
Fungi	4–8	0.75
Bacteria	0.5	2–3

FOCUS ON . . . PHYSIOLOGY

1. What is the fundamental difference between eukaryotic and prokaryotic cells?

2. How do phototrophic and chemotrophic (or heterotrophic) soil organisms differ in the way they get their energy for growth?

3. What does it mean for a soil organism to be autotrophic?

4. What does a saprophyte use as food?

5. What does the Gram stain tell you?

ENVIRONMENTAL EFFECTS AND SOIL ORGANISMS

pH, temperature, aeration, and water are the most important environmental factors affecting soil organisms.

Environmental factors have a tremendous effect on the type, number, and activity of soil organisms. The most important environmental factors affecting soil organisms are pH, temperature, aeration, and available water. They most directly affect soil organisms, particularly microorganisms, which are permanent soil residents.

pH

Soil acidity and alkalinity are described by pH, the negative log of the H+ ion (technically speaking, the hydronium ion, H_3O^+) concentration. Plant growth in most soils is optimal between a pH of 6.0 to 8.0. Soil organisms will also be most abundant in this neutral (slightly acid to slightly basic) pH range. Once you leave this optimal range the number and diversity of soil organisms begins to decline. Soils with a pH < 6.0 (acidic) or > 8.0 (alkaline) will have dramatically reduced biological populations for two reasons. First, as plant growth declines there will be progressively fewer large soil organisms, such as earthworms and insects, supported by the available plant growth. Second, microscopic soil organisms like protozoa and bacteria will be affected by the pH of the soil solution and its influence on

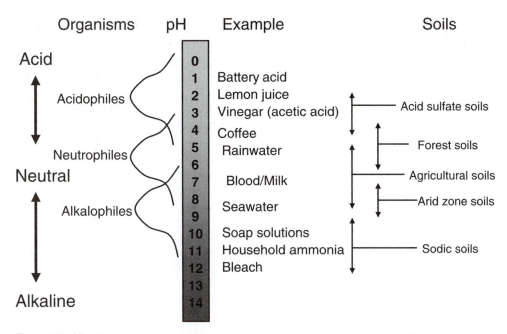

FIGURE 9-11 The pH scale relative to some common materials and growth of representative soil organisms.

the availability of essential nutrients such as phosphorus or the toxicity of elements such as aluminum and manganese (Figure 9-11). Third, fewer plants means less carbon for microbial growth.

As a rule, only microorganisms such as bacteria and fungi predominate in acid or alkaline soils. In extremely acid (pH 1.0) or alkaline (pH 13.0) soils, bacteria alone can survive, although from the human perspective these soils would appear to be wastelands. In slightly acidic soil (pH 4.0 to 6.0) fungi, which are somewhat acid tolerant (acidophilic), will dominate the soil population. At alkaline pH (8.0 to 9.0) filamentous bacteria called actinomycetes will start to dominate the soil population because they are somewhat alkalophilic, or tolerant of basic conditions.

Some microbial processes such as NH_4^+ or S oxidation will decrease the pH of the soil environment. This can affect nutrient transformations. For example, nitrification practically ceases at a soil pH < 4.0. Soil pH can also be managed to control soil organisms to some extent. For example, plant pathogenic actinomycetes, such as *Streptomyces scabies,* can be inhibited by acidifying the soils in which they grow. Some pathogenic fungi such as *Fusarium oxysporum* grow more poorly in well-buffered, neutral soils than acid soils.

> Fungi dominate the soil population in mildly acidic conditions while actinomycetes do so in mildly alkaline conditions.

Temperature

The temperature of the soil environment influences the growth and activity of soil organisms. The boundaries of biological activity in a soil environment range from 0 to 50°C (32 to 122°F). Within that range there are organisms that grow best when the temperature is < 10°C (< 50°F; cryophiles or psychrophiles), 10 to 40°C (50 to 104°F; mesophiles), and > 40°C (> 104°F; thermophiles). Compost piles are good examples of thermophilic environments in which the organisms responsible for decaying the organic material thrive where the temperature rises to 50 or 70°C (158°F).

> Soil organisms can thrive in hot (thermophylic), temperate (mesophylic), or cold (cryophilic) temperature regimes.

Most soil organisms would be considered mesophiles. Small arthropods such as collembola, for example, will move deeper into the soil profile to avoid a source of light and heat at the soil surface. Although microorganisms, typically bacteria, can grow at temperatures below 0°C and close to 100°C (21°F) these would obviously not be environments permitting plant growth.

As the soil temperature rises, the activity of soil organisms increases until an optimum temperature is reached. One reason soil organic matter is lower in the tropics than the tundra is because the decomposition rate is much higher. As a general rule for biological process in soil, the activity increases twofold for every 10°C rise in temperature until the optimum is reached. The optimum temperature varies depending on the typical soil temperature. Within the mesophilic range, for example, the optimum temperature for soil organisms that live in cool soils will be lower than the optimum temperature for soil organisms growing in warmer soils. These optimum temperatures can change on a seasonal basis for several nutrient transformations. For example, the optimum temperature for nitrification is higher in summer than it is in the spring.

> Microbial metabolism increases as the temperature rises, but only to a point or limit.

Available Water

Water can help to moderate changes in soil temperature because it has a high heat capacity, which is one of the reasons that moist soils tend to be cooler than dry soils. Water is also important for permanent soil residents because they require moist environments in which to live, either because they are susceptible to desiccation, as are springtails and isopods, or because they depend on water films in the soil environment in which to move, or through which nutrients can diffuse to them. The latter case is particularly important for nonfilamentous organisms such as nematodes, protozoa, and bacteria. If the water content of the soil is excessive and the soil is flooded, it can drive soil organisms such as mites and springtails out of the soil pores. If the water content of the soil is limited, water films in the soil become discontinuous and isolate soil organisms.

> One of the most important features of water in soil is whether there is enough to make continuous water films for microbial transport.

EGGS AND GELATIN

Why do high and low temperatures kill soil organisms? At high temperatures the principle mechanisms are desiccation and denaturation. At low temperatures, freezing is a problem. Denaturation and freezing disrupt the lipids and proteins that are essential for cells to function. This is best illustrated by what happens to gelatin when it is frozen and eggs when they are boiled. Ice crystals that form in the gelatin as it freezes prevent the orderly arrangement of the proteins in the gelatin. Hardboiled eggs are a good example of how heat causes cell proteins and lipids to coagulate (denature). Because most soil organisms reflect the temperature of their environment (unlike mammals, which maintain a relatively constant body temperature regardless of the external temperature) they are susceptible to temperature extremes. Unless they have time to adapt, drastic temperature changes will kill them. That is one reason why you never freeze a soil sample that you are going to use to examine biological activity.

Water availability, or potential, is more important than water content in determining biological activity of microorganisms in soil. Because water flows from high (e.g., −1 kPa) to low (e.g., −100 kPa) potential, microorganisms must maintain an internal water potential the same as or slightly lower than their external environment or they risk becoming desiccated. Depending on the water content of the environment in which they are found, microorganisms use several strategies to ensure that this happens. Organisms that grow in relatively wet environments will take up solutes from the environment in response to decreased water availability. These organisms are susceptible to further drying because if they accumulate too many external solutes they begin to impair their own metabolism. This response is typical of most protozoa.

In progressively drier environments typical of most soils, some microorganisms can respond to reduced water availability by synthesizing solutes to lower their internal water potential. This response is typical of most Gram-negative bacteria. In the driest or most saline environments the microorganisms have adapted by continuously producing internal solutes. This response is typical of Gram-positive bacteria and many fungi.

Biological activity in soil is highest when the water potential is between -0.1 and -1.0 Mpa (or -1.0 to -10.0 atmospheres tension). Activity declines steeply when the water potential drops below this optimum, and becomes negligible below -6.0 Mpa, although some fungi can maintain activity at water potentials as low as -400 Mpa (Table 9-3).

When soils dry many soil organisms can persist because they enter a dormant state or because they encyst—surround themselves with an impermeable coating and become inactive until water is available again. Protozoa typically encyst. Bacteria and fungi can also produce spores that can persist for long periods.

> Soil organisms have many different approaches to deal with water stress.

Oxygen and Aeration

In addition to moderating soil temperatures, serving as a substrate in enzyme reactions, acting as a solvent, and facilitating transport through soil, water plays a critical role by affecting soil aeration. Because O_2 diffusion is 10,000 times slower through water than air, and because water can only hold a small amount of dissolved O_2 (approximately 8 mg/L at room temperature), it does not take long in a saturated soil for all of the available O_2 to be consumed, to the detriment of those organisms that require O_2 for growth. This is one important reason why, as Figure 9-12 demonstrates, soil activity can actually decline when the water potential is greater than -0.1 Mpa.

> Water is also important because it slows the diffusion of O2.

TABLE 9-3 Influence of water potential on soil organisms. *(Adapted from Harris, 1981)*

Water Potential (Ψ)			
MPa	**Bars**	**Example**	**Activity in Soil**
-0.03	-0.3	Ciliates	Movement of protozoa, bacteria
-0.1	-1.0	Flagellates	Movement of protozoa, bacteria
-0.5	-5.0	Naked amoeba *Spirillum*	Movement of protozoa, bacteria
-1.5	-15.0	*Nitrosomonas*	NH_4^+ and S oxidation affected Wilting point of many plants affected Gram-negative bacteria affected
-2.5	-25.0	*Pseudomonas*	
-4.0	-40.0	Basidiomycete yeasts	Phycomycete fungal growth NH_4^+ and S oxidation cease
		Archrobacter Streptococcus	Gram-positive bacteria affected
-6.0	-60.0		Fungal growth
-10	-100	Saccharomyces	Fungal growth

FIGURE 9-12 Relationship between water potential and respiration. *(Adapted from Paul and Clark, 1996)*

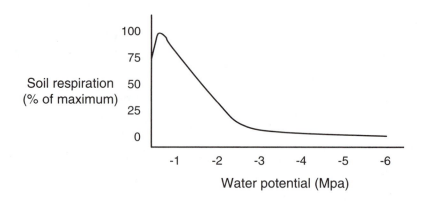

FIGURE 9-13 Relative sequence of reduced compound formation in a flooded soil.

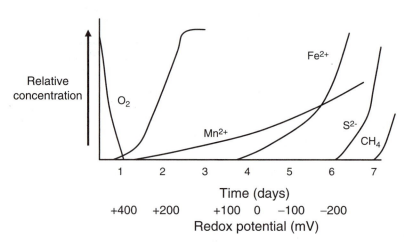

In the absence of O2 some soil organisms can use other compounds for respiration.

With the exception of fungi, some of which are capable of fermentation, all eukaryotic organisms are obligate **aerobes**—they require O_2 for growth and metabolism. Many bacteria are obligate anaerobes—they do not require O_2 and are actually killed by O_2. Between these two extremes are facultative anaerobes. They will preferentially grow when O_2 is present but have the ability, or faculty, to grow in the absence of O_2, often by using other compounds to take the place of O_2 during anaerobic metabolism. This is very important because as a soil changes from being well-aerated to poorly aerated, from aerobic to anaerobic, the products of metabolizing these alternatives to O_2 start to be produced and can be used to estimate the **redox potential** of the soil environment (Figure 9-13). Furthermore, as these products are produced they can affect the chemistry of the soil environment and can result in the loss of gaseous C and N from the soil (see Chapter 10).

FOCUS ON . . . ENVIRONMENTAL EFFECTS ON SOIL ORGANISMS

1. What are some of the consequences if soil pH falls outside a neutral range?

2. In what kinds of soil environments would you expect thermophiles to live?

3. Oxygen is important for soil organisms, so how does good or poor soil aeration affect different groups of soil organisms?

4. Why is diffusion important for soil organisms?

5. Will fertilizing a soil make water more available or less available to soil organisms?

SOIL ORGANISMS AND THEIR DISTRIBUTION

The distribution of soil organisms is not uniform. It is affected by the soil structure, to which the soil organisms contribute, and it is affected by the distribution of available carbon and energy that the soil organisms require for growth.

Distribution and Porosity

Soil is a porous media composed of various-sized sand, silt, and clay particles. Within these different sized aggregates are innumerable pores forming a continuum from the very small (< 20 μm) to the very large (> 250 μm). The size of the aggregates and the size of the pores determine the kind of organisms that can be found there. The smallest pores have few if any organisms and they will be microscopic. One advantage to inhabiting small pores is that larger predatory soil organisms are unable to get in. One disadvantage is that the smallest aggregates with the smallest pores have the least available food for growth and the pores tend to be water-filled and poorly aerated. Consequently, most soil organisms are found in larger soil aggregates that have a mixture of water- and air-filled pores and abundant food sources.

> Pore size is an important determinant of predation and food availability in soil.

Distribution and Depth

The soil is a thin, biologically active skin surrounding Earth. Even in that skin, the zone where soil organisms are found is quite limited, usually in the first meter (40 inches) or so. If you confine yourself to the nonburrowing organisms, then the biologically active layer shrinks even further to approximately 30 cm (12 inches), or the approximate plant rooting depth. Photosynthetic soil organisms will be active only at the soil surface where there is enough light to generate energy. As the soil depth increases, the number of bacteria significantly decreases, and this is true for most soils. For example, Table 9-4 shows results of a study in Nebraska that examined the distribution of soil bacteria by depth. If you look at the column for nutrient agar medium, 77 percent of the bacteria were in the first 30 cm of soil.

Soil organisms accumulate at the soil surface because most soil organisms are saprophytes and only grow where there is available organic matter, which tends to accumulate at the soil surface where plants occur. Tillage management practices, such as no-tillage or conservation tillage, that do not return organic matter into the soil accentuate this stratification (Table 9-5).

Occasionally microbial populations can be observed to increase within the soil profile. This is usually because there is an intervening layer such as clay or organic matter that has more nutrients, more water, or more

> Soil organisms can be found throughout the soil profile, but most are found close to the soil surface.

TABLE 9-4 Number of bacteria in air-dried soils from Nebraska sampled by soil depth. *(Data from Putnam, 1913)*

Depth		Growth Media[a]	
		Nutrient Agar	Ashby's Medium
5.1 cm	2 inches	2,500,000	610,000
10.2 cm	4 inches	660,000	458,000
30.5 cm	1 foot	290,000	417,000
61.0 cm	2 feet	282,000	250,000
91.4 cm	3 feet	69,000	185,000
1.2 meter	4 feet	77,000	210,000
1.5 m	5 feet	56,000	114,000
1.8 m	6 feet	66,000	47,000
2.2 m	7 feet	11,000	2,000
2.4 m	8 feet	7,400	7,100
2.7 m	9 feet	700	300
3.0 m	10 feet	1,200	1,000
3.3 m	11 feet	4,700	2,700
3.6 m	12 feet	1,200	2,600
4.0 m	13 feet	26,500	116,000
4.2 m	14 feet	50	0
4.6 m	15 feet	0	0
4.9 m	16 feet	0	0
5.5 m	18 feet	0	0
5.8 m	19 feet	0	0
6.1 m	20 feet	0	0

[a]Nutrient agar is a complex, protein-rich medium. Ashby's medium is an example of a minimal medium, it contains mannitol (a sugar) as a carbon source and other essential growth elements but no growth factors, so the bacteria that grow must be able to synthesize everything for themselves.

TABLE 9-5 The ratio of various soil biological parameters in untilled and tilled soils. *(Adapted from Doran and Linn, 1994)*

	Untilled/Tilled by Depth	
	0–7.5 cm	7.5–15.0 cm
Organic C	1.4	1.0
Mineralizable N	1.4	1.0
Microbial biomass	1.5	1.0
Fungi	1.4	0.6
Aerobic bacteria	1.4	0.7
Obligate anaerobic bacteria	1.3	1.1

available carbon for growth than soil immediately above or below it, thus resulting in a more favorable environment in which organisms can grow. Soil organisms can be found at great depths in soil but this does not mean that they are active. Their presence could simply reflect their transport through soil channels from the soil surface to deeper depths.

TABLE 9-6 The effect of the rhizosphere on bacteria populations and diversity. *(Adapted from Paul and Clark, 1996)*

Distance from Root (mm)	Population ($\times 10^9$ cm^{-3})	Distinct Morphological Types
0–1	120	11
1–5	96	12
5–10	41	5
10–15	34	2
15–20	13	2

Distribution and Plant Roots

One of the most dramatic influences on the distribution of soil organisms is the presence of plant roots. The soil zone immediately around the plant root and influenced by the growth and exudates of the plant root is called the **rhizosphere.** Because plant roots represent an abundant food source for saprophytes and parasites and the organisms that prey on them, the population of soil organisms immediately adjacent to the plant root, in the rhizosphere, is much greater than the surrounding soil, as Table 9-6 illustrates. Notice how quickly the number and diversity of bacteria decreases in this example. As you move away from the plant root the population of soil organisms exponentially declines and does so at a much faster rate than it does as you move through the soil by depth.

Soil organisms can be characterized by whether they grow preferentially in the root zone or bulk soil. The R/S (rhizosphere/soil) ratio is the ratio of the mass or number of organisms in the rhizosphere relative to that in the bulk soil several cm away. Organisms with a high R/S ratio grow preferentially in the vicinity of the rhizosphere. This includes many Gram-negative bacteria and the protozoa and nematodes that feed on them. Organisms with a low R/S ratio are not influenced by the rhizosphere or grow preferentially in the bulk soil. This includes lithotrophic organisms that get their energy from oxidizing inorganic compounds rather than organic carbon. It also includes many Gram-positive bacteria that can't compete for readily available carbon in the root zone but can successfully compete for less-available carbon in the bulk soil.

> The rhizosphere, which surrounds plant roots, is an important microbial habitat in soil.

FOCUS ON . . . THE HABITATS OF SOIL ORGANISMS

1. What are three ways to look at the distribution of soil organisms?

2. How does porosity affect the distribution of soil organisms?

3. Where in soil would you expect to find the most soil organisms?

4. What is the rhizosphere?

5. How do you calculate the R/S ratio and what does it mean?

NUMBERS AND MASS OF SOIL ORGANISMS

Soil organisms make up part of the solid fraction of soil, the organic matter. The living organisms in soil are generally referred to as biomass. As a general rule, the greater the biomass, the greater the activity and productivity of soil.

Mass versus Numbers

> Bacteria may be more numerous than soil fungi, but fungi make up most of the soil biomass.

There is a direct relationship between the numbers and mass of soil organisms (Table 9-7). Sheer numerical superiority does not necessarily mean that an organism will play a significant role in soil. Two earthworms may have a greater impact on soil structure and function than all the viruses combined; bacteria may outnumber fungi 100 to 1, but it is the fungi that make up the greatest fraction of biomass in most soils.

Enumeration and Estimation

> A variety of techniques are used to extract or trap macrofauna in soil.

The size and mobility of soil organisms has a tremendous influence on the techniques used to enumerate them. The very largest soil organisms, such as insects and earthworms, can be directly extracted by digging up a known soil volume and physically removing each specimen. Alternately, noxious compounds can be added to the soil, such as a weak formalin solution, which will drive these macrofauna to the soil surface, where they can be collected.

Small insects and mesofauna that are too small to be individually picked can be harvested in traps. Typically, a volume of soil is placed on top of a

TABLE 9-7 Size, mass, and numerical relationships among soil organisms.
(Adapted from Coleman and Crossley, 1996)

Organisms	Approximate Length (mm)	Approximate Biomass (kg dw ha^{-1})	Individuals Per 1000 Cubic cm^{3a}
Earthworm	15–85	25–50	2
Potworm	1.0–60.0	1–8	50
Arthropods	1–30		100
Nematodes	0.2–2.0	1.5–4.0	30,000
Mites	1.0	2–8	2,000
Collembola	0.5	0.2–0.5	1,000
Protozoa			1,000,000,000
Naked amoeba	0.03	47.5	
Flagellates	0.01	2.5	
Ciliates	0.08	< 0.5	
Fungi		700–2700	Billions
Bacteria		500–750	Billions

[a]1000 cm^3 has the same volume as a cube of soil 4 inches on each side.

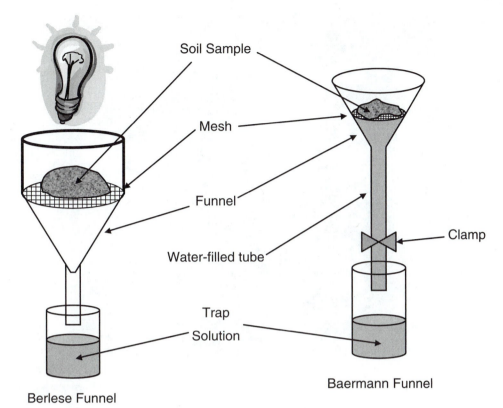

Soil Sample

Mesh

Funnel

Water-filled tube

Clamp

Trap

Solution

Berlese Funnel

Baermann Funnel

FIGURE 9-14   Simple devices used to trap mesofauna.

trap and some stimulus, such as a heat and light source, is used to drive the soil organisms through the soil and into the traps below (Figure 9-14).

Isolating smaller soil organisms most frequently makes use of their growth on specific prey or substrates. Because these organisms are so numerous in soil, one of the first procedures is to dilute the soil sample so that only a portion of the total population is estimated. Thereafter, the organisms are grown in a selective culture medium that only supports the growth of some organisms. This makes use of physiological specificity that increasingly becomes important as soil organisms get smaller. For bacteria and fungi, each individual cell, spore, or fragment can multiply and rapidly form a mass of cells called a colony. The individual colonies that develop on selective media, therefore, tell what the population of the organism is and the morphological differences between colonies tell something about the individual species present in the soil sample (Figure 9-15).

Viruses are particularly difficult to enumerate because of their size. The most common strategy is to grow their specific host and infect them with a solution containing the virus. For bacteriophage, the host cells are distributed over the surface of a selective media that permits the host growth. As the host grows, however, the viruses replicate, are released, and spread to surrounding hosts, infecting and killing them as well. Eventually you are left with a lawn of cells pocketed with cleared zones, called plaques, indicative of where individual viruses developed and were released.

> Soil microorganisms are so numerous that they first have to be diluted before they can be counted.

FIGURE 9-15 Bacterial colonies (colony-forming units) on agar plate.

A CLOSER LOOK AT SERIAL DILUTION

Serial dilution is one of the most important techniques in soil biology. It is a process of reproducibly diluting soil or water samples containing millions of organisms so that one examines only a few representative organisms. In serial dilution a soil or water sample is dispensed into a buffered solution called a diluent and agitated to release as many organisms as possible. A known volume of this agitated sample is collected and added to more diluent (usually in a 1 to 10 ratio). Further dilution is repeated until the organisms present are diluted almost to extinction. The diluted sample is then evenly dispensed on a solid growth medium (plating) or added to broth in a flask that permits growth. Based on the number of colonies that form on the medium, or the evidence of growth in broth and the extent to which the original sample was diluted, you can calculate how many organisms were originally present.

FOCUS ON . . . THE NUMBERS OF SOIL ORGANISMS

1. What is the relationship between size and numbers of organisms in soil?

2. What organism represents the greatest fraction of biomass in soil?

3. Why can you use traps to collect some soil organisms?

4. Which kinds of protozoa are most important in terms of biomass?

5. Why is serial dilution important?

SOIL ORGANISMS AND THEIR ECOLOGICAL ROLES

Everything you have read about in terms of soil organisms pales in comparison to what these organisms do in soil, and how they are essential to the structure and function of soil systems. Soil organisms have an active role in developing soil architecture and an even more important role in decomposing, transforming, and distributing nutrients in the soil environment (Hendrix et al., 1990).

Developing Soil Structure

In many soils, aggregation of minerals and organic matter into clay-, silt-, and sand-sized particles depends on biological activity. At this scale it is the microbial populations that are paramount. Microbial exudates and decomposing cells help to bind particles into packets > 2 μm in diameter. Living bacteria and fungal hyphae help to bind these packets into larger aggregates > 20 μm in diameter. Fungal hyphae and plant roots help to bind these small aggregates into even larger aggregates > 200 to 2000 μm in diameter (Tisdale and Oades, 1982). Mycorrhizal fungi produce a compound called glomalin, which is thought to have significant soil aggregation properties. In no-tillage soils much of the aggregation is due to fungal growth and activity, because networks of fungal hyphae are not disturbed by soil tillage (Beare et al., 1994a; Beare et al., 1994b). In tilled soils bacteria appear to be the most significant influence on soil aggregates.

Soil organisms play crucial roles in developing soil structure.

The larger soil organisms influence soil structure through processes collectively called **bioturbation.** Burrowing animals make large openings in the soil while small insects make small cavities. Termites and carpenter ants create large mounds of excavated soil material. Earthworms, ants, and termites all burrow vertically and laterally in soil, which affects water infiltration. Earthworms consume soil for its organic material and produce wastes called casts, helping to improve soil stability because these casts, in addition to being a rich source of soil enzymes and nutrients, are also more water stable than surrounding soil aggregates. Burrowing earthworms also line the channels they form with a layer of polysaccharides that can influence the surrounding soil. These earthworm-influenced zones in soil are called the **drilosphere.**

Fenestration and Soil Mixing

There is a direct relationship between the surface area of organic debris and the rate at which it will decompose. Before plant material and other organic debris is readily available to microorganisms and other saprophytes it must first be broken down. **Fenestration,** which is carried out by soil insects, isopods, and mollusks, is a process in which plant material is fenestrated or opened to reveal the interior and allow soft tissues to be attacked by bacteria and fungi. In addition, the burrowing activity of insects and other macrofauna helps to bring the organic material into the soil, where it is mixed with other soil material and becomes even more accessible to decomposition and promotes **humification.** If macrofauna are excluded from the process of fenestration and mixing, the overall decomposition rates slow, as illustrated in Figure 9-16. This figure shows that when mesh bags fine enough to exclude most macrofauna are used, the overall leaf decomposition is retarded.

Without the activity of macrofauna, rates of plant litter incorporation into soil would be very slow.

FIGURE 9-16 Decomposition as a function of mesh size in leaf litter bags. *(Phillipson, 1961)*

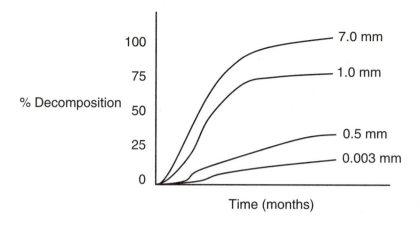

Mineralization and Decay

Algae are a small but significant contributor to the primary productivity of many soils, particularly soils that are just in the process of forming and that do not yet have more developed plant life. In most soils, higher plants supply the organic carbon and nitrogen to soil, and it is the activity of saprophytic fungi and bacteria that makes this material available to other soil organisms: first as it is converted to new microbial growth, then as that new microbial growth is eaten or dies and releases the carbon and nitrogen to the soil environment. Bacteria rapidly metabolize much of the readily available carbon and nitrogen in organic debris, compounds such as sugars, proteins, and lipids. Much of this material, however, is composed of resistant compounds such as cellulose and lignin, which give plants their structural strength. These compounds are best decomposed by saprophytic lignin and cellulose-digesting fungi. Fungi, in fact, are the main agents responsible for decomposing fallen trees and limbs.

> Mineralization by soil organisms recycles organic matter back into plant nutrients.

Population Control

Unrestricted growth of soil organisms rarely if ever occurs because of the limitation of necessary growth factors, or because predator populations keep growth in control. One of the principal ecological roles of the mesofauna and microfauna—the nematodes, protozoa, and microarthropods—is to consume the bacteria and fungi that grow as a result of organic matter decomposition. The same is also true of some of the insects, which serve to control the populations of herbivorous and saprophytic microfauna such as collembolas and rotifers.

> Macro-, meso-, and microfauna all help keep microbial populations under control.

Nutrient Supply

Decomposing microbial biomass represents the largest and most readily available source of nutrients in soil. Because fungi make up the largest fraction of the biomass they also represent the largest single source of readily available nutrients. Protozoa also represent a large supply of readily available nutrients since the protozoa populations in soil are thought to turnover as much as ten times a year (Beare et al., 1992). Dung-burying beetles in environments with considerable animal waste play a significant role in adding organic nitrogen to soil. As much as 75 percent of cattle feces in some instances is buried by beetles, representing an annual nutrient addition of approximately 175 kg N per hectare (Fincher et al., 1981).

Bacteria also contribute to nutrient cycling through symbiotic and asymbiotic nitrogen fixation, and enhanced P uptake by mycorrhizal fungi. Bacteria along with fungi also contribute to the steady dissolution of soil minerals, which releases elements, particularly Ca and P, that are otherwise slowly available.

Symbiotic bacteria and fungi help acquire plant-available N and P.

Parasitism

Viruses are the arch parasites in the soil environment. Through their activity they help keep other soil populations in control. Some bacteria are also parasites. There are several nematode-trapping fungi in soil that increase and develop specialized structure to ensnare, trap, and otherwise infect nematodes (Figure 9-17).

FOCUS ON . . . ECOLOGICAL ROLES

1. How do macrofauna and microorganisms contribute to soil structure?
2. What evidence suggests that macrofauna contribute to organic matter decomposition?
3. What is probably the most important ecological role of fungi?
4. Although protozoa aren't the largest fraction of the soil biomass, they make a significant contribution to nutrient cycling. How?
5. What is the major ecological role of viruses in soil?

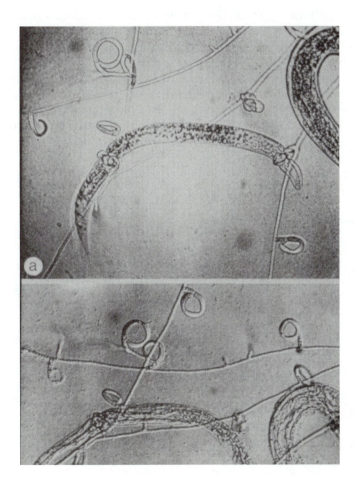

FIGURE 9-17 Nematode trapping fungi.

ECOLOGICAL INTERACTIONS

There are several ways in which soil organisms interact. *Neutralism* occurs when soil organisms have little influence on one another because they are separated spatially, temporally, or occupy different ecological niches. *Competition* is when two organisms directly compete for the same scarce resources to the detriment of both. *Commensalism* occurs when the activities of one organism are beneficial to the growth of another. **Ammensalism** occurs when the activities of one organism are detrimental to the growth of another. *Predation* and *parasitism* happen when one organism is the prey or host of another. *Mutualism* is the condition when the growth of two organisms is mutually beneficial. Mutualism takes several forms. One is **synergism,** in which two organisms growing together will grow better than either organism alone. Another is **symbiosis,** in which the union of two organisms is essential for either the survival or certain physiological processes to occur.

Symbioses

Three types of symbioses stand out as contributing significantly either to the fertility of soil systems, and consequently to their productivity, or to the initial development of soil in harsh environments: the legume/rhizobia and woody shrub/*Frankia* N_2-fixing symbiosis, the mycorrhizal symbiosis, and the symbiosis between algae and fungi to produce lichens.

Legume/Rhizobia

Because atmospheric N_2 is relatively inert, and soil nitrogen is the most limiting nutrient in most soil systems, symbiotic N_2-fixation is a critical path by which plant-available N can enter the soil. The symbiotic association is relatively specific. Forage and seed-bearing legumes form symbiotic associations with bacteria from the genera *Rhizobium, Bradyrhizobium,* and *Azorhizobium.* *Rhizobium* and *Bradyrhizobium* form root nodules while *Azorhizobium* forms stem nodules in which the fixation occurs. Bacteria in the soil invade root or stem tissue and the plant host, in turn, creates a specialized structure in which the bacteria are housed. In exchange for converting N_2 into NH_3, which plants can use, the bacteria are provided with shelter, nutrients, and carbon (Figure 9-18). Only bacteria can fix N_2, and although they can be made to do so apart from the host, they normally do not.

FIGURE 9-18 A well-nodulated plant.

Woody Shrub/*Frankia*

Members of the genus *Frankia* are actinomycetes that also have the ability to infect plant tissue and fix N$_2$ in a symbiotic relationship with their host. *Frankia* infect woody shrubs such as *Alnus, Alder,* and *Ceanothus.* They form nodules that can be the size of baseballs. This association, the actinorhizal association, is important in cool, moist environments, forest soils, and soils developing in adverse conditions. *Frankia* can be grown apart from their host only with great difficulty.

Mycorrhizae

Mycorrhizae are fungi that infect the plant roots of nearly all plants. The symbiosis is almost as ancient as are terrestrial plants. The most significant role associated with mycorrhizae is in plant nutrition, particularly phosphorus nutrition, although additional roles have been ascribed to enhanced drought tolerance, improved disease resistance, and improved soil structure (Figure 9-19). There are two broad categories of mycorrhizae: **ectomycorrhizae** and **endomycorrhizae**.

> Mycorrhizae means "fungus root," and refers to the symbiosis between plant roots and certain fungi.

Ectomycorrhizae

Ectomycorrhizae infect most deciduous plants and shrubs. They form a characteristic structure called the "Hartig net" in which the fungal mycelium surrounds the plant root and penetrates between plant cells. The roots themselves develop a stunted and stubby appearance (Figure 9-20). Ectomycorrhizal species are numerous and the fungus can grow apart from the host, but it is usually a poor competitor with other soil fungi.

Endomycorrhizae

Endomycorrhizae, characterized by the genus *Glomus,* are critical symbionts with most agriculturally important plants. The endomycorrhizae are obligate symbionts and cannot grow apart from their host. Endomycorrhizae usually form two distinct structures that develop in the plant root: an arbuscle, which is a very fine, highly branched invagination into the cell, and a vesicle, which is a spherical structure forming between cells (Figure 9-21). The arbuscles are the site of nutrient exchange between plant and mycorrhizae and the vesicle is a fungal storage organ.

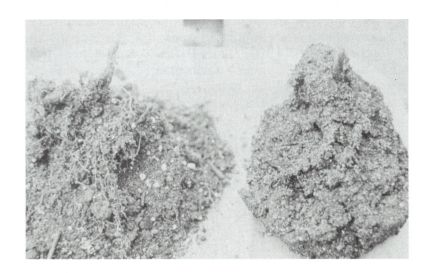

FIGURE 9-19 Mycorrhizae help soil aggregation. The root ball on the left is non-mycorrhizal while that on the right is mycorrhizal. *(Photograph courtesy of J. W. Hendrix)*

FIGURE 9-20 Ectomycorrhizae form a mantle around
the plant root and also cause
the roots to become stubby and
branched. *(Photograph courtesy of J. W. Hendrix)*

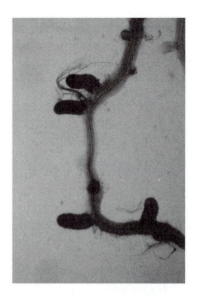

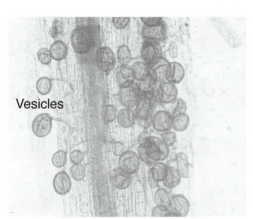

FIGURE 9-21 Endomycorrhizae form characteristic structures inside plant roots called vesicles (a) and arbuscles (b). The endomycorrhizae are sometimes called VA mycorrhizae because of these structures. *(Photograph courtesy of J. W. Hendrix)*

Lichens

Lichens help to initiate the physical and chemical weathering of rocks, the first step in soil formation.

Lichens are obligate and transitory symbioses between algae and fungi or cyanobacteria (photosynthetic, N_2-fixing bacteria) and fungi. Lichens appear in deserts, on rocks, and in soils undergoing the first stages of development. Crustose lichens are small and colorful and frequently appear on bare rock (Paul and Clark, 1996; Figure 9-22). Foliose lichens form colorful ruffled mats, usually in moist environments or on trees. Fruticose lichens are stalked and bushy in appearance.

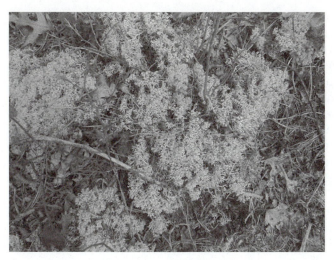

FIGURE 9-22 Crustose lichen growing on rock.

FOCUS ON . . . SYMBIOSES

1. What are several types of ecological interactions in soil?
2. What are the three types of nodule-forming bacteria?
3. What are mycorrhizae?
4. What is the difference between ectomycorrhizae and endomycor-rhizae?
5. What are lichens composed of?

SUMMARY

In this chapter you examined eight key features of soil organisms. The soil organisms can be divided into basic groups by size: macrofauna (insects, arachnids, earthworms, and potworms), mesofauna (nematodes and microarthropods), microfauna (protozoa), and microorganisms (algae, fungi, bacteria, and viruses). Soil organisms can be characterized as prokaryotes or eukaryotes and by where they get their energy and carbon for growth. They can be grouped based on how they are affected by environmental factors such as temperature and pH.

The nutritional requirements of soil organisms and the most important chemical elements required for their growth were discussed. The enormous diversity of soil organisms was examined. You looked at how the soil structure affects the habitats and distribution of soil organisms, and also how it influences their biomass and numbers. You also looked at the various ways in which soil organisms reproduce and are distributed. Finally, you surveyed important ecological roles of soil organisms and how they interact in relationships, like symbioses, with higher plants and animals.

In the next chapter you will take a closer look at how soil organisms are involved in important biogeochemical cycles in soil and the transformation and availability of important soil elements.

END OF CHAPTER QUESTIONS

1. Draw a scale showing the relative sizes of soil organisms. How does cell complexity change as the size of the organism increases or decreases?

2. How does size affect the ecological roles of soil organisms and the locations in soil where they can exist?

3. In terms of C and energy requirements, how would you classify an organism that grew on CO_2 as its C source while metabolizing iron for energy?

4. Draw a graph showing the change in population of an obligate anaerobe as the O_2 content in soil increases.

5. What is the H+ ion concentration when the pH is 4, and what soil organisms are most likely to compete at this pH?

6. Draw a graph that illustrates the change in population of cryophilic, mesophilic, and thermophilic bacteria as the temperature rises from 5 to 75°C.

7. If a soil sample is serially diluted 1000-fold before bacteria are enumerated and 1.6×10^2 bacteria are subsequently observed, what was the starting population?

8. Using the data in Table 9-4, what depth would account for approximately 50 percent of the bacteria?

9. What is the R/S ratio for a ciliated protozoa if its population in the bulk soil is 1.6×10^3 g^{-1} and its population in the rhizosphere is 3.0×10^4 g^{-1}?

10. Draw a graph showing what would happen to the population of a soil organism if an ammensal organism were introduced into its environment.

Chapter 10

SOIL BIOGEOCHEMICAL CYCLES

*"Land, then, is not merely soil; it is a fountain of energy flowing
through a circuit of soils, plants, and animals."*
Aldo Leopold, *A Sand County Almanac*, 1949

OVERVIEW

Chapter 9 introduced various organisms in soil, what they do, and why
they are important. This chapter continues that discussion by examining
how the soil community, and the soil environment itself, contributes to
transforming minerals and chemical compounds. You looked at part
of that activity when you examined weathering (Chapter 2 and Chapter 7).
In this chapter your focus will narrow to just a few key elements that are
transformed chemically and biochemically. Cycling these elements
through air, soil, and water is essential to life and soil's ability to sustain
living things.

OBJECTIVES

After reading this chapter, you should be able to:

✔ Distinguish between organic and inorganic compounds.

✔ Draw biogeochemical cycles for the most important elements: C, N, P,
S, Fe, and Mn.

✔ Recognize oxidation and reduction reactions, and where they occur.

✔ Understand how environmental conditions such as pH, temperature,
and aeration affect the form and rate at which element transformations occur.

KEY TERMS

ammonification	assimilation	chelate
anaerobic	biogeochemistry	chemodenitrification

denitrification	mineralization	phosphatases
disproportionation	nitrification	redox reactions
fermentation	nitrogenase	reduction
heterotrophic	orthophosphate	saprophytic
nitrification	oxidation	siderophore
immobilization		

BIOGEOCHEMICAL TRANSFORMATIONS

The topic of this section is the elements involved in the biochemistry of soil and the organisms that live in soil. Biochemical transformations are often inseparable from geochemical transformations, hence the term biogeochemistry. Although there may appear to be countless different transformations that can occur, these transformations can usually be grouped into one of four types:

❖ **mineralization**—transformation from an organic form to an inorganic form

❖ **assimilation**—transformation of an inorganic form to an organic form (also called *immobilization*)

❖ **oxidation**—removing electrons from a compound

❖ **reduction**—adding electrons to a compound

Biochemical transformations affect the chemistry, availability, and behavior of compounds. For the remainder of the chapter you will examine the most significant transformations that affect soil systems.

The Elements of Soils and Life

> Six elements (C, H, O, N, P, and S) make up 95 percent of living things.

Biogeochemistry is the chemistry of the "light" elements that compose the Earth and its inhabitants. All of the biologically relevant elements have an atomic weight less than iodine (atomic weight, 126.9; atomic number, 53) (Schlesinger, 1997). Conveniently, just as there are twenty-six letters in the English alphabet, there are twenty-six elements that play a role in biogeochemistry (Figure 10-1). Of these twenty-six elements, just six—C, H, O, N, P, and S—make up 95 percent of living things. Most of the living mass is actually water (H_2O). Many other elements are found in soil, and many other elements are found in living organisms and their by-products. But those elements do not play a key role in life processes. Some of these nonessential elements are accidentally transformed because of their chemical similarity to essential elements, while others are just adsorbed or absorbed during growth.

Mineralization

> Compounds in soil are either C-containing organic compounds or inorganic compounds.

Mineralization transforms organic compounds into inorganic compounds. Organic compounds contain C and usually H. The sugar glucose ($C_6H_{12}O_6$), the organic acid acetic acid (CH_3COOH), the gas methane (CH_4), and the amino acid glycine (NH_2CH_2COOH) are all organic compounds. However, there are several exceptions to this general rule. For example,

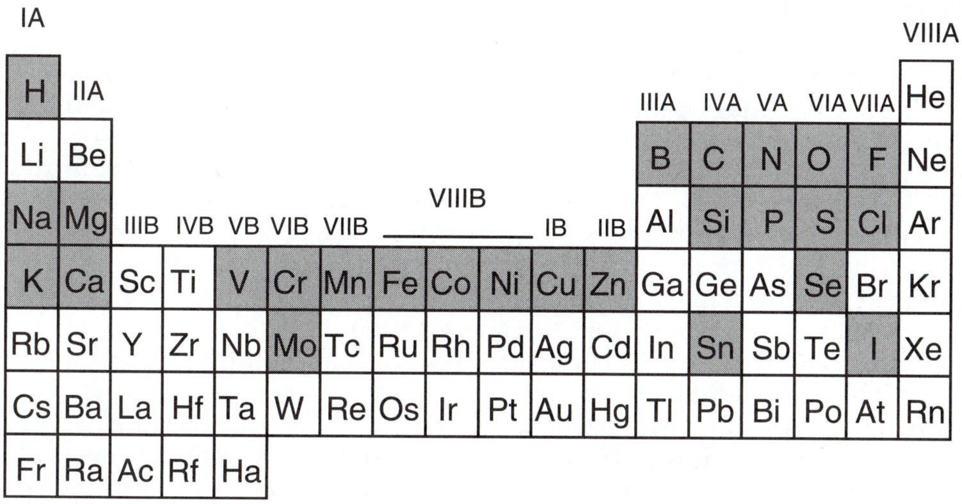

FIGURE 10-1 Modified periodic table of the elements showing the key elements involved in life processes in soil (Lanthinide and Actinide series omitted for simplicity).

carbon dioxide (CO_2) and carbonate ($CO_3{}^{2-}$) contain C, and bicarbonate ($HCO_3{}^-$) and carbonic acid (H_2CO_3) contain C and H, but none of these compounds are considered to be organic.

When an organic compound is completely mineralized it is decomposed into just a few basic inorganic constituents, such as H_2O, CO_2, $NH_4{}^+$, $NO_3{}^-$, $NO_2{}^-$, $HPO_4{}^{2-}$, $H_2PO_4{}^-$, $SO_4{}^{2-}$, or H_2S, and some trace elements. These compounds can be lost as gas to the atmosphere, dissolved in the soil solution, or adsorbed to the cation and anion exchange complex of soil. Because the compounds are now in their inorganic forms, they can be taken up during plant and soil organism growth.

Immobilization/Assimilation

When a plant or a soil organism takes up an inorganic molecule from soil, the compound is said to have been **assimilated**. **Immobilization** is another term describing this process because if the molecule is taken up and incorporated it is no longer free to be used elsewhere (it is immobile). When plants and soil organisms take up $NH_4{}^+$ and $NO_3{}^-$, for example, they typically incorporate the N as amino acids. The first amino acids that the N appears in are glutamate and glutamine. Sulfate is also incorporated into amino acids, and is an active element in several proteins.

Phosphate is converted into ATP (Adenosine triphosphate), the energy currency of cells. Phosphate is also incorporated into electron carriers, and into phospholipids, which are important parts of cell membranes. Iron and Mn are taken up and become important components of electron carriers and cell enzymes. Many of the trace metals, such as Zn and Mo, are also important in enzymes, where they function as cofactors. Cofactors are metal ions that assist the activity of enzymes. Calcium and Si are assimilated and converted to structural elements such as the shells of invertebrates, in snails and some protozoa, and of algae such as diatoms.

> Immobilization makes nutrients unavailable.

Oxidation

Oxidation is the loss of electrons by a compound. A compound becomes *oxidized* when this occurs. Oxygen is often, but not always, involved in this reaction. For example, CH_4 is oxidized to CO_2, NH_4^+ is oxidized to NO_3^-, and H_2S is oxidized to SO_4^{2-}, but ferrous iron (Fe^{2+}) is oxidized to ferric iron (Fe^{3+}). Oxidizing agents, such as O_2, have a very strong tendency to oxidize other compounds by removing electrons from them. Because the Earth's atmosphere is approximately 21 percent O_2, it creates an oxidizing environment in which most compounds that are reduced (have lots of electrons) will lose them to O_2 and become oxidized.

The oxidation state of a compound is a very important concept to understand. The oxidation state tells you whether a compound is likely to gain or lose electrons in a particular environment. As a general rule, oxidized compounds gain electrons and become reduced in reducing environments. Reduced compounds lose electrons and become oxidized in oxidizing environments. The following box gives some guidelines for determining the oxidation state of a compound.

GENERAL RULES FOR DETERMINING OXIDATION STATE

1. The oxidation state of any free element is zero; for example, elemental S has an oxidation state of zero and O_2 has an oxidation state of zero.

2. Elemental ions have an oxidation state equal to their charge; for example, Ca^{2+} and Mg^{2+} each have an oxidation state of +2.

3. Hydrogen and oxygen have oxidation states of +1 and −2, respectively, when they are combined with other elements.

4. The sum of oxidation states must equal the net charge of the compound.

5. Hydrogen and oxygen excepted, if these rules are applied and a compound has a positive value it is said to be oxidized; if the value is negative the compound is reduced.

Reduction

Reduction is the opposite of oxidation. If a compound *is reduced* it has electrons to spare, and if it *becomes reduced* it gains electrons. For example, CH_4, NH_4^+, and H_2S are all reduced compounds. Hydrogen is often transferred to a compound when it gets reduced, so looking at the number of H in a compound can suggest how much it is reduced. Reducing environments are electron rich. These environments are typically **anaerobic** (O_2-free) because, as you have seen, O_2 is a strong oxidizing agent and will capture electrons when possible.

Oxidation and reduction reactions are never wholly separate from one another in the environment. If one compound gets oxidized another compound must get reduced. These interactions are called **redox reactions.** The scheme is illustrated below:

$$A_{reduced} \longrightarrow A_{oxidized} + e^-$$
$$B_{oxidized} + e^- \longrightarrow B_{reduced}$$
$$\overline{A_{reduced} + B_{oxidized} \longrightarrow A_{oxidized} + B_{reduced}}$$

A good example of a redox reaction is H_2 oxidation by O_2:

$$2H_2 + O_2 \longrightarrow 2H_2O$$

Hydrogen, which is reduced, donates electrons to O_2, which is oxidized. The hydrogen gas with an oxidation state of zero goes to hydrogen in water with an oxidation state of $+1$ while the oxygen in oxygen gas (with an oxidation state of zero) goes to oxygen in water with an oxidation state of -2.

A Model for Viewing Biogeochemistry

If you look at biogeochemistry in terms of these four basic reactions (mineralization, assimilation, oxidation, and reduction) you can develop a simple model (Figure 10-2) for soil transformations that will let you evaluate nutrient element cycling from a common perspective. In this model, the element in question (within the central circle) can travel between inorganic or organic forms, and oxidized (aerobic) or reduced (anaerobic) environments. Depending on where it is, the element can take many forms: organic, inorganic, oxidized, or reduced. This model will help to guide you through some of the most significant biogeochemical transformations in the next few sections.

FOCUS ON . . . BIOGEOCHEMICAL TRANSFORMATIONS

1. Which key elements comprise 95 percent of the mass of most living organisms?
2. What four processes underlie most biogeochemical processes?
3. What transformation occurs during mineralization?
4. Why is assimilation also called immobilization?
5. How are oxidation and reduction reactions linked?

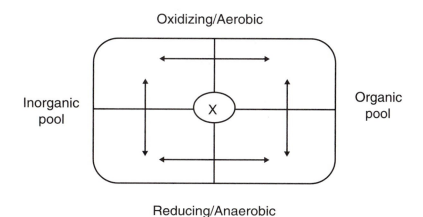

FIGURE 10-2 Biogeochemical model for element transformations in soil. X is the element of interest.

THE CARBON CYCLE

Earth is a world dominated by carbon (C). It is an element that is literally the "stuff of life." Carbon is not only a source of energy for most living things, it is also the element that makes up most of the dry mass of living things. To understand the cycling of C in the soil, therefore, is to understand much about how life in soil is possible.

Organic and Inorganic Forms

There are four major active pools of C on Earth: in soil (1500×10^{12} kg), in water ($38,000 \times 10^{12}$ kg), in air (750×10^{12} kg), and in biomass (560×10^{12} kg)[2]. The Earth itself contains about 10^{20} kg C in the form of sedimentary rocks and carbonate. Limestone ($CaCO_3$) is an example of one such sedimentary rock. Extractable fossil fuels represent about 4×10^{15} kg C (Figure 10-3).

| Carbon is found in active and passive pools. |

Carbon takes several inorganic and organic forms within these active pools. The most common inorganic forms are carbon dioxide (CO_2), carbonic acid (H_2CO_3), bicarbonate (HCO_3^-), and carbonate (CO_3^{2-}). Note that these are all oxidized forms of C. If you want to look at reduced forms of C you must look at organic compounds. These organic compounds can be solids, liquids, or gases. The two most important groups are the carbohydrates (general formula $[CH_2O]_n$) and hydrocarbons (general formula $[C_nH_{2n+2}]$). Carbohydrates make up the bulk of living and dead plant tissue, while hydrocarbons are a major form of extractable fossil fuel (Figure 10-4).

Mineralization

The products of complete C mineralization are methane (CH_4), bicarbonate (HCO_3^-), and carbon dioxide (CO_2) in an anaerobic environment and

FIGURE 10-3   The soil C cycle.

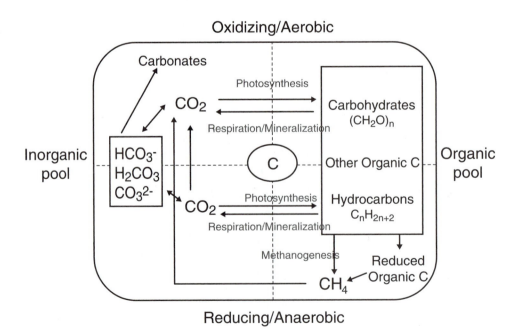

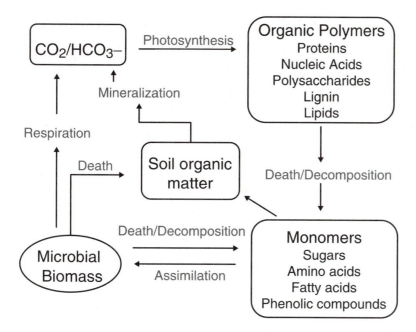

FIGURE 10-4 Organic compounds and transformations in the soil C cycle. *(Adapted from Stevenson and Cole, 1999)*

HCO_3^- and CO_2 in an aerobic environment. In both cases the gases are the products of cell metabolism in which the C is oxidized to release energy. Intermediate compounds, such as organic acids (acetic and butyric acid, for example) can also be released in each environment because many organic C compounds cannot be completely mineralized by any one organism. Ultimately, however, these intermediates will be completely metabolized. In an aerobic environment, for C to be mineralized is for it to be oxidized to CO_2.

Assimilation/Immobilization

With the exception of invertebrates such as snails, which precipitate C in their shells as carbonate, for C to be assimilated it is reduced. Assimilation wholly depends on the ability of a few microorganisms, but mostly green plants and algae, to convert CO_2 in the atmosphere into carbohydrates during photosynthesis. The photosynthetic organisms thereby meet all of their C requirements for growth, and synthesize many other complex organic molecules. All other living organisms, humans included, rely on consuming these photosynthetic organisms or their products to meet requirements for C.

Higher organisms such as cows and people can consume plant life directly, but most organisms in soil are **saprophytic,** meaning they consume the plant tissue once it has died and fallen onto the soil. Assimilation, like mineralization, involves breaking down the polymers that make up plants into smaller units (monomers) such as simple sugars. These simple compounds are then reconstructed, in the case of assimilation, into new polymers such as proteins and cell walls, or the monomers are oxidized biologically during respiration and fermentation to generate energy (see the following box). Although soil represents the biggest pool of active soil C apart from the ocean, virtually all of it at one time or another originally derived from photosynthetic organisms.

Saprophytes consume dead and decaying organic matter.

WHAT'S THE DIFFERENCE BETWEEN RESPIRATION AND FERMENTATION?

Living organisms generate energy for growth by three major processes: photosynthesis, respiration, and fermentation. Photosynthesis uses light energy to generate ATP. Respiration uses chemical energy to generate ATP. The essence of both pathways is similar. Electrons gain high potential energy because they have been energized by light or come from reduced compounds. Like water flowing down a stream, these high-energy electrons flow through a series of electron carriers that are analogous to water wheels. As the wheels turn they generate ATP if the potential energy of the electron is great enough. With enough potential energy you can generate many ATP from each electron transfer. Eventually the electrons are accepted by a terminal electron acceptor such as O_2.

Fermentation occurs in the cytoplasm of cells. In fermentation, electrons are moved between compounds in a cell so that some of them become oxidized and others become reduced. During that transfer, intermediate compounds are formed that have high-energy phosphate bonds. This process is known as *substrate-level phosphorylation*. The high-energy bonds can be used to do work in the cell. Fermentation can occur in the absence of O_2 because it doesn't require a terminal electron acceptor, but fermentation doesn't yield nearly as much energy as respiration or photosynthesis.

Environmental Influences

Global warming stimulates C fixation but also C mineralization.

Because the principal source of C for life processes is photosynthesis, any environmental factors that limit or enhance this process will subsequently affect mineralization and assimilation. Thus, global warming and increased atmospheric CO_2 levels may stimulate greater C fixation through longer growing seasons and greater plant growth. Forests in temperate environments such as New England may be actively accumulating CO_2 in biomass as a result of a warming climate (Birdsey et al., 1993). Carbon accumulation, or sequestration, in soil is offset by higher soil temperatures if global warming occurs (Schimel, 1995), leading to fears that large deposits of soil C in cool, moist environments such as the tundra will begin mineralizing and producing CO_2.

FOCUS ON . . . THE CARBON CYCLE

1. Four sources make up most of the active C on Earth; what are they?
2. What are examples of organic C?
3. What is produced when organic C is completely mineralized?
4. Why do most other organisms rely on plants for C?
5. How can global warming contribute to C fixation?

THE NITROGEN CYCLE

Nitrogen (N) plays a special role in life. It is a fundamental component of protein. In terms of plant production it is often the most limiting of all elements. The atmosphere is predominantly nitrogen gas (N_2). But, perversely, N_2 is one of the least available N forms. Hence, the focus of much of soil fertility is on how to increase N availability in soil.

TABLE 10-1 The N in various pools of the soil ecosystem. *(Adapted from Stevenson and Cole, 1999)*

Pool	kg N	% of Total Soil N
Biomass		
plants	1.4×10^{13}	4.2%
animals	2.0×10^{11}	0.1%
microorganisms	5.0×10^{11}	0.2%
Litter	3.3×10^{12}	1.0%
Soil organic matter	3.0×10^{14}	89.5%
Fixed NH_4^+	1.6×10^{13}	4.8%
Soluble inorganic N	1.0×10^{12}	0.3%

Organic and Inorganic Forms

Excluding N in the lithosphere (rocks and minerals) there is estimated to be about 388.4×10^{16} kg of N actively circulating on Earth (Burns and Hardy, 1975). In marked contrast to C, 99.4 percent (386×10^{16} kg) of the N is in the form of relatively inert N_2 gas in the atmosphere. The remaining N is partitioned between the oceans (0.59 percent), biomass (0.02 percent), and soil (0.006 percent) (Stevenson, 1982).

If you take a closer look at the soil you can place N into several groups, as shown in Table 10-1. Most of the N in soil is stored in soil organic matter (about 90 percent) with a relatively small amount of inorganic forms such as NO_3^-, NO_2^-, and NH_4^+. Most of the organic N in soil originated from decomposing plant tissue and microbial cells. Much of this organic N is unidentifiable, but it is possible to detect organic N compounds such as proteins, amino acids, nucleic acids, and amino sugars.

Because N is one of the most significant elements in living things, and because so much of it is unavailable as N_2 gas, it has one of the most active of all cycles, readily changing between assimilated and mineralized forms, and between oxidized and reduced states (Figure 10-5).

Most of the plant available N in soil is organic.

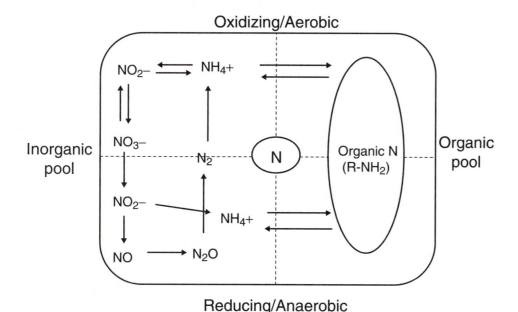

FIGURE 10-5 The soil nitrogen cycle.

Mineralization

Nitrogen mineralization transforms organic N to NH_4^+, a process also called **ammonification**.

$$\text{Organic N} \xrightarrow{\text{ammonification}} NH_4^+$$

The soil organisms most responsible for this step are heterotrophic bacteria. The substrates for mineralization are varied, and include such compounds as proteins, peptides, nucleic acids, aminopolysaccharides, urate, and ureides. Because the substrates are varied, the enzymes that are used to mineralize organic N are also varied. Urease, for example, is the enzyme that decomposes urea (NH_2CONH_2) to NH_4^+. Urea is an organic N compound of great significance because it is one of the dominant organic N wastes excreted by animals, including humans. It is also among the most common solid N fertilizer sources in the world. Two molecules of NH_4^+ and one molecule of CO_2 are released for every molecule of urea during urea mineralization.

Immobilization/Assimilation

Nitrogen assimilation transforms inorganic N, NH_4^+, NO_3^-, and NO_2^-, into organic N.

$$\text{Organic N} \underset{\text{assimilation}}{\overset{\text{mineralization}}{\rightleftarrows}} NH_4^+, NO_3^-, NO_2^-$$

Ammonium is first assimilated as glutamate and glutamine. Later, the N is incorporated into other amino acids and N-containing compounds. When NO_3^- is assimilated it is first reduced to NO_2^-, and then NO_2^- is reduced to NH_4^+, which is assimilated as previously described.

Mineralization and assimilation of N occur simultaneously in soil. Whether the balance of these processes leads to inorganic N accumulation or depletion in soil depends on the C:N ratio of the soil and the organic compounds being mineralized. Typically, compounds with C:N ratios less than 30 will cause net production of inorganic N in soil. Compounds with C:N ratios greater than 30 will lead to net immobilization.

> The C:N ratio determines whether N will be released or immobilized.

Oxidation

The NH_4^+ released into the environment through either fertilization or mineralization will be oxidized in aerobic environments in a process called **nitrification**. Nitrification is a two-step process in which NH_4^+ is first oxidized to NO_2^- by chemoautotrophic bacteria such as *Nitrosomonas* and the NO_2^- is oxidized to NO_3^- by NO_2^--oxidizing bacteria such as *Nitrobacter.*

$$NH_4^+ \longrightarrow NO_2^- \longrightarrow NO_3^-$$

The NO_2^- rarely accumulates in soils because its rate of oxidation is faster than the rate of NH_4^+ oxidation.

Nitrate and NO_2^- can also be formed by a process called **heterotrophic nitrification**. Reduced N in organic compounds is oxidized by heterotrophic bacteria and fungi, which metabolize the organic compounds for energy. Unlike the chemoautotrophic nitrifiers, which do not grow below pH 5, the heterotrophic nitrifiers are active in acid soils, particularly forest soils. Nitrate can also be formed abiologically, although the amounts

> Nitrate is formed in soil by autotrophic and heterotrophic nitrification.

are much less than through the combined action of mineralization and nitrification (Hutchinson, 1944). Atmospheric N can be oxidized to NO_3^- during electrical discharge by lightning, photochemically, and in the trail of meteorites. Nitrate can also be formed by burning fossil fuels, which is a major contribution to acid rain.

Reduction

Inorganic NO_3^- and NO_2^- suffer several fates in soil. These anions can be assimilated by plants and microorganisms after first being reduced to NH_4^+. This process is called assimilatory reduction. In environments that have lots of C and remain persistently anaerobic, another fate of NO_3^- and NO_2^- is dissimilatory reduction to NH_4^+. The NH_4^+ that is produced is not used to develop new organic N compounds but to generate energy through substrate-level phosphorylation (see box on page 212). A third fate of NO_3^- and NO_2^- in anaerobic and reducing environments is conversion to N_2 gas in a process called **denitrification** (see the sequence below). In denitrification, the NO_3^- and NO_2^- are used for anaerobic respiration in place of O_2 by a diverse group of bacteria called denitrifiers.

$$NO_3^- \longrightarrow NO_2^- \longrightarrow NO \longrightarrow N_2O \longrightarrow N_2$$

Denitrification generally takes place only in anaerobic, O_2-free environments. Two of the intermediates that it produces, nitric oxide (NO) and nitrous oxide (N_2O), have been implicated in global warming and destruction of the ozone layer (Conrad, 1990).

A chemical process, **chemodenitrification**, also reduces N. In chemodenitrification, NO_2^- reacts with soil organic matter and amino acids, usually in acid conditions, to produce N_2, NO, and N_2O. This process increases when any phenomenon interferes with the further oxidation of NO_2^- during nitrification.

This description of N reductions in soil is incomplete because it's obvious that it has not represented a true N cycle. If all reactions proceeded as described, then all of the available N would accumulate as N_2 gas in the atmosphere over geologic time. The last, and perhaps the most important, reduction in the N cycle is the reduction of N_2 (N oxidation state = 0) to NH_3 (N oxidation state = −3). This reduction is performed industrially by the Haber-Bosch process. Biological nitrogen fixation is carried out by various asymbiotic and symbiotic bacteria that share a common enzyme complex called **nitrogenase**. Nitrogenase enables these bacteria to convert (fix) atmospheric N_2 into NH_3, which is subsequently incorporated into organic N compounds.

Nitrogen-fixing bacteria can be free-living, photosynthetic organisms such as cyanobacteria (blue-green algae). Examples of cyanobacteria are *Anabaena*, *Nostoc*, and *Calothrix*. Nitrogen-fixing bacteria can be free-living (asymbiotic), heterotrophic bacteria such as *Azotobacter* and *Clostridium*. The most important N-fixing bacteria in soil are those that form symbiotic relations with leguminous plants and a select number of woody shrubs. These are the rhizobia and *Frankia*, respectively. The symbiotic nitrogen-fixing bacteria form special structures, principally on the roots of their host, called nodules (Figure 10-6). In exchange for a home and a guaranteed supply of C, the N-fixing bacteria supply the host plant with NH_3 that the plant can use for growth in N-limited soils. Until industrial fixation was developed in the twentieth century, biological nitrogen fixation was the principal mechanism by which the N fertility of soils was maintained.

NO and N_2O are gases formed during denitrification that are harmful to the ozone layer.

Nitrogenase is the enzyme complex that performs N_2 fixation.

FIGURE 10-6 A well-nodulated legume infected by N_2-fixing rhizobia.

TABLE **10-2** Environmental influences on the N cycle.

Change	Environmental Factor		
	pH	Temperature	Aeration
Increasing	NH_3 volatilization ↑ Nitrification ↑ NO_2^- accumulation ↑ (at high pH) Ammonification ↓ (> pH 8)	Mineralization ↑ Nitrification ↑ Volatilization of NH_3 ↑	Nitrification ↑ Dentrification stops Mineralization ↑ Assimilation ↑ NO_3^- accumulates
Decreasing	Chemodentrification ↑ (at low pH) Nitrification inhibited NH_4^+ oxidation stopped	Mineralization ↓ Dentrification ↓ N_2 fixation ↓	Nitrification ↓ Dentrification ↑ Mineralization produces odorous organic N intermediates Assimilation slows NH_4^+ accumulates NO_3^- leaches or denitrifies

Environmental Influences

Environmental factors have a tremendous influence on the various transformations in the N cycle. The soil pH determines whether some processes such as nitrification and chemodenitrification can occur, and whether NH_4^+ can be lost as volatile NH_3 gas. Whether an environment is or becomes aerobic or anaerobic also determines whether a particular microbially driven process can occur. And it determines the further transformation of the inorganic N that is already present. Because these reactions are all biologically driven, too low a temperature results in little or no activity. Too high a temperature results in organism death. At moderate temperatures between 10 and 40°C the transformation rates tend to double for every 10°C rise in temperature. Table 10-2 illustrates how the environment can influence various parts of the N cycle.

Fates of Inorganic N in the Soil

In terms of managing soil, and in terms of plant growth in soil, it is the fate of inorganic N such as NH_4^+ and NO_3^- that is most important. Sustainable agriculture depends on the ability of soil to release these nutrients to plants in a timely manner during plant growth. Sustainable agriculture also depends on preserving soil N, because only 5 percent of the soil organic N mineralizes each year (Stevenson and Cole, 1999). In lieu of biologically fixed N, which adds NH_4^+ to soil once the roots and tissue of the plants have decomposed, fertilizer N is added to soil as manure or inorganic fertilizer. In cases where plant uptake and immobilization in soil organic matter are too slow to remove all of the available N, some of this inorganic N can be lost from the soil system to the detriment of the environment. Figure 10-7 illustrates some of the potential fates of inorganic N.

About 5 percent of the soil N mineralizes each year.

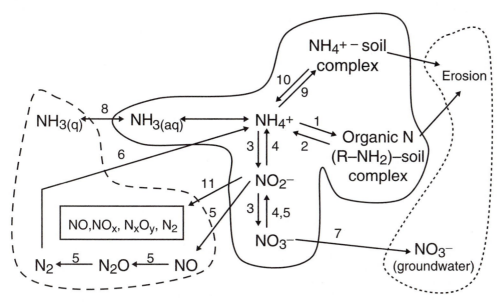

FIGURE 10-7 N losses from soil. Major pathways are numbered: 1) Assimilation; 2) Ammonification; 3) Nitrification; 4) Assimilatory and Dissimilatory NO_3^- reduction to NH_4^+; 5) Denitrification; 6) Nitrogen fixation; 7) NO_3^- leaching; 8) NH_3 volatilization; 9 & 10) Cation exchange reactions; 11) Chemodenitrification. Transformations that lead to N presence in soil, water, and air are partitioned accordingly. *(Adapted from Stevenson and Cole, 1993)*

FOCUS ON . . . THE NITROGEN CYCLE

1. What are the major reservoirs of N on Earth?
2. What are examples of organic N compounds in soil?
3. What kinds of compounds does oxidation of inorganic N produce?
4. Why do reducing environments lead to N loss from soil?
5. What occurs during biological N_2 fixation?

THE PHOSPHORUS CYCLE

In addition to nitrogen, the other element most often limiting plant productivity in soil is phosphorus (P). In aquatic systems it is often the most limiting nutrient. Phosphorus availability suffers because of its tendency to form relatively insoluble compounds throughout the range of soil pH, particularly in acid and alkaline soils. It is little wonder that a neutral soil pH optimizes most plant growth, because this is also the pH at which P is most available in soil solution.

FIGURE **10-8**   The phosphorus cycle in soil.

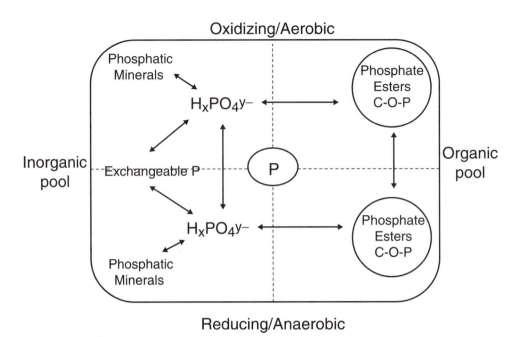

FIGURE **10-8**   The phosphorus cycle in soil.

Organic and Inorganic Forms

Phosphorus is among the most essential of plant nutrients, second per-haps only to N in terms of importance as a growth-limiting factor. In con-trast to the C, N, and S cycles, the P cycle is confined to soil and water. There are no transfers between these environments and the atmosphere. In addition, P typically undergoes no changes in oxidation state regardless of the environment in which it occurs, whether that environment be oxi-dizing or reducing. In both environments the P will typically occur as PO_4^{3-}. As a consequence, our model for the biogeochemical cycle of P in soil is somewhat simplified (Figure 10-8). There are no transformations of P between oxidized and reduced states. Rather, the P transformations are between mineralization and assimilation reactions, and between precipita-tion and weathering reactions.

The major reservoirs of P on Earth reflect the fact that over geologic time erosion and weathering of P from soil has transported most of the P to ocean sediments, where it is unavailable (Table 10-3). The P in soil can be grouped into six categories:

> Phosphorus has no gaseous transfers to the atmosphere.

❖ soluble inorganic and organic compounds
❖ weakly adsorbed inorganic P

TABLE 10-3 Major reservoirs of P. *(Adapted from Richey, 1983)*

Reservoir	Estimated P Content (kg)
Marine sediments	$840{,}000 \times 10^{12}$
Dissolved oceanic P	80×10^{12}
Particulates and biomass	$0.70 - 0.77 \times 10^{12}$
Soil	$96 - 160 \times 10^{12}$
Mineable P	19×10^{12}
Biomass	2.6×10^{12}
Dissolved	0.09×10^{12}

❖ sparingly soluble Ca-, Fe-, and Al-phosphates
❖ insoluble organic P in microbial biomass, humus, and plant material
❖ strongly adsorbed Fe and Al hydrous oxides
❖ phosphate fixed in silicate minerals

The soluble form of inorganic P is **orthophosphate**, which takes several forms depending on the pH of the environment:

> Orthophosphate is the most soluble form of P.

$$\text{Acidic } H_3PO_4 \longleftrightarrow H_2PO_4^- \longleftrightarrow HPO_4^{2-} \longleftrightarrow PO_4^{3-} \text{ Alkaline}$$

In most slightly acid to slightly alkaline soils the dominant soluble P forms are $H_2PO_4^-$ and HPO_4^{2-}. The minerals from which this P derived are called *apatites*, with the general formula $3[Ca_3(PO_4)_2] \cdot CaX_2$. The type of apatite depends on whether X is Cl^-, F^-, OH^-, or CO_3^-. Calcium phosphate, $Ca(H_2PO_4)_2$, is one of the precipitates of P that form in soil. Secondary hydrous oxide minerals such as stengite ($Fe(PO_4) \cdot H_2O$) and variscite ($Al(PO_4) \cdot 2H_2O$) are extremely insoluble P forms (Table 10-4).

The most common identifiable forms of organic P in soil are inositol phosphates, phospholipids, nucleic acids, phosphoproteins, metabolic P such as ATP, ADP, and NADP, and sugar phosphates. However, as much as a third or more of the organic P in soil is unrecognizable. Inositol phosphate comprises 10 to 50 percent of the P in most soils (Stevenson and Cole, 1999). Inositol phosphates are esters of hexahydroxy cyclohexane. The hexaphosphate member of this group is called phytic acid or phytin. Phytin is produced by plants and soil microorganisms, and is the dominant form of inositol phosphate in soil.

> Phytin is one of the most important forms of organic P in soil.

Mineralization

Most if not all of the organic P in soil consists of orthophosphate esters. These chemical bonds are broken by enzymes that are collectively called **phosphatases**. Phosphatases hydrolyze the ester bonds with the subsequent formation of an alcohol and inorganic P.

$$\underset{\underset{O^-}{\overset{O}{\|}}}{\overset{O}{\underset{\|}{R-O-P-O^-}}} + H_2O \xrightarrow{\text{phosphatase}} ROH + HPO_4^{2-}$$

There are several phosphatases depending on the substrate present and the pH of the environment. Acid phosphatases have an optimum pH range of 4–6, while alkaline phosphatases have an optimum pH range of

TABLE 10-4 Solubility of P-containing minerals in soil. *(From Lindsay, 1979)*

Mineral	Formula	Solubility Product (M)
Fluoroapatite	$Ca_5(PO_4)_3F$	10^{-59}
Hydroxyapatite	$Ca_5(PO_4)_3OH$	10^{-57}
Tricalcium phosphate	$Ca_3(PO_4)_2$	10^{-29}
Strengite	$FePO_4 \cdot 2H_2O$	10^{-26}
Variscite	$AlPO_4 \cdot 2H_2O$	10^{-21}

9–11. Phosphatases can be monoesterases, which break single phosphate ester bonds in compounds such as in inositol phosphate. Phosphatases can also be diesterases, which break phosphate ester bonds linking two adjacent groups, such as the phosphate bonds holding nucleic acids together. Phytase is a specific phosphatase that removes P from phytin.

Phosphatases are sensitive to dissolved P in the environment. When the dissolved P in soil exceeds 260 to 350 mg kg^{-1} P, phosphatase activity decreases dramatically (Speir and Ross, 1978).

> Phosphatases decompose organic P by hydrolyzing P bonds.

Immobilization/Assimilation

Phosphorus is essential to plant and microbial life. The P assimilated from the environment is incorporated into such essential compounds as nucleic acids, phospholipids, phosphorylated sugars, phytin, ATP, and phosphopyridine nucleotides such as NAD and NADP, which serve as electron carriers. As a result, much of the mass of microorganisms (2 to 5 percent) and plants (0.1 to 0.5 percent) is composed of P. The P is incorporated directly as orthophosphate without any change in oxidation state.

Immobilization and mineralization of P occur simultaneously. Whether there will be a net increase of P in the soil from mineralization or a net decrease due to immobilization and P fixation depends a great deal on the C:P ratio of soils. If the C:P ratio in soils is < 200 there is usually a net increase in soluble P in soil. There is essentially enough P in the mineralizing organic P to sustain the P requirements of the soil biomass. If the C:P ratio is > 300, soluble P disappears from the environment into soil biomass because there is inadequate P in the mineralizing tissue to sustain growth and decomposition.

Fixation and Solubilization

As previously noted, P undergoes neither oxidation nor reduction reactions. Yet, there is very little soluble P in the soil solution. This is because it is in great demand as a plant and microbial nutrient, which results in considerable assimilation, and because P readily undergoes fixation and precipitation reactions. Chemical fixation of P occurs in all pH ranges but is most significant when soil pH becomes acid or alkaline. This causes the formation of Fe- and Al-phosphates and Ca-phosphates, respectively (Figure 10-9).

Microorganisms play a significant role in solubilizing P and making it available through three key mechanisms:

1. Solubilizing Ca- and Mg-phosphates by producing weak (H_2CO_3) and strong (H_2SO_4) acids.
2. Destabilizing Fe-phosphate complexes by reducing Fe to the soluble ferrous form.
3. Producing organic acids (citric acid, oxalic acid, 2-ketogluconic acid) and other chelating compounds that form metal **chelate** complexes and soluble P.

Mycorrhizal fungi associations with plant roots are particularly important in P nutrition. In addition to the mechanisms for solubilization, the mycorrhizae increase P uptake by infected plants through increasing the volume of soil that is exploited for P, increasing the surface area available for P uptake from solution, and increasing the longevity of roots for P uptake.

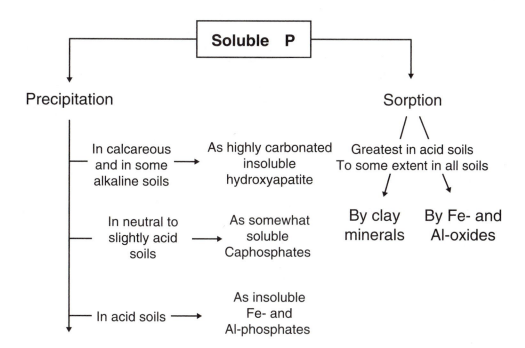

FIGURE 10-9 Phosphorus fixation in soil. *(From Sauchelli, 1951)*

FOCUS ON . . . THE PHOSPHORUS CYCLE

1. What are the principal organic forms of P in soil?
2. What form does soluble P usually take in neutral soil?
3. Why is P such a limiting element in soil?
4. What is the role of phosphatase in soil?
5. By what mechanisms do microbes solubilize P?

THE SULFUR CYCLE

Sulfur (S), like N, can be found in the atmosphere, lithosphere, hydrosphere, and biosphere. Relative to C, N, and P, it is required in much lower amounts. What S lacks in its contribution to soil fertility, however, it makes up for in terms of its effect on soil environmental quality through such activities as S oxidation, reduction, and acid rain.

Organic and Inorganic Forms

Sulfur, like N, has extremely dynamic soil transformations that involve conversion to organic and inorganic forms, oxidized and reduced forms, and gaseous forms (Figure 10-10). Unlike N, there is no great reservoir of S compounds in the atmosphere. Rather, the S is located almost exclusively in sedimentary rocks in forms such as gypsum ($CaSO_4$) and pyrite (FeS_2), sea water, and soil (Table 10-5).

FIGURE 10-10 The sulfur cycle in soil.

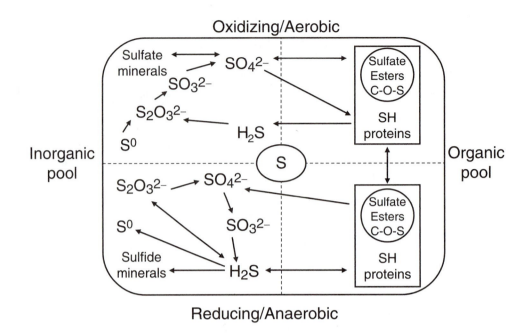

TABLE 10-5 Active S reservoirs. *(Adapted from Pierzynski et al., 1994; Schlesinger, 1997)*

Reservoir	kg of S ($\times 10^{15}$)
Sedimentary Rocks	
Shales	4970
Evaporites	2470
Sandstone	300
Limestones	100
Sea water	1280
Soil organic matter	0.0155
Land plants	0.0085
Atmosphere	0.0000028

The principal organic S compounds in the environment are so-called Fe-S proteins in which S groups are present at reactive sites. Two amino acids, cysteine and methionine, contain S. Cysteine is essential to the way proteins and enzymes contort themselves. Cysteine, for example, forms interchain and intermolecular disulfide (S-S) bonds. Additional S-containing compounds are things such as reducing agents (lipoic acid and glutathione), antibiotics (penicillin), sugar sulfates (glucose sulfates), sulfamates, and adenosine phosphosulfates. Most of these compounds have sulfate ester bonds (C-O-S or C-N-S) or carbon-sulfur bonds (C-S).

The major inorganic S compounds in soil and their oxidation states are listed in Table 10-6. They range from the most oxidized form of S (SO_4^{2-}) with an oxidation state of +6 to the most reduced form (S^{2-}) with an oxidation state of −2.

TABLE **10-6** Common inorganic S compounds and their oxidation states.

Compound	Formula	Oxidation State
Sulfate	SO_4^{2-}	+6
Tetrathionate[a]	$S_4O_6^{2-}$	-2, +6
Dithionate	$S_2O_6^{2-}$	+6
Thiosulfate	$S_2O_3^{2-}$	-1, +5
Sulfite	SO_3^{2-}	+4
Dithionite	$S_2O_4^{2-}$	+2
Sulfur	S_8^0	0
Sulfide	S^{2-}	-2

[a]S compounds like tri-, tetra-, and pentathionate have S atoms with different oxidation states and so can undergo an interesting chemical reaction called disproportionation.

SULFUR AND DISPROPORTIONATION

Inorganic S compounds such as thiosulfate and tri-, tetra-, and pentathionate have S that exists in two oxidation states. This allows them to undergo an interesting chemical reaction called **disproportionation** in which part of the molecule becomes oxidized and part of the molecule becomes reduced. It is particularly important in anaerobic environments where these compounds may be formed by the incomplete oxidation of organic S.

Disproportionation also occurs when S^{2-} is oxidized chemically by Fe^{3+} to form $S_2O_3^{2-}$ (thiosulfate). The thiosulfate disproportions to either SO_4^{2-} or S^{2-}

$$S_2O_3^{2-} + H_2O \rightarrow SO_4^{2-} + HS^- + H^+$$

Disproportionation is not a redox reaction in the classic sense because the transformations are accomplished within the same compound. It may be a significant way in which inorganic S cycles in sediments.

Mineralization

The amount of S in soil varies widely, from as little as 25 mg/kg S in agricultural soils of Nigeria to greater than 35,000 mg/kg in tidal marshes in the eastern United States (Germida, 1998). Typical agricultural soils in the United States will have several hundred ppm S. More than 90 percent of the S in soil is organic. Mineralization constantly occurs, but whether there is a net increase in inorganic S in the environment depends on the C:S ratio of the decomposing material and the soil biomass. If the C:S ratio is < 200:1 inorganic S accumulates; if the C:S ratio is > 400:1 immobilization of the mineralized S quickly occurs.

The types of mineralization that occur depend on the environment of the mineralizing organisms and the substrate. These pathways are illustrated in Figure 10-11. Like P, sulfate esters are cleaved by hydrolytic enzymes called sulfatases to release SO_4^{2-}. Desulfurization is the process by which C-S bonds are broken to release sulfhydryl (−SH) groups, which are converted to H_2S anaerobically and rapidly oxidize to SO_4^{2-} aerobically.

The S content of soils varies widely.

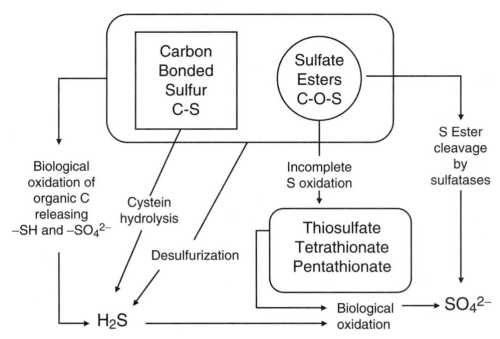

FIGURE 10-11   Pathways of organic S mineralization. *(Adapted from Germida, 1998)*

Immobilization/Assimilation

Plants and microorganisms assimilate S as SO_4^{2-}, although some microorganisms in anaerobic environments can assimilate H_2S. Most of the SO_4^{2-} is incorporated into sulfate esters (30 to 70 percent of the organic S in soil; Germida, 1998), sulfonates, vitamins, and cofactors. Assimilatory SO_4^{2-} reduction converts the S to the level of HS^-, which is then incorporated into amino acids.

Oxidation

Reduced inorganic S compounds can be oxidized in the environment by several S-oxidizing bacteria. These bacteria are best characterized by the genus *Thiobacillus*. Thiobacilli grow in a range of soil pH, and many species in this genus are able to grow as chemolithotrophs, which oxidize the reduced sulfur compounds for energy (Table 10-7). Some of the reactions they perform are as follows:

$$H_2O + S^0 + \frac{1}{2} O_2 \longrightarrow H_2SO_4$$

$$Na_2S_2O_3 + 2O_2 + H_2O \longrightarrow 2NaHSO_4$$

$$FeS_2 + 3\frac{1}{2}O_2 + 2H_2O \longrightarrow FeSO_4 + H_2SO_4$$

The last reaction is one of the main reasons exposure of pyrites (FeS_2) to air during mining operations contributes to acid mine drainage. During S oxidation a considerable amount of acidity develops. In anaerobic conditions there is also photosynthetic S oxidation, although this is less likely to occur in soils than it is to occur in sediments. This process, carried out by anaerobic photosynthetic bacteria, is analogous to the photosynthesis

TABLE 10-7 pH range and energy-yielding substrates of thiobacilli. *(From Coyne, 1999)*

Thiobacillus Species	pH Range	pH Optimum	Substrates
T. thiooxidans	1.5–5.0	2.0–3.5	S^{2-}, S^0, $S_2O_3^{2-}$, $S_4O_6^{2-}$
T. ferroxidans	1.5–5.0	2.0–3.5	S^0, $S_2O_3^{2-}$, Fe^{2+}
T. thioparus	5.0–8.0	7.0	S^{2-}, S^0, $S_2O_3^{2-}$, $S_4O_6^{2-}$
T. novalis	5.0–8.0	7.0	$S_2O_3^{2-}$, organic S
T. denitrificans	5.0–8.0	7.0	S^{2-}, S^0, $S_2O_3^{2-}$, $S_4O_6^{2-}$

performed by plants and algae, but photosynthetically oxidizes H_2S rather than water, and produces elemental S rather than O_2:

$$6CO_2 + 6H_2O \longrightarrow C_6H_{12}O_6 + 6O_2 \text{ (oxygenic photosynthesis)}$$

$$6CO_2 + 6H_2S \longrightarrow C_6H_{12}O_6 + 6S \text{ (anaerobic photosynthesis)}$$

Reduction

In anaerobic conditions, oxidized S compounds are reduced to H_2S and either volatilize to the atmosphere or combine with reduced Fe and Mn to form FeS and MnS.

$$SO_4^{2-}, SO_3^{2-} \longrightarrow H_2S$$

$$Fe^{2+,} Mn^{2+} + H_2S \longrightarrow FeS, MnS$$

Sulfate and sulfite are generally reduced directly to S^{2-} whereas compounds like thiosulfate are disproportionated to SO_4^{2-} and S^{2-} (see box on page 223). The organisms that reduce SO_4^{2-} and SO_3^{2-} are termed SO_4^{2-}-reducing bacteria. They are strict anaerobes that grow in neutral environments and utilize the oxidized S compounds as electron acceptors much like NO_3^- is used as an electron acceptor during denitrification. Some examples of the S-reducing bacteria are *Desulfovibrio* and *Desulfobacter*. Although S oxidation commences almost as soon as O_2 returns to the soil, S reduction does not begin until all of the substances more highly oxidized than SO_4^{2-} have been reduced. This means that S reduction does not really begin until the soil redox potential is < -200 mV and most if not all the Mn and Fe is in the reduced form.

> S begins to be reduced only at very low redox potentials.

Environmental Effects of S Transformations

There are a variety of volatile S compounds such as H_2S, dimethyl sulfide, carbonyl sulfide, and carbon disulfide that are emitted from soil, but these compounds are rapidly oxidized in the atmosphere. Likewise, ore smelting and combustion of fossil fuels and organic matter releases sulfur dioxide (SO_2) into the atmosphere, where it is oxidized and falls as wet or dry deposits that contribute to soil acidification. One of the first international treaties regulating air pollution was an agreement between Canada and the United States to remediate the environmental damage in Washington caused by SO_2 from the nearby smelter in Trail, British Columbia.

S oxidation and reduction have several undesirable effects.

Sulfur reduction has several undesirable effects. It leads to the corrosion of buried iron and concrete pipes. Sulfate reducers in golf greens can produce impermeable black layers that retard water infiltration and ultimately kill the overlying sod. Hydrogen sulfide is a toxic gas and stains the materials it touches.

Sulfur oxidation likewise has negative effects, the most significant of which is acid mine drainage. Soil acidification during S-oxidation can also occur in other environments, such as marine sediments rich in FeS and MnS, that are drained for agriculture. Metal and concrete pipes likewise corrode in soils with active S oxidation.

Focus on . . . The Sulfur Cycle

1. What are examples of organic S?
2. What is the range of oxidation state in inorganic S compounds?
3. What process do sulfatases carry out?
4. What environments support H_2S production?
5. How does S oxidation contribute to soil acidity?

The Iron and Manganese Cycle

Iron (Fe) and manganese (Mn) are micronutrients. Iron, of course, plays a pivotal role in photosynthesis, as does Mn to a certain extent. Both elements are also involved in enzyme function. Manganese, for example, is essential in the enzymes (ligninases) that fungi secrete to decompose lignin in the environment. Without ligninases, C cycling in the environment would be dramatically slowed. Both elements also have significant environmental effects.

Organic and Inorganic Forms

Iron and Mn are essential metal cofactors in numerous enzymes, electron carriers, and other organic compounds. Iron, for example plays a pivotal role in chlorophyll and cytochromes, and Mn is critical to the enzyme splitting H_2O to produce O_2 during photosynthesis and the Mn peroxidases released by fungi to degrade lignin.

The principal inorganic forms of Fe are ferric iron (Fe^{3+}) and ferrous iron (Fe^{2+}). Ferric iron is usually precipitated in the soil environment as ferric hydroxide ($Fe[OH]_3$). Manganese exists as manganate (Mn^{4+}) and manganous (Mn^{2+}) ion. The manganate usually exists as manganese oxide (MnO_2).

Immobilization/Assimilation

Iron and Mn are released during the decomposition of organic compounds that contain them. There are no special enzymes involved in their transformations back to inorganic forms. When plants and microorganisms assimilate Fe and Mn, they invariably do so by taking up the reduced forms,

Fe^{2+} and Mn^{2+}. In oxidized and alkaline environments this may not be possible because the Fe and Mn will be in the relatively insoluble Fe^{3+} and Mn^{4+} states. At pH 7, for example, the solubility of $Fe(OH)_3$ is only 10^{-17} M and this can lead to problems with Fe and Mn deficiency. Soybeans (*Glycine max*) are particularly susceptible, and in well-drained calcareous soils with high pH they suffer from a symptom called *iron chlorosis* in which the leaves turn light green or yellow due to the lack of Fe for chlorophyll. Maples in clay soils can also show signs of leaf chlorosis due to Fe deficiency. A corresponding deficiency of Mn in oats is called *grey speck disease*. These deficiencies can be solved by adding Fe^{2+} or Mn^{2+} as foliar fertilizers or by acidifying the soil environment.

Microorganisms can increase Fe availability by releasing special chelating compounds called **siderophores**. These are low-molecular-weight organic compounds with a high affinity for Fe^{3+}. Other chelating compounds such as citric and oxalic acid are also released to increase Fe solubility. The Fe is tightly bound to the siderophores until they attach to receptors in the microbial cell membrane. There the Fe is reduced, taken into the cell, and the siderophore released. The transported iron is then incorporated into organic compounds.

> Oxidation and reduction of Fe and Mn are very important processes in freshwater systems.

Oxidation

Like P, Fe and Mn do not undergo transfers to a gaseous phase. The principal reactions that they undergo are oxidations and reductions in the soil environment depending on whether the environment is aerobic and oxidizing or anaerobic and reducing (Figure 10-12).

In an aerobic environment Fe and Mn will become oxidized. This can occur chemically and biologically. At least 3.5 billion years ago huge accumulations of Fe_2O_3 called "Banded Iron Formations" were deposited when the Earth's atmosphere, as a result of photosynthesis, changed from one that was predominantly reducing to one that was oxidizing (Schlesinger, 1997).

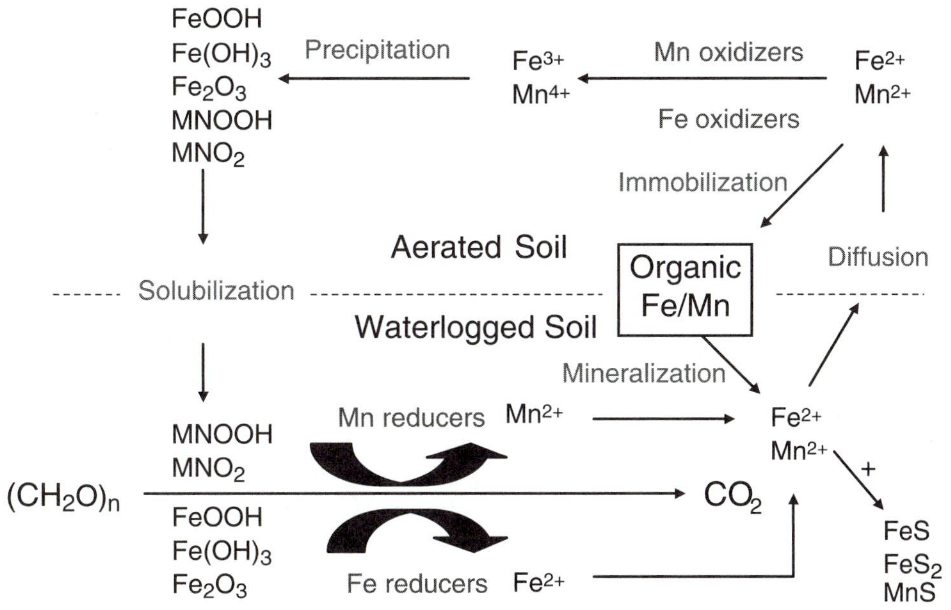

FIGURE 10-12 The Fe and Mn cycles in soil. *(Adapted from Nealson and Meyers, 1992)*

During Fe and Mn oxidation some lithotrophic microorganisms get energy much the same way that nitrifying bacteria gain energy by oxidizing inorganic NH_4^+ and NO_2^-.

Reaction	Representative microbe
$2Fe^{2+} \longrightarrow 2Fe^{3+} + 2e^-$	*Thiobacillus*
$Mn^{2+} \longrightarrow Mn^{4+} + 2e^-$	*Arthrobacter, Pseudomonas*

Other organisms can biologically oxidize Fe and Mn without gaining energy. For example, proteins in the cell surface of bacteria such as *Gallionella* and *Crenothrix* will oxidize Fe^{2+} and deposit a sheath or ribbon of precipitated $Fe(OH)_2$.

Chemical oxidation of Fe also occurs in aerobic environments at pH > 3. Elemental Fe (Fe^0) undergoes spontaneous oxidation to Fe^{2+} in neutral aqueous environments. This spontaneous chemical oxidation (termed cathodic depolarization) is one of the reasons buried iron pipes develop pitting. As the soluble Fe^{2+} is formed, some of it diffuses away from the pipe. Mn oxidation occurs spontaneously at pH > 5:

$$Mn^{2+} + 2OH^- \longrightarrow MnO_2 + H_2O$$

Manganese is also oxidized by H_2O_2 released by microorganisms during the activity of catalase.

$$Mn^{2+} + H_2O_2 \longrightarrow MnO_2 + 2H^+$$

Reduction

In anaerobic and reducing environments Fe^{3+} and Mn^{4+} will be reduced chemically and biologically.

Reaction	Representative microbe
$2Fe^{3+} + 2e^- \longrightarrow 2Fe^{2+}$	*Geobacter, Pseudomonas*
$Mn^{4+} + 2e^- \longrightarrow Mn^{2+}$	*Bacillus, Geobacter*

The characteristic grey-green color of some waterlogged soils is called *gleying* and reflects environmental conditions suitable for Fe and Mn reduction. Ferric iron can be used as a terminal electron acceptor during the oxidation of organic C by several obligately anaerobic bacteria, such as *Geobacter metallireducens* (Ghiorse, 1994). The process is illustrated as follows:

$$C_6H_{12}O_6 \longrightarrow CH_3COOH + CH_3CH_2COOH + CO_2 + H_2$$
$$\text{(fermentation step)}$$

$$2Fe^{3+} \text{ or } Mn^{4+} + H_2 \longrightarrow 2Fe^{2+} \text{ or } Mn^{2+} + 2H^+ \text{ (respiration step)}$$

or

$$CH_3COO^- + 8Fe^{3+} + 4H_2O \longrightarrow 2HCO_3^- + 8Fe^{2+} + 9H^+$$
$$\text{(respiration step)}$$

Environmental Consequences

The evidence of Fe and Mn biogeochemistry is everywhere. The bright red of Oxisols and Ultisols reveals the intense weathering that has left these soils enriched in Fe oxides. Exposed steel girders convert to ferric oxide, better known as rust. Manganese concretions accumulate in soils where the water table fluctuates. Sediment in ponds and streams develops dark black layers indicative of reduced Fe and Mn sulfides. In the west, sheer rock walls are coated with black, brown, and red bands called *rock varnish,* which are layers of oxidized Fe and Mn precipitated on the rock walls (Dorn, 1991).

One of the most significant consequences of Fe and Mn oxidation in the environment is acid production. When reduced Fe- and Mn-sulfides are exposed, often during mining operations, a combination of chemical and biochemical oxidations creates runoff or drainage water with a pH that can be below 2. You have already seen the consequences of S oxidation to soil acidification. Iron oxidation contributes to acidification in the following way:

$$12FeSO_4 + 3O_2 + 6H_2O \longrightarrow 4Fe_2(SO_4)_3 + 4Fe(OH)_3$$

The ferric sulfate that forms can chemically oxidize more reduced compounds, releasing even greater amounts of acidity. The ferric hydroxide that forms stains the surrounding environment a bright red, which characterizes environments affected by acid mine drainage.

FOCUS ON . . . THE IRON AND MANGANESE CYCLE

1. What kinds of organic compounds contain Fe and Mn?

2. What is an example of an oxidized Mn compound?

3. How are Fe and Mn involved in soil respiration?

4. What does "mottling" in soil tell you?

5. What are some adverse consequences of Fe oxidation in the environment?

BIOGEOCHEMICAL TRANSFORMATIONS OF HEAVY AND TRACE METALS

Heavy metals generally have molecular weights greater than Fe (Table 10-8). Several, such as Cu, Zn, Mo, and Se, are vital to the function of cell metabolites. These heavy metals are referred to as the *trace elements* because a trace of them is essential for life. However, trace elements and heavy metals in high quantities are toxic. Some soils, as a result of parent material and development, have high concentrations of heavy metals. Serpentine soils, for example, have high concentrations of Cr, whereas many soils that developed from sediment in the western United States have high Se concentrations. Some soils near Cu, Pb, and Zn smelting facilities are contaminated as the result of industrial activity.

> Heavy metals have molecular weights greater than iron.

There are several ways that heavy metals are transformed in the soil environment, usually leading to situations that reduce their toxicity:

❖ *Precipitation as a metal sulfide.* As, Cd, Cu, Hg, Pb, Se, and Zn can all be precipitated as metal sulfides, which reduces their biological availability. These reductions are microbially driven, usually in anaerobic environments, by the activity of S-reducing bacteria.

❖ *Volatilization.* As, Hg, and Se all form volatile compounds that will disperse through the atmosphere.

❖ *Methylation.* As, Hg, and Se also form methylated compounds that are more soluble and more volatile than the parent metal. Methylation occurs aerobically and anaerobically.

❖ *Precipitation.* Many metals are precipitated as metal carbonates (e.g., $CuCO_3$, $PbCO_3$, $CdCO_3$), hydroxides (e.g., $Sb[OH]_3$, $Cr[OH]_3$, $Cu[OH]_2$, $Fe[OH]_3$), elemental metal (Cu, Ag, Au), or oxides (MnO_2).

❖ *Photolytic reduction of oxyanions.* Several oxyanions such as TeO_3^{2-}, WO_4^{3-}, SeO_3^{2-}, and AsO_4^{3-} can be reduced during photosynthesis by anaerobic bacteria to leave insoluble elemental metals.

❖ *Sequestration in organic matter.* Many of the most toxic heavy metals such as Cu, Pb, and Zn can be immobilized by adsorption to reactive groups in soil organic matter such as hydroxyls and carboxyls. This is a temporary immobilization, however, because as soon as the organic matter decomposes the heavy metals are free to move again.

TABLE 10-8 Heavy metals and trace elements of environmental concern.

Metal	Symbol	Atomic Mass	Sample Metabolic Role(s)
Vanadium	V	50.9	Metal cofactor in some nitrogenases
Chromium	Cr	52.0	Metal cofactor
Manganese	Mn	54.9	Metal cofactor in enzymes, electron transport
Iron	Fe	55.8	Fe-S proteins, electron transport
Cobalt	Co	58.9	Metal cofactor in nitrogenase
Nickel	Ni	58.7	Metal cofactor in urease
Copper	Cu	63.5	Metal cofactor in nitrite reductase
Zinc	Zn	65.4	Metal cofactor in DNA polymerase
Arsenic	As	74.9	No metabolic role, biocidal
Selenium	Se	79.0	Metal cofactor in vitamin E
Molybdenum	Mo	95.9	Metal cofactor enzymes including nitrogenase
Silver	Ag	107.9	Precious metal
Cadmium	Cd	112.4	No metabolic role
Tin	Sn	118.7	Metabolic role unknown
Iodine	I	126.9	Metabolic role unknown
Platinum	Pt	195.0	Precious metal
Gold	Au	197.0	Precious metal
Mercury	Hg	200.6	No metabolic role, biocidal
Lead	Pb	207.2	No metabolic role, biocidal

FOCUS ON . . . TRANSFORMATIONS OF OTHER ELEMENTS

1. What are trace elements?

2. Why are trace elements needed?

3. What are examples of heavy metals?

4. What is the problem with heavy metals?

5. How are heavy metals detoxified in the environment?

SUMMARY

This chapter summarized the most important elements that constitute living things in soil and are involved in biochemical transformations. You could describe most of these biochemical reactions as processes of mineralization and immobilization, which moves the elements back and forth between inorganic and organic forms. You could also describe many of the reactions that occur as oxidations or reductions. For the most important elements, C, N, P, S, Fe, and Mn, you took a more detailed look at the specific transformations involved in the cycling of these elements, the environments in which these transformations take place, and the environmental consequences of these reactions. Finally, you examined some of the ways in which heavy metals and trace elements are transformed in the environment.

END OF CHAPTER QUESTIONS

1. Distinguish between mineralization and immobilization.

2. The chemistry of life has been described as the chemistry of "light" elements. What does that mean?

3. What is the oxidation state of C or N in the following compounds: CH_4, CO_2, NH_4^+, and NO_2^-?

4. Can reductions in soil take place without corresponding oxidations?

5. Could life exist on Earth without C fixation?

6. Why is nitrification an important process in soil?

7. What sort of processes will cause inorganic N to be lost from soil?

8. Apatites may be common P-containing minerals in soil, yet the P level remains low; why is this true?

9. Why are mycorrhizal fungi important in P nutrition in soil?

10. Flour of S (elemental S) is often used to treat soils that show iron chlorosis. Why?

11. What sort of reductions occur in a *gleyed* soil?

12. Should you install a septic tank and leach field in a soil horizon that contains mottling?

13. How is Fe related to the color of Oxisols and Ultisols?

14. What are examples of volatile and precipitated forms of heavy metals in the environment?

FURTHER READING

There are many books addressing biogeochemistry in soil environments, but the best is probably *Biogeochemistry: An Analysis of Global Change,* 2nd ed. by W. H. Schlesinger (1997, San Diego, CA: Academic Press). This book takes a broad global look at the key nutrient cycles in terms of how they influence air, water, and terrestrial systems.

Cycles of Soil by F. J. Stevenson and M. A. Cole (1999, New York: John Wiley & Sons, Inc.) gives a particularly good overview of the C, N, and P cycles and to a lesser extent the S cycle. There is also some treatment of the trace elements in soil. One advantage of this text is that it contains specific examples of how the different components of each cycle are examined experimentally.

REFERENCES

Birdsey, R. A., A. J. Plantinga, and L. S. Heath. 1993. Past and prospective carbon storage in United States forests. *Forest Ecology and Management* 58: 33–40.

Burns, R. C., and R. W. F. Hardy. 1975. *Nitrogen fixation in bacteria and higher plants.* New York: Springer-Verlag.

Chapelle, F. H. 1993. *Ground-water microbiology and geochemistry.* New York: John Wiley & Sons, Inc.

Conrad, R. 1990. Flux of NOx between soil and atmosphere: Importance and soil microbial metabolism. In N. P. Revsbech and J. Sorensen (eds.), *Denitrification in soil and sediment.* New York: Plenum Press, pp. 105–128.

Coyne, M. S. 1999. *Soil microbiology: An exploratory approach.* Clifton Park, NY: Thomson Delmar Learning.

Dorn, R. I. 1991. Rock varnish. *Scientific American* 79: 542–553.

Germida, J. J. 1998. Transformations of sulfur. In D. M. Sylvia et al. (eds.), *Principles and applications of soil microbiology.* Upper Saddle River, NJ: Prentice Hall, pp. 346–368.

Ghiorse, W. C. 1994. Iron and manganese oxidation and reduction. In R. W. Weaver et al. (eds.), *Methods of soil analysis, part II: Microbiological and biochemical properties.* Madison, WI: Soil Science Society of America, pp. 1079–1096.

Hutchinson, G. E. 1944. *American Scientist* 32: 178.

Lindsay, W. L. 1979. *Chemical equilibria in soils.* New York: John Wiley & Sons, Inc.

Nealson, K. H., and C. M. Meyers. 1992. Microbial reduction of manganese and iron: New approaches to carbon cycling. *Applied and Environmental Microbiology* 33: 279–318.

Pierzynski, G. M., J. T. Sims, and G. F. Vance. 1994. *Soils and environmental quality.* Boca Raton, FL: CRC Press.

Richey, J. E. 1983. The phosphorus cycle. In B. Bolin and R. B. Cook (eds.), *The major biogeochemical cycles and their interactions.* New York: John Wiley & Sons, Inc., pp. 51–56.

Sauchelli, V. 1951. *Manual on phosphates in agriculture.* Baltimore, MD: Davidson Chemical Corp.

Schimel, D. S. 1995. Terrestrial ecosystems and the carbon cycle. *Global Change Biology* 1: 77–91.

Schlesinger, W. H. 1997. *Biogeochemistry.* San Diego, CA: Academic Press.

Speir, T. W., and D. J. Ross. 1978. Soil phosphatase and sulphatase. In R. G. Burns (ed.), *Soil enzymes.* New York: Academic Press, pp. 197–250.

Stevenson, F. J. 1982. Origin and distribution of nitrogen in soils. In F. J. Stevenson (ed.), *Nitrogen in agricultural soils.* Madison, WI: American Society of Agronomy, pp. 1–42.

Stevenson, F. J., and M. A. Cole. 1999. *Cycles of soil,* 2nd ed. New York: John Wiley & Sons, Inc.

Chapter 11

SOIL ORGANIC MATTER

*"We are part of the earth and it is part of us. . . . What
befalls the earth befalls all the sons of the earth."*
Chief Seattle, 1852

OVERVIEW

In Chapter 10 you read about the various biogeochemical cycles of important nutrients in soil. In this chapter you will read about soil carbon, more specifically, soil carbon in the organic matter of soil. **Soil organic matter, (SOM)**, is the debris of biological activity in soil. But unlike most debris it plays a critical role, influencing the chemical and physical properties of soil and the extent of biological activity that occurs therein. The amount and depth of the SOM in soils is one of the most important properties that influences soil productivity and its ability to sustain plant life and carry out other important ecological roles such as buffering and moderating hydrological cycles, physically supporting plants, retaining and delivering nutrients to plants, disposing of wastes and dead organic material, renewing soil fertility, and regulating major element cycles (Daily et al., 1997).

OBJECTIVES

After reading this chapter, you should be able to:

✔ Describe the basic composition of soil organic matter.

✔ Identify five or six key roles that soil organic matter plays in soil systems.

✔ Understand the processes by which soil organic matter forms.

✔ Evaluate how soil-forming processes and human activity influence the amount of soil organic matter.

KEY TERMS

active and passive
   SOM
chelate
fulvic acid
glomalin

humic acid
humin
humus
lignin

nonhumic compounds
soil organic matter
   (SOM)
subsidence

WHAT IS SOIL ORGANIC MATTER?

In its broadest sense SOM is every organic compound that is in soil, which includes roots, animals, and microorganisms. Broadly classified, this means living organisms, fresh residues, and **humus** (the finely divided, nonrecognizable fraction of organic matter). In typical grassland soil the distribution of organic material would look something like that shown in Table 11-1. Humus, the nonrecognizable fraction of SOM, is by far the most abundant component.

Excluding living organisms and plants for the moment, the SOM can be divided into two categories: **nonhumic compounds** and humic substances. The relative contribution of each category to soil is shown in Figure 11-1.

Nonhumic substances have identifiable chemical composition and are readily available for microbial decomposition, so they are still in the process of decay. This fraction is often called the *active* component of SOM. Humic substances can be chemically analyzed but don't have a consistent, identifiable chemical composition. They are a dark-colored collection of compounds that resist further decomposition. One of the characteristics of humic substances is that these resistant materials are polymerized into new compounds such as **fulvic acid** and **humic acid**. Soil organic matter is unique in the sense that it reflects both degradative and constructive processes.

> SOM is usually partitioned into humic and nonhumic materials.

TABLE 11-1 Average contents of organic materials in a temperate grassland soil. *(Adapted from Troeh and Thompson, 1993)*

Organic Component	kg/ha	%
Plant roots	16,800	8.4
Living macrofauna	1,826	1.0
Identifiable dead remains	4,480	2.2
Living microbial biomass	8,008	4.0
Humus	168,000	84.4

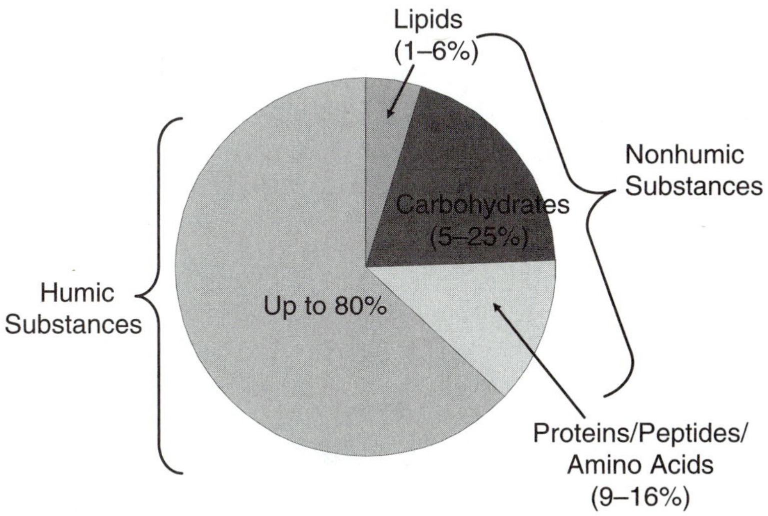

FIGURE 11-1 Contribution of different organic fractions to SOM. *(Adapted from Stevenson and Cole, 1999)*

FOCUS ON . . .THE IDENTITY OF SOM

1. What are the types of organic matter in soil?
2. What are examples of nonhumic substances?
3. How do humic and nonhumic substances differ?
4. What fraction of organic matter makes the greatest contribution to SOM?
5. What property of SOM makes it unique?

WHAT IS THE COMPOSITION OF SOM?

It is much easier to ask this question than to answer it because SOM has neither an identifiable structure nor a uniform composition. For that matter, it is simultaneously forming and degrading, so the amount and distribution of SOM also continuously fluctuates.

| Fresh plant residues are about 10 percent of SOM. |

Fresh plant and microbial residues make up about 10 percent of SOM. The most abundant and persistent microbial contributions are the cell wall polymers such as peptidoglycan, cellulose, and chitin. These polymers break down to release simple molecules like aminosugars and glucose. Plants, likewise, have structural polymers like cellulose and hemicellulose that are their principal contribution to SOM. Other compounds in fresh plant residue are sugars and starches, proteins and amino acids, waxes and pigments, pectin, fats, oils, lipids, organic acids, hydrocarbons, and lignin (Figure 11-2).

| Lignin is a key compound involved in SOM formation. |

Lignin is perhaps second to cellulose in terms of the plant biomass returned to soil in residues and it is one of the key ingredients involved in forming soil humus. Lignins provide structural rigidity in plants, helping to glue cellulose polymers together and provide resistance to compression and bending. Lignin is a three-dimensional and highly branched molecule that forms by random condensation and polymerization reactions of just a few basic subunits called phenyl propanoids. The structure of lignin and its building blocks are illustrated in Figure 11-3.

FIGURE 11-2 Typical composition of green plant residues. *(Adapted from Stevenson and Cole, 1999)*

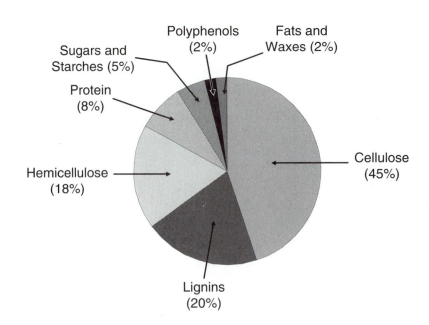

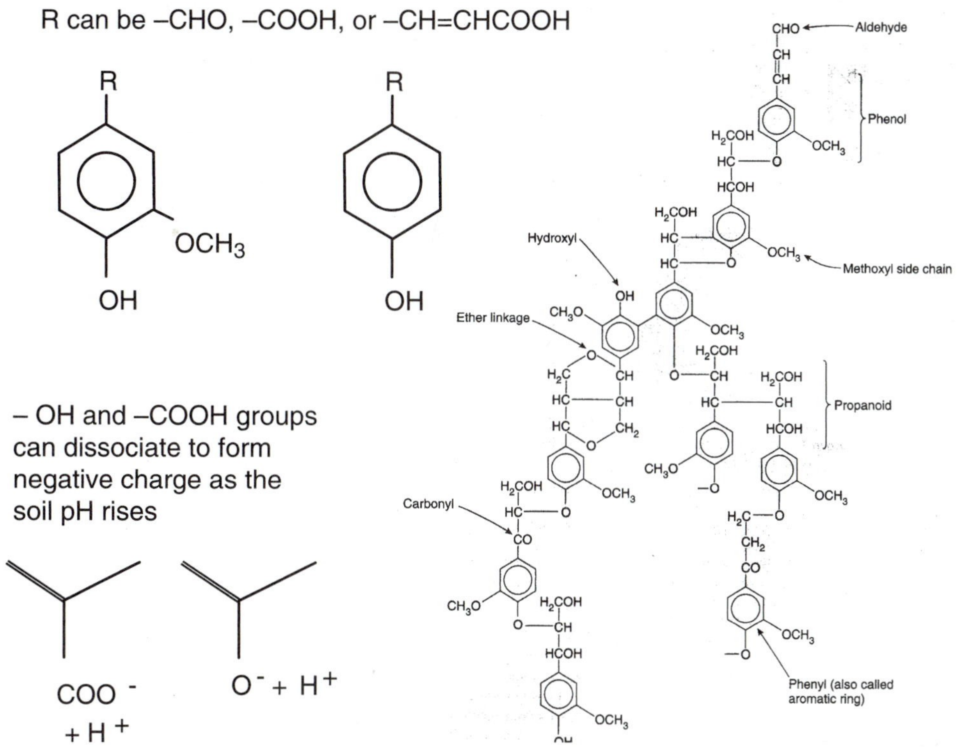

R can be –CHO, –COOH, or –CH=CHCOOH

R

OCH₃

OH

R

OH

– OH and –COOH groups
can dissociate to form
negative charge as the
soil pH rises

COO ⁻
+ H ⁺

O⁻ + H⁺

CHO ← Aldehyde

Phenol

Methoxyl side chain

Hydroxyl

Ether linkage

Propanoid

Carbonyl

Phenyl (also called
aromatic ring)

FIGURE 11-3 Representative structure of lignin and its building blocks. *(Adapted from Paul and Clark, 1996)*

During plant residue decomposition, a series of chemical, biochemical, and biological transformations takes place that returns CO_2 to the atmosphere, converts identifiable plant material into unidentifiable humic materials, and promotes the ecological succession of various microbial communities (Figure 11-4). Humus is generally fractionated into three components following extraction in NaOH: humin, fulvic acid, and humic

Humus is fractionated based on its solubility in acid and alkaline solutions.

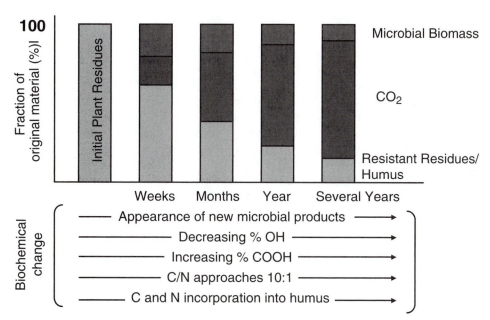

FIGURE 11-4 Physical and biochemical succession during plant residue decomposition in soil. During decomposition the majority of the organic carbon is mineralized to carbon dioxide, 10-15% is incorporated into microbial biomass, and approximately 10% is incorporated in soil organic matter.

EXTRACTING FULVIC AND HUMIC ACID FROM SOIL

Extracting fulvic and humic acids from soil is one of the easiest procedures a soil scientist can perform. It requires a strong base, a strong acid, and a little patience. The procedure is diagrammed below.

The humic acid can be collected by centrifugation after settling for twenty-four hours. The fulvic acids still in solution, along with other soluble compounds such as sugars and small polysaccharides, can be collected by freeze-drying. The procedure should be conducted in a nitrogen gas atmosphere to minimize the oxidation that takes place. Even so, it is impossible to extract humic and fulvic acids from soils without in some way changing the chemical and certainly the physical composition of the material.

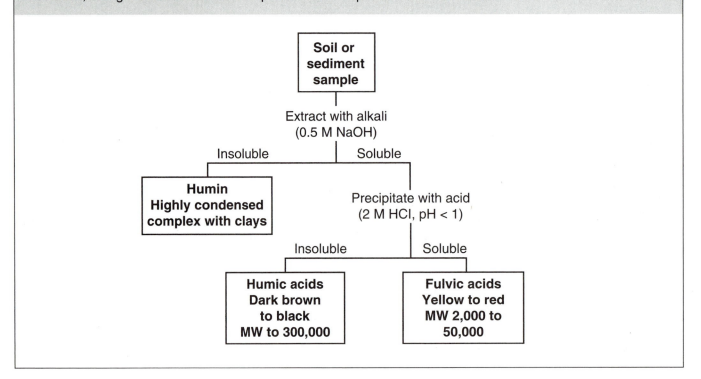

acid. **Humin** is the non-NaOH-dispersible fraction. Humic acid is a NaOH soluble fraction that is insoluble at pH 2. The molecular weight of humic acid varies from 10,000 to 100,000 and it is composed of aromatic rings, cyclic nitrogen compounds, and peptide chains. The important functional groups of humic acids, those groups that give it pH-dependent charge and help bind soil elements and other organic compounds, are carboxyls, phenolic OH, alcoholic OH, and ketones. Fulvic acids are soluble in NaOH and at pH 2. They are smaller than humic acids, with molecular weights ranging from 1,000 to 30,000. Fulvic acids have the same functional groups as humic acids, but they tend to be more oxidized and more mobile. On the basis of chemical analysis, the composition of humic acid and fulvic acid will look something like Figure 11-5. However, the actual chemical composition of humic materials in soils will vary depending on the type of soil and climate.

Another constituent of SOM, and one which has only recently been recognized, is a glycoprotein secreted by mycorrhizal fungi called **glomalin.** This compound is not extracted from soil during routine humus fractionation, but it can be recovered by vigorously autoclaving the soil in buffer. When this procedure is carried out, it reveals that glomalin constitutes considerable quantities of the organic material in soil.

Glomalin is a fungal product contributing to SOM.

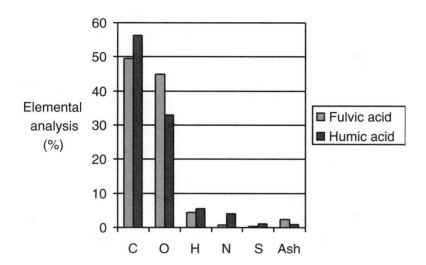

FOCUS ON . . . SOM COMPOSITION

1. What are the major plant components entering soil?
2. What is the basic chemical structure of lignins?
3. What happens to plant material as it gets incorporated into SOM?
4. What is the difference between fulvic and humic acid?
5. Why is glomalin important in soil?

WHY IS SOM IMPORTANT?

Although you've just read about some of the details of SOM composition, what's really important is not what SOM is made of, but what it does and how it contributes to the formation, function, and fertility of soil. That will be the topic of this section.

What are the functions or roles of SOM?

❖ Soil organic matter is the major source of nitrogen and an important source of phosphorus, sulfur, and trace nutrients.

❖ Soil organic matter contributes to the cation exchange capacity (CEC) and anion exchange capacity (AEC) of soil, which makes it a major storehouse of exchangeable nutrients.

❖ Soil organic matter binds organic chemicals and pesticides, which reduces their movement and activity, and lessens their effect on the environment.

❖ Soil organic matter is a type of glue that binds clay and silt particles into larger structural units and contributes to desirable soil structure with improved percolation and tilth. The binding makes these aggregates water stable and gives them improved resistance to destructive forces such as water erosion.

❖ Soil organic matter helps move tightly bound and normally insoluble metals by forming water-soluble metal–organic matter complexes. It helps to **chelate** micronutrients and helps create spodic horizons in some soils, so it contributes to forming diagnostic soil horizons in soil taxonomy.

❖ Soil organic matter contributes to the water-holding capacity of soil—directly because it holds five times more water by weight than clay minerals and indirectly by its contribution to developing soil structure and porosity.

❖ Soil organic matter affects soil temperature directly and indirectly. Soil organic matter, mulch, and litter at the soil surface buffer the soil from extreme temperature changes, keeping soils cooler in summer and warmer in winter. The dark color of organic matter in soil adsorbs more solar radiation than lighter-colored soils with lower SOM contents; this leads to warmer soils. Indirectly, because SOM has a high water-holding capacity, it buffers the soil from temperature changes because the water has a very high heat capacity.

❖ Soil organic matter provides carbon and energy for the growth of heterotrophic microorganisms in soil and larger organisms such as earthworms.

❖ Organic acids in SOM contribute to mineral weathering.

FOCUS ON . . . ROLES OF SOM

1. How does SOM contribute to plant and microbial nutrition?
2. What is a dark-colored soil likely to mean?
3. How does SOM contribute to improved soil structure?
4. What are direct and indirect interactions of SOM and water?
5. Why are pesticide spills less damaging in soils with high SOM?

HOW DOES SOM FORM?

There are many steps in SOM formation. As noted before, SOM develops during simultaneous decomposition of nonhumic organic carbon entering soil, and formation of the decomposition products into humic materials such as humic acid and fulvic acid. The overall process is illustrated in Figure 11-6.

FIGURE 11-6 Steps in developing SOM. Plant polymers are decomposed into individual subunits during microbial decomposition and then resynthesized by polymerization.

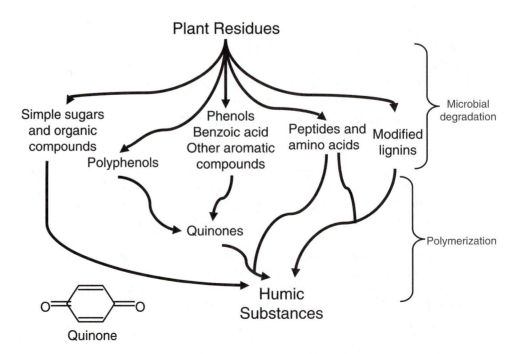

Peptides, amino acids, and aromatic compounds from lignin decomposition are the key ingredients that form the humic materials. Ultimately this results in the structures illustrated in Figure 11-7. It would be misleading to imply that humic material is small, linear and two-dimensional as Figure 11-7 suggests. Humic materials are large, shapeless, three-dimensional structures coating soil minerals. You can get a sense of the true nature of SOM from examining the complexity of the compound illustrated in Figure 11-8.

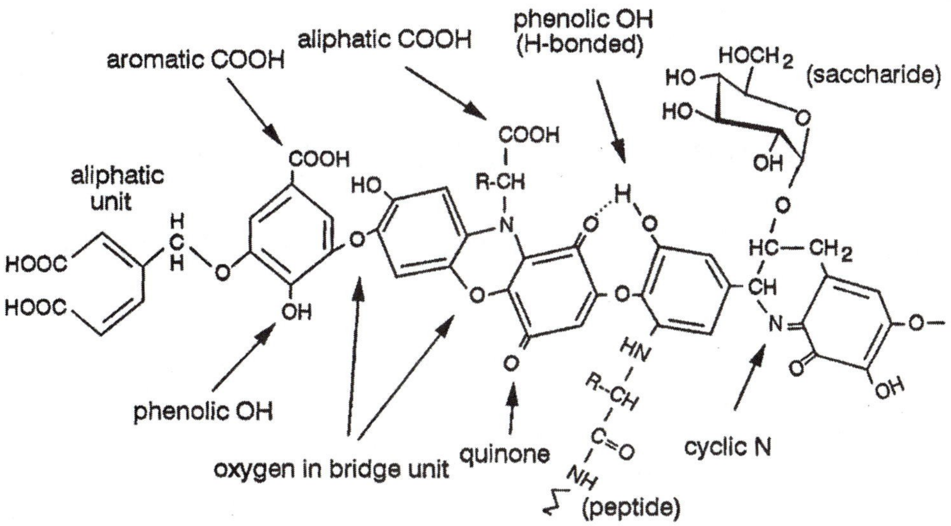

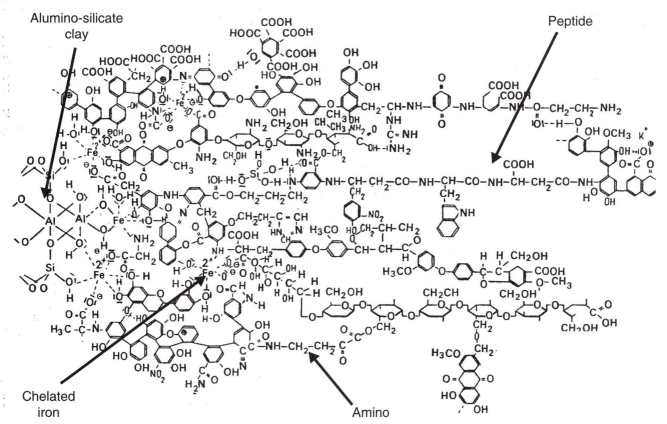

FIGURE 11-7 Representative structure of humic material showing various linkages to peptides and sugars and reactive hydroxyl and carboxyl groups. *(Adapted from Stevenson and Cole, 1999)*

FIGURE 11-8 Representation of the complexity of humic materials in soil. *(Adapted from Orlov, 1985)*

The complexity of SOM gives it some of its beneficial properties.

It is the complexity of humic materials, the bulk of SOM, that give them some of their most beneficial properties. As you can see from Figure 11-8, the random polymerization of peptides and sugars into humic acids traps important plant nutrients like N and S in soil, which is why SOM is considered a great storehouse of plant nutrients. You can also see the humic materials binding silicate clay in Figure 11-8. This binding is the first step in developing soil aggregation and also in creating stable aggregates. Finally, because the material is complex, it resists decomposition. So the beneficial effects of SOM can be long lasting. However, SOM will decompose over time, and that is the subject of the next section.

FOCUS ON . . . SOM FORMATION

1. Why does SOM formation require simultaneous decomposition and constructive processes?
2. What are the key building blocks of humic materials in soil?
3. What are some of the reactive groups in humic materials?
4. What are unique characteristics of humic materials that contribute to their beneficial functions?
5. How does SOM contribute to soil structure?

SOM DECOMPOSITION

SOM forms active and passive pools that differ in terms of how readily they decompose.

Soil organic matter decomposition is perhaps the major contributor to nutrient availability in soil, so it is vitally important to have a sense of how rapidly it decomposes. Soil scientists typically think of SOM in terms of **active and passive SOM** pools. The active pool is the fraction of soil SOM that is readily available (or labile) for microbial growth and activity. Fresh residues and soluble materials fall into this category, and their decomposition is quite rapid, occurring within days or weeks. This pool is characterized by materials that have simple structures and low C:N ratios. Low C:N ratios are important because they mean there is adequate N in the material to support growth of the microbial community on the available C; decomposition will readily occur and decomposition will not tie up or immobilize inorganic N in soil, preventing it from use by plants.

The passive pool of SOM resists decomposition because it is either physically or chemically protected. Physical protection means that during soil aggregate formation the SOM (either labile or nonlabile) is trapped in the developing aggregate and shielded from microbial attack, either because it is armored in clay minerals or, more likely, because it is located in pores too small to permit microbial access. Chemically protected SOM, however, is material that resists decomposition because of its chemical composition and high C:N ratio. Humin and humic acids fall in this category. They are large, complex, insoluble materials that are both physically and chemically difficult to degrade. The relative turnover times of the various SOM fractions are illustrated in Figure 11-9.

About 2 to 5 percent of SOM decomposes yearly.

All told, about 2 to 5 percent of the SOM decomposes on a yearly basis. This varies from environment to environment, with warmer and drier soils having greater SOM decomposition rates than cooler and wetter soils (Table 11-2).

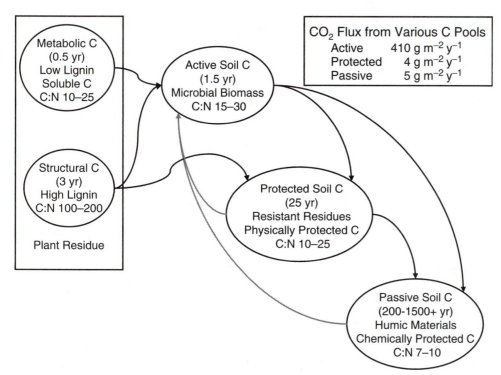

FIGURE 11-9 Soil C flow in the CENTURY model. Residence time in parentheses. *(Adapted from Parton et al., 1975; Paustian et al., 1992; Schlesinger, 1997)*

TABLE 11-2 Turnover time of soil C in various ecosystems. *(Adapted from Raich and Schlesinger, 1992)*

Vegetation Type	Soil C (Mg/ha)	Soil Respiration (Mg/ha)	Turnover Time (years)
Swamps and marshes	723	2.0	520
Tundra	204	0.6	490
Boreal forest	206	3.2	91
Temperate grassland	189	4.4	61
Tropical lowland forest	287	10.9	38
Desert scrub	58	2.2	37
Temperate forest	134	6.6	29
Cultivated soil	79	5.4	21
Tropical grassland	42	6.3	10

FOCUS ON . . . SOM DECOMPOSITION

1. What are the active, protected, and passive components of SOM?

2. What mechanisms protect SOM from decomposition in soil?

3. Why do the different fractions of SOM decompose at different rates?

4. What are approximate turnover times for different pools of SOM?

WHERE IS SOM FOUND AND HOW QUICKLY DOES IT ACCUMULATE?

If SOM is so important, then one of the central questions you have to ask yourself is where it is most likely to be found. Furthermore, if SOM accumulation is beneficial to soil function, then an equally important question is how quickly it forms, and whether there is anything that can accelerate that accumulation.

Amounts of Soil Organic Matter

The surface horizons of most well-drained mineral soils contain anywhere from 1 to 6 percent SOM depending on the climate and their management history. The O horizon of forest soils and poorly drained soils, such as Histosols, will frequently have higher SOM contents, ranging up to 20 to 30 percent C.

Distribution by Depth

Most SOM is in the upper soil surface.

Most of the SOM is in the surface horizons of soil and progressively and sometimes rapidly decreases with soil depth. as Figure 11-10 illustrates. Spodosols, soils that form beneath cool, wet, coniferous forests, have a unique feature called a spodic horizon in which SOM increases. This horizon is formed by leaching of soluble humic compounds from the soil surface.

Some soils have buried A horizons, caused when wind- or water-borne soil is deposited over an existing soil. In this case, there is also an increase in SOM content at lower depths.

Effects of Soil-Forming Factors

You are already familiar with the five soil-forming factors. How does each of them, in turn, affect SOM accumulation?

FIGURE 11-10 Vertical distribution of organic C in various soils. To convert between % total C and % OM in soil use the following equation: OM = TC × 1.72. (*Adapted from Brady and Weil, 2001*)

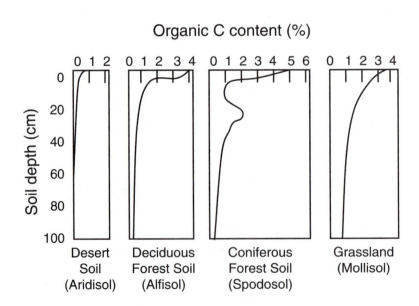

TABLE 11-3 Distribution of soil organic matter in various ecosystems. *(Adapted from Schlesinger, 1997)*

Ecosystem Type	Mean Soil Organic Matter Content (kg C/m^2)	Total World Soil Organic C (Mg C $\times$ 10^9)
Forests		
Tropical forests	10.4	255
Temperate forests	11.8	142
Boreal forests	14.9	179
Woodlands and shrublands	6.9	59
Grasslands		
Tropical savannas	3.7	56
Temperate grasslands	19.2	173
Cultivated	12.7	178
Tundra and alpine	21.6	173
Deserts		
Desert scrub	5.6	101
Extreme deserts, rocks, ice	0.1	3
Swamps and marshes	68.6	137

Organisms (Vegetation)

The type of vegetation, or lack thereof, has a tremendous influence on SOM content. The extremes run from swamps and marshes, which contain soils that are almost entirely organic matter, to warm and cold deserts that have little if any organic matter. Table 11-3 illustrates how SOM content varies depending on the different plant communities present in the ecosystem.

Climate

Climate naturally influences the type of vegetation that will be present in an environment. The climate, particularly temperature and annual precipitation, also has a significant effect on SOM contents through the dual and opposing processes of residue return to soil and SOM decomposition. At low temperatures and high precipitation SOM accumulates faster than it decomposes. Even though plant productivity is lower in colder environments, for much of the year the soils may be frozen, thus preserving the SOM content. This is one of the reasons why tundra and alpine regions (see Table 11-3) have such high SOM contents. However, if you compare temperate and tropical grasslands in terms of SOM content, tropical grasslands typically have much lower SOM contents, in part because they are hotter on a year-round basis, which promotes decomposition (Figure 11-11). Deforestation in tropical rainforests is particularly harmful because without the extremely high productivity of the overlying forest returning residues to soil, the high temperatures and rainfall in the tropics lead to extremely rapid loss of SOM from the underlying soil. As the temperature rises much above 35° C, SOM decomposition rates exceed organic residue deposition and SOM content decreases.

SOM accumulates in cool and moist environments.

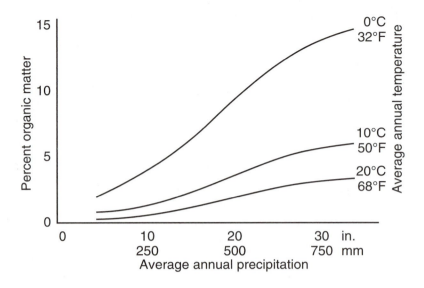

Topography and Aspect (Relief)

> SOM accumulation is influenced by microtopography, relief, and aspect.

Water, soil, and SOM are subject to gravity, so it is not surprising that topography has a significant effect on SOM deposition in the environment. Soils on steep slopes have greater erosion and more water runoff than soils on gentler slopes. Consequently, they tend to have lower SOM contents. Soils in slope positions that receive colluvial deposits tend to be wetter and have higher SOM contents. The distribution of SOM in a typical catena is illustrated in Figure 11-12. Convex positions on a slope tend to shed water and are drier than concave positions, which tend to accumulate water and consequently tend to have slightly higher SOM contents. A depression is a good example of a concave position in soil. Depressions modify the microenvironment of the soil by accumulating water, and so will tend to accumulate SOM.

Soil aspect is also important in determining the SOM content. Slopes that receive sunlight will tend to be several degrees warmer than opposing slopes that receive less sunlight. In the northern hemisphere, this means that south-facing slopes will be warmer and drier than north-facing slopes and consequently have less SOM. This is particularly applicable to forest

FIGURE **11-12** Soil organic matter content and relative A horizon thickness as related in topographic position. Where the potential for erosion exists the soil organic matter decreases, as it does on shoulders. In depressions, however, where soil organic matter can accumulate, the depth of the A horizon increases. *(Adapted from Troeh and Thompson , 1993)*

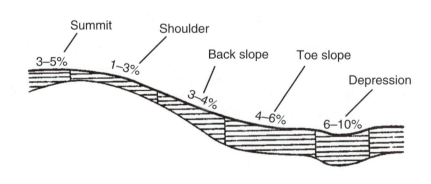

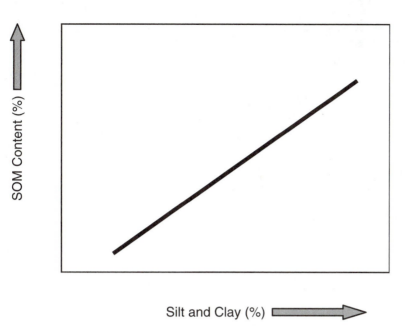

soils. The cause is twofold. First, there is the increase in SOM decomposition due to higher temperature on south-facing slopes. Second, there is less water available in south-facing slopes because of the higher temperatures. This causes differences in the plant community, which in turn influence the amount and availability of the litter that is deposited.

Parent Material

Parent materials that provide adequate nutrients for plant growth will tend to promote higher SOM contents. Parent materials will also influence the dominant texture of a soil. So, parent materials that lead to sandy soil formation will have relatively lower SOM contents than soils producing silty and clayey soils. Sandy soils are better drained and drier than other soils; they have fewer-plant available nutrients and typically support vegetation such as trees, which do not contribute as much available residue to SOM as grasses. For similar reasons parent material that produces finer-textured soils will promote SOM accumulation.

Clays play another important role in preserving SOM. The reactive groups of SOM, which are among the first groups subject to biological decomposition, are attracted to the charged groups on clay particles. This helps protect the SOM from decomposition. Furthermore, clay/OM interactions will lead to aggregation, which in turn physically protects SOM from decomposition. Consequently there is usually a positive relationship between the SOM content and the percentage of silt and clay in a soil (Figure 11-13).

Clay helps to preserve SOM by forming crypts.

Time

How fast does SOM accumulate? It may take hundreds to thousands of years for the SOM to accumulate to relatively steady-state conditions in a mature soil. Chronological sequences in various soils suggest that the rates range from 1 to 12 g m^{-2} per year (Schlesinger, 1997) depending on the environment. The SOM accumulation rates can be significantly higher, however, particularly in the first decades after the effects of human influence are removed. Table 11-4 illustrates the profound recovery of SOM content in abandoned soils with time.

TABLE 11-4 SOM accumulation in abandoned agricultural and disturbed soils. *(Adapted from Schlesinger, 1997)*

Ecosystem	Previous Land Use	Years Since Abandonment	Rate of Accumulation (g C m^{-2} yr^{-1})
Subtropical forest	Agriculture	40	30–50
Temperate deciduous forest	Agriculture	100	45
	Mining	50	55
Temperate coniferous forest	Agriculture	50	21–26
	Diked	100	26
Temperate grassland	Agriculture	5	110
	Agriculture	53	1.6
	Mining	28–40	28

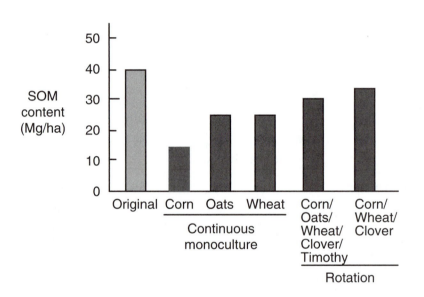

FIGURE 11-14 Effect of various cropping systems on soil organic matter content. Although cultivation will decrease soil organic matter content, different rotations will have different success at maintaining soil organic matter levels. *(Adapted from Frye et al., 1985)*

Human Activities

Cultivation has a dramatic effect on SOM (Figure 11-14). From a geological perspective, human activity can remove in an instant the SOM that had taken millennia to produce. With continuous cultivation more than half of the C in soil can be lost in a short period.

In the long-term Morrow plots at the University of Illinois, over 125 years of continuous corn cultivation has reduced the SOM content by more than 60 percent. The SOM content decreases for several reasons. Cultivation aerates the soil briefly, which stimulates aerobic decomposition. It buries residue in the soil, which puts organic matter in closer proximity to decomposing microbes. Generally speaking, cultivated crops do not add as much organic residues to soil as do perennial grasses, and in many cases little residue is returned at all because it is harvested for forage or burned to fa-

> Cultivation helps to accelerate SOM disappearance.

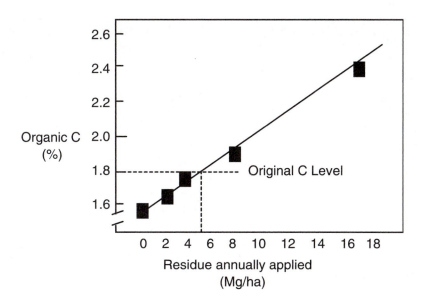

FIGURE 11-15 SOM accumulation in cultivated soil with increasing residue application. *(Adapted from Larson et al., 1978)*

cilitate planting. Tillage also serves to break up aggregates that have formed, and to some extent expose labile but physically protected soil C to microbial decomposition. Nevertheless, the SOM content in cultivated soils does not decline to zero regardless of the length of cultivation. Why? Because there remains the large pool of passive SOM, which is resistant to decomposition.

As a rough guide, approximately 5,000–8,000 Mg ha^{-1} of residue are required to maintain the SOM of a soil once it has gone into cultivation (Larson et al., 1978). It is also possible to increase SOM content, though not usually to the levels previously found in undisturbed soil. Studies have shown an approximately linear increase in SOM content with increase in crop residues applied to soil (Figure 11-15).

Drainage also has a dramatic effect on SOM content. In organic soils that contain 20 to 30 percent organic matter, SOM decomposition is retarded by wet conditions, and organic matter tends to accumulate. Once the soils are drained and aerated, decomposition of the accumulated SOM occurs at a much faster rate than its accumulation and a process called **subsidence** occurs. Subsidence is nothing more than a rapid decrease in the soil depth, sometimes as much as 2.5 to 5 cm (1 to 2 inches) per year. It is a major problem in organic soils in Florida, where subsidence has decreased the soil depth more than 3 m (10 ft) in the past 100 years in some locations.

> Subsidence due to loss of SOM is a major problem in organic soils that are drained and aerated.

FOCUS ON . . . THE DISTRIBUTION OF SOM

1. How is most SOM distributed in soil?

2. How does topography influence the distribution of SOM?

3. How quickly does SOM accumulate in undisturbed soil?

4. Why does the SOM content of soil never go to zero?

5. What is subsidence and where does it occur?

SUMMARY

Soil organic matter is one of the critical components contributing to a soil's ability to carry out necessary environmental functions such as supplying plant nutrients and facilitating soil aggregation. The SOM is composed of nonhumic and humic materials, basically representing plant and microbial residues that are in the process of decay, and organic materials that are unrecognizable and are in the process of polymerizing into compounds like humic and fulvic acid. One of the key ingredients in developing SOM is the polymerization of lignin and lignin decomposition products with peptides. The SOM can also be divided into active and passive fractions, which represents the SOM that is readily decomposed and that which is physically and chemically protected. Many soil-forming factors influence SOM distribution, and quantity in soil and human activity can both reduce and build up SOM content in soil.

END OF CHAPTER QUESTIONS

1. What are major ecological roles of soil to which SOM contributes?
2. What percent of SOM is typically composed of humic materials?
3. Give some examples of typical plant materials that enter soil as residues.
4. What is lignin and what are the major components of lignin?
5. Diagram how lignin is involved in SOM formation.
6. Give two or three specific reasons why SOM is important to preserve.
7. Distinguish between humic and fulvic acids.
8. What is the relative difference in decomposition rate between the active and passive fractions of SOM?
9. What is the difference between chemical and physical protection of SOM?
10. How can human activity accelerate SOM decomposition?

FURTHER READING

The Plowman's Folly by Edward E. Faulkner (1977, Washington, DC: Island Press, http://www.islandpress.org) gives an interesting alternative perspective on how conventional agricultural activities have contributed to the loss of SOM and consequently the loss of soil fertility and soil quality. For a more technical treatment of soil C and soil C cycles there are two important references that are invaluable: *Cycles of Soil*, 2nd ed., by F. J. Stevenson and M. A. Cole (1999, New York: John Wiley & Sons), which gives a thorough biochemical treatment of SOM, and *Biogeochemistry: An Analysis of Global Change*, 2nd ed., by W. H. Schlesinger (1997, San Diego, CA: Academic Press), which treats SOM in the context of soil as an ecosystem.

REFERENCES

Daily, G., P. A. Matson, and P. M. Vitousek. 1997. Ecosystem services supplied by soil. In G. C. Daily (ed.), *Nature's services: Societal dependence on natural ecosystems.* Washington, DC: Island Press, pp. 113–132.

Larson, W. E., R. F. Holt, and C. W. Carlson. 1978. Residues for soil conservation. In W. R. Oschwald (ed.), *Crop residue management systems.* Spec. Pub. 31. Madison, WI: American Society of Agronomy.

Raich, J. W., and W. H. Schlesinger. 1992. The global carbon dioxide in soil respiration and its relationship to vegetation and climate. *Tellus* 44B: 81–99.

Schlesinger, W. H. 1997. *Biogeochemistry: An analysis of global change.* San Diego, CA: Academic Press.

Troeh, F. R., and L. M. Thompson. 1993. *Soils and soil fertility.* New York: Oxford University Press.

SECTION 5

SOIL MANAGEMENT

Chapter 12

PHYSICAL MANAGEMENT

"History is largely a record of human struggle to wrest the land from nature, because man relies for sustenance on the products of the soil. So direct is the relationship between soil erosion, the productivity of the land, and the prosperity of people, that the history of mankind, to a considerable degree at least, may be interpreted in terms of the soil and what has happened to it as the result of human use."

Hugh H. Bennett and W. C. Lowdermilk, circa 1930s

OVERVIEW

Soil has been described as the thinnest of skins covering the Earth and permitting life to exist on land. Its physical management is of the utmost importance. Management typically means its uses for agriculture. However, you can easily extend those uses to other purposes, such as treating wastes deposited on soil. In addition, humanity has a long history of using soil to build the foundations of its civilizations from pottery to shelter. Today, the physical properties of soil and their management are critical to such things as land value; road, dam, and pond construction; building placement; and mitigating environmental contamination. In this chapter you will evaluate some of the principles underlying how physical management of soil affects agricultural productivity and environmental quality.

> Soil has many other uses besides agriculture.

OBJECTIVES

After reading this chapter, you should be able to:

- ✔ Identify various tillage systems and how tillage can promote or retard soil erosion.
- ✔ Describe how compaction occurs in soil environments.
- ✔ Identify mechanisms promoting soil erosion and conservation practices that reduce erosion.
- ✔ Detail how residue management controls soil erosion.
- ✔ Suggest how soil properties influence nonagricultural land uses.

KEY TERMS

allelopathic
compaction
conservation tillage
desertification
ephemeral gully
    erosion
eutrophication
fallow
grassed waterway
no-tillage

Proctor density test
rill erosion
salinization
saltation
sheet erosion
splash erosion
surface creep
tillage
Universal Soil Loss
    Equation (ULSE)

Water Erosion
    Prediction Project
    (WEPP)
Wind Erosion
    Equation (WEQ)
Wind Erosion
    Prediction System
    (WEPS)

TILLAGE AND AGRICULTURE

Humanity's ties to soil can be epitomized by the plow. The beating of swords into plowshares is just one image capturing that relationship. Yet, this relationship has been equivocal—at once providing the basis for feeding millions, and now billions, but also creating conditions that lead to environmental disaster.

Purposes of Tillage

Tillage intentionally disturbs the soil.

Tillage is soil disturbance with an underlying reason. Tillage has three basic purposes:

- ❖ weed control
- ❖ altering soil physical and chemical properties
- ❖ residue management

Before extensive herbicide use for weed control, tillage was the main way cropland was kept weed free before and after planting. Weed control is a necessary feature of high-production agriculture because weeds compete with crops for available nutrients and water, intercept the light necessary for photosynthesis, and in some instances are **allelopathic**. Summer **fallow** cropping systems, for example, use tillage to keep soil weed free during the fallow period and maximize the amount of water available to subsequent small grain crops.

Beneficial aspects of tillage are weed control, aeration, preparing good seedbeds, and incorporating soil amendments.

In terms of physical and chemical properties, tillage briefly aerates the soil, prepares a seedbed that improves seed-to-soil contact and emergence, and temporarily destroys surface crusts and compacted layers that might retard shoot emergence and root penetration. Tilled soils are drier and warmer in spring and it is possible to begin planting earlier on tilled soils. Surface-applied fertilizers, liming agents, and other soil amendments are mixed with the upper soil profile by tillage, which promotes more rapid dissolution and thus more uniform availability to subsequent crops. Residue that would interfere with uniform planting and seed placement and harbor plant diseases and insect pests is buried by tillage. This has the additional benefit of returning some of the plant-extracted soil nutrients back to the soil where they can be recycled.

Tillage Methods

In mechanized agriculture, tillage can be divided into two categories: conventional or plow tillage and **conservation tillage.** Plow tillage has two stages, a primary tillage, which inverts the soil surface and buries residue, and a secondary tillage that breaks up large clods and produces a uniform seedbed. Conservation tillage, in contrast, minimizes the amount of residue that is buried. Conservation tillage ranges all the way from mulch till or chisel-plow tillage, which loosens the soil but does not invert it, to **no-tillage,** in which nothing but a small slit in the soil is prepared for seed placement.

Consequences of Tillage

What is good for crops in terms of tillage is not necessarily beneficial to the soil and its overall properties. As you read in Chapter 11, plowing soils can reduce the SOM content by over 50 percent (Figure 12-1). Repeated tillage operations crush and destroy soil aggregates and can cause soil **compaction.** Without overlying plant residue and with minimal amounts of organic matter holding them together, aggregates in tilled soils are exposed to the full brunt of rainfall and are easily dispersed, which can cause soil crusting, blocked soil pores, and subsequently increased runoff. Increased runoff has the inevitable consequence of removing topsoil. Where sloping soil occurs, heavy rain on an exposed soil can erode 112 Mg/ha (50 tons of soil/acre), removing in minutes the product of hundreds of years of soil formation (Rasnake, 1983).

> Tillage can cause compaction, soil crusting, and erosion.

FOCUS ON . . . TILLAGE

1. What are the two basic types of tillage practices?
2. What are three purposes of tillage?
3. Why might you want to use tillage in a fertilized field?
4. What are some of the advantages of tillage?
5. What are two major consequences of tillage?

FIGURE 12-1 Effects of plowing compared to no-tillage after 30 years of continuous corn cultivation. The difference in elevation between the two tillage practices is primarily due to the steady loss of soil organic matter.

CONSTRUCTION AND NONAGRICULTURAL USES OF SOIL

Nonagricultural uses of soil include sports, construction, and insulation.

Soil can be used as a structural material to build dams and seal ponds. It can be used to build the foundation of roads and houses, and in some cases it can be used to build the houses themselves. Soil has also been used as a home insulator. Perhaps the most exacting demands on soil come from sporting industries such as golf and baseball, which require carefully constructed soils for optimum performance of the turf and player, respectively. The construction and engineering industries likewise have a vested interest in managing soil and there are numerous soil properties that must be considered for any project (Table 12-1).

Roads and buildings require a firm foundation, which is usually provided by sandy or gravelly soils. These materials frequently have to be placed over clayey soils to provide adequate support. Clays, however, form the compacted core of earthen dams and levees. This core is overlain by sand and loam to permit vegetation to develop. Ponds are likewise sealed against water loss by a compacted layer of clay that is kept continually wet. Closed landfills and dumps, in contrast, are covered by a clay cap to keep water from infiltrating into the buried material.

Septic systems require soils with particular properties to function well.

Soil properties are extremely important in installing septic systems in rural areas. In this case soil has a role in secondary waste treatment. To protect groundwater and surface water from contamination, septic systems require a minimum depth of soil through which raw septage must percolate as a treatment process. Many states require a minimum of 50 cm (about 18 inches) between the bottom of a septic system drain field (absorption field) and any water table or impermeable layer. If the soil is too shallow inadequate filtration occurs, and if the soil has a seasonally high

TABLE 12-1 Engineering and construction considerations for soil. *(Adapted from Harpstead et al., 1997)*

Parameter of Interest	Engineering or Construction Concerns
Soil texture	Unconsolidated soils are subject to liquefaction during earthquakes.
Clay type	The shrink-swell potential of different clay types is a consideration.
Depth to bedrock	Shallow soils are unsuitable for excavating basements, ditches, and utility lines. Installation of on-site septic systems is also affected.
Type of bedrock	Bedrock can be hard (granite), porous (sandstone), or soluble (limestone), which affects the support of major structures and creates the potential for sinkholes to develop.
Density	Soils may be porous or cemented, which affects water movement.
Content of rock fragments	Soils with many rock fragments may be difficult to excavate.
Erosion potential	Sandy soils are subject to wind erosion, silty soils to water erosion, and certain clays to piping, or subsurface erosion by water, which was the cause of the Grand Teton Dam failure.
Surface geology	Homes and structures built on hillslopes can be lost due to mass wasting.
Soil pH and corrosivity	Wet, acid soils are corrosive and can corrode metal and concrete pipes laid within them.
Salinity	Saline soils can have very poor structure.
Depth to seasonal water table	Frost action is severe in wet soils. Seasonally high water tables provide inadequate support for roads and buildings. High water tables also influence the placement of septic systems.
Plasticity	Clay soils are plastic and become fluid with wetting.
Organic matter content	Soils with high organic content are subject to subsidence once drainage occurs and provide poor support.

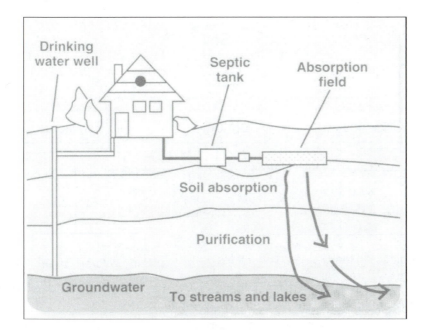

FIGURE 12-2 Schematic design of a septic system for on-site waste treatment. Although home occupancy determines the size of the system, soil properties determine whether it can be installed. *(Diagram courtesy of the National Smallflows Clearinghouse, Morgantown, West Virginia)*

water table the homeowner runs the risk of having sewage rise to the soil surface. For lack of taking soil properties into consideration, it is frequently the case that homeowners building $200,000 homes in rural sites discover, too late, that they are unable to install a required septic system at 1/100th the cost of the home and consequently are unable to occupy their new homes (Figure 12-2).

PHYSICAL DEGRADATION OF SOIL

Soil degradation can be expressed through chemical, biological, and physical processes that are inextricably linked (Figure 12-3). Unfortunately, nonsustainable human agricultural and industrial activity has caused severe soil degradation in many parts of the world. The effects of soil degradation can be measured on-site through such factors as the reduction in soil organic matter, reduced yields, increased **salinization,** and diminished soil depths. Soil degradation can also be measured at a distance in terms of wind-blown sediments, siltation in ponds and ditches, and increased nutrient contents in surface waters, which leads to **eutrophication.** In this section you will look specifically at some of the indicators of physical soil degradation.

> Soil degradation can be seen in chemical, biological, and physical properties.

Desertification

Desertification is primarily a problem in arid and semiarid regions. As the name implies, it is a process in which arable land is replaced by progressively desertlike conditions. Several factors contribute to desertification: prolonged drought, vegetation removal, and irrigation with low-quality water, which over time leads to soil salinization. Many once-fertile regions in the Middle East have become effectively desertified due to the salinization of land.

The Great Plains was the scene of desertification in the 1930s, the so-called "Dust Bowl" era, when long-term drought, excessive tillage, and naturally high winds caused heavy soil losses through wind erosion and

> Desertification replaces arable land with unproductive soil.

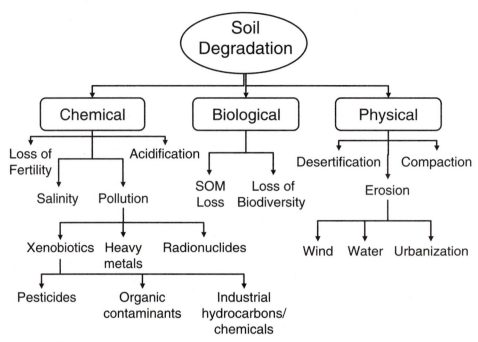

FIGURE 12-3   Forms of soil degradation. *(Adapted from Dubbin, 2001)*

left many farms covered in dunes of windborne soil. (See Chapter 16 for illustrations.)

Compaction

Soil compaction occurs when a given mass of soil is pressed into a smaller volume, which increases a soil's bulk density; because the mass of soil is not changing, it means that the soil porosity must decrease. This is obviously detrimental. If the soil becomes compacted and the bulk density increases it becomes increasingly difficult for plant roots to penetrate the soil. If the plant roots are able to penetrate the soil they are liable to experience an environment in which aeration is decreased because more of the pore space is filled by water than air for a given water content. Compaction also has undesirable effects with respect to microbial activity. Soil respiration and enzyme activity decrease in compacted soils (Nannipieri, 1994). Because microbial activity is decreased, nutrient cycling from fresh residues and SOM is decreased, and plants will ultimately have fewer nutrients on which to grow. Compaction also has unfavorable effects on water relations in soil. Compacted soils at the soil surface are less permeable to infiltration and promote runoff. Compacted layers in the soil can cause perched water tables or otherwise divert infiltrating water from uniformly wetting a soil.

Soil compaction does not necessarily occur at the soil surface. It typically occurs in a layer 8 to 14 inches below the soil surface, which would be a pattern typical of conventional tillage systems with a chisel or moldboard plow, although this depth can greatly vary. Compaction at the soil surface is not apparent because of the action of the tillage instruments, root growth, and animal burrowing. Once formed, the compacted layers persist for considerable periods.

What are the conditions that lead to soil compaction? The most common cause of soil compaction is vehicular traffic with heavy farm machinery on wet soils. This occurs in tilled soils and in soils where no-tillage is practiced.

> Compaction decreases porosity, root penetration, and biological activity and increases the potential for erosion.

MEASURING COMPACTION

The most common method of evaluating soil compaction is by means of the *soil penetrometer,* a device that measures the resistance of soil to the penetration by a conical steel probe (Figure 12-4). Soil resistance is actually a measure of soil strength, which depends on the type of soil particles and their architecture. However, when the soil is neither too wet nor too dry (close to field capacity is ideal), soil penetration resistance is a good surrogate measure for bulk density, which is directly related to compaction. Penetrometers typically measure soil resistance in terms of lbs/in^2 (psi). The metric equivalent is kilopascal/cm^2 (kPa/cm^2). Soils with resistance measures > 300 psi are considered to be compacted.

(a)

(b)

FIGURE 12-4 A soil penetrometer (a) and student measuring soil resistance to penetration (b).

In conventional tillage systems multiple passes are made through a field to prepare a clean seedbed. In no-tillage systems compaction occurs, in part, because the same wheel tracks are followed repeatedly in an effort to minimally disturb the soil.

Wet soils will compact more than dry soils. Each soil has an optimum water content at which maximum compaction occurs. Thus, the density to which a soil compacts increases as the soil water content increases, but declines for the same compactive force as the soil becomes saturated. This is the basis of the **Proctor density test.**

Soil compaction is not restricted to agricultural activities. Anyone who has enjoyed a footpath along a recreational trail has unwittingly contributed to compaction in that environment. One of the reasons these trails

> Traffic on wet soils is a major cause of soil compaction.

THE PROCTOR DENSITY TEST

The Proctor density test is typically used as a measure in engineering construction, but it provides a useful illustration of how increasing soil moisture can lead to increased compaction by farm machinery. The Proctor density is the maximum density that a soil can achieve for a given water content and a given amount of compactive energy. When a uniform compactive force is applied (usually by dropping a standard weight from a standard height a set number of times) the soil will continue to compact up to an optimum water content; thereafter, the soil density will decrease.

Obviously a soil will not become less compacted if weights are continually dropped on it. Therefore the Proctor test is conducted in two directions: for a soil being progressively wet and for the same soil being progressively dried. As Figure 12-5 illustrates, lines can be drawn through wetting and drying regimes to find a point of intersection, which represents the maximum density that can be achieved.

FIGURE 12-5 The Proctor density test.

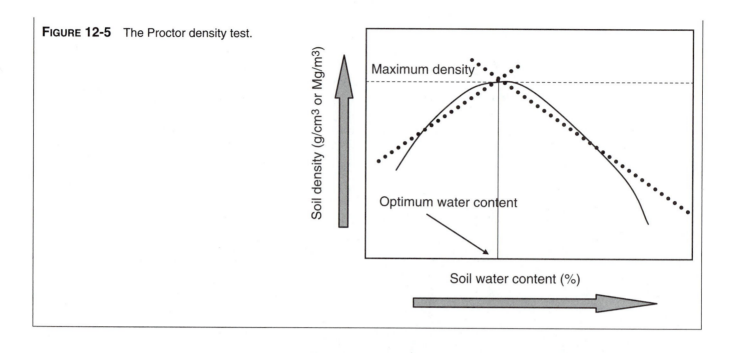

Maximum density

Soil density (g/cm³ or Mg/m³)

Optimum water content

Soil water content (%)

FIGURE 12-6 Mulch is often spread over a foot trail to reduce compaction. Another approach in high-traffic or sensitive areas is to build raised walkways. Here, a district conservationist and member of the Lamar County Exposition Authority tour the Barnsville Recreation Center in Georgia. The trail and center serves as an outdoor classroom and has a nature trail through natural and constructed wetlands. *(Photograph courtesy of the USDA-NRCS)*

Mulch helps prevent soil compaction on hiking trails.

become apparent, in addition to the physical removal of vegetation, is that the soils are too compacted to permit plant root growth. A simple solution to the problem of recreational compaction is distributing coarse mulch over the trail, which disperses the compactive pressure of hikers over a greater surface area (Figure 12-6).

FOCUS ON . . . COMPACTION

1. What is compaction and how is it measured?
2. Why does compaction inhibit plant growth?
3. What conditions favor compaction?
4. What does the Proctor density test measure?
5. In tilled soils, where does most compaction occur?

Erosion

Erosion is a natural process on a dynamic planet such as Earth. The density of meteor craters that pockmark the moon are repeated on Earth, but their evidence has been erased by hundreds of thousands of years of erosion. Compare the height and shape of the newly forming Rockies in the American west with the ancient Appalachian Mountains in the American east. All mountains eventually erode to the sea. Erosion in Africa was responsible for the rise of civilizations. If not for the annual flooding of the Nile, which carried eroded soil from the Ethiopian highlands to the Nile Delta, the fertility of the soil would not have been replenished and the great civilization of ancient Egypt would have been impossible. Nevertheless, erosion is a slow process, and in most places soils will form at a faster rate than naturally occurring erosion—1 cm or so every 100 to 400 years, depending on the amount of water available to support plant growth.

> Erosion is a natural but slow process that human intervention can greatly accelerate.

Human and animal activities greatly accelerate naturally occurring erosion, sometimes by more than a hundredfold or greater. Deforestation, development, overgrazing, and farming hillslopes are all factors that contribute to erosion. Mass wasting is a type of erosion in which large masses of soil move by the force of gravity. Slumps along riverbanks and gullies exemplify mass wasting (Figure 12-7). Mudslides are an even more dramatic example of mass wasting. In soils subject to freezing and thawing a type of erosion called soil creep occurs, in which the soil develops a ripple appearance due to vertical heaving in winter.

The effects of erosion are damaging in two respects. First, by removing the top layer of soil, which has the preponderance of SOM and nutrients, the overall soil quality is decreased. Water-holding capacity, cation exchange capacity, and biological activity are all diminished when topsoil is lost. Second, when severe erosion occurs and exposes the underlying B and C horizons of soil, plants are exposed to an environment in which root penetration is more difficult because pores are smaller, the soil is typically more acid, cation exchange capacity is reduced, and infiltration is poor, which ultimately leads to greater amounts of surface runoff. In this section you will examine the two principal forces causing erosion: wind and water.

> Erosion removes nutrients and SOM. It also exposes less-fertile subsoil.

Wind Erosion

Wind erosion, which is responsible for 30 to 40 percent of the erosion that occurs annually, is a global problem, occurring whenever wind velocities

FIGURE 12-7 Mass wasting along a stream bank in Warren Co., IA. *(Photograph courtesy of USDA-NRCS)*

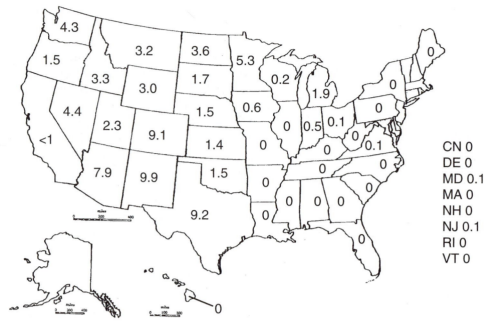

FIGURE 12-8 Average annual wind erosion, by state, on nonfederal cultivated land in tons/acre/year. *(Adapted from the USDA National Resources Inventory, 2002)*

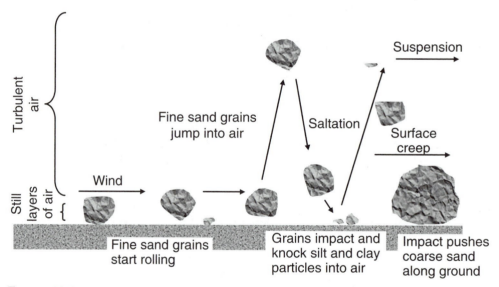

FIGURE 12-9 Saltation triggers wind erosion. At sufficient wind velocity, fine sand grains become airborne in a process called saltation. As the windborne sand periodically strikes the soil, the impact knocks silt and clay particles into the air where they remain suspended. Saltation can also impact coarse sand particles and push them along the soil in a process called surface creep.

> The most important effects of wind erosion occur with 1 foot of the soil surface during saltation.

increase above approximately 25 km/h (16 miles/h). In the United States, wind erosion is the dominant form of erosion in the plains states, which run in a swath from Texas to Washington (Figure 12-8).

The mechanism of wind erosion is illustrated in Figure 12-9. When the wind velocity approximately 30 cm (1 ft) above bare soil approaches 25 km/h, fine (0.1 to 0.5 mm) sand grains begin to roll along the soil surface. Some of these sand grains eventually jump into the turbulent stream of air. When the sand grains fall back and strike the soil they knock other particles loose in a process called **saltation.** Saltation suspends fine soil particles like silt and clay in the turbulent air, which can carry them great distances. Suspended soil from Africa, for example, can be collected on the east coast of North America. Larger sand particles 0.5 to 1.0 mm in diameter are too big to be suspended, but the impact of the fine grains causes them to roll along the soil surface in a process called **surface creep.** Surface creep is a major factor in the movement of dunes.

Several factors determine the amount of wind erosion that occurs:

❖ *Texture*—soils predominating in fine sand are very susceptible to wind erosion, as are organic soils. Soils high in clay or well-aggregated soils are less prone to erosion. Because tillage physically destroys aggregates, it tends to make soils more susceptible to wind erosion.

❖ *Surface roughness*—clods and ridges act as obstacles to wind movement and decease the wind velocity.

❖ *Climate*—low rainfall, high temperature, and high winds can all contribute to wind erosion.

❖ *Field length*—the longer the field, or length of unimpeded wind movement, the greater the extent of wind erosion. As more soil becomes airborne, the effect of saltation increases down the length of a field. As Figure 12-8 shows, the eastern United States is little prone to wind erosion because of its higher rainfall than the Coast Plains and because the distances between obstacles in a field, which reduce wind velocity, are much shorter than in the plains states.

❖ *Vegetative cover*—bare soil is fully exposed to the force of wind. Therefore, any type of ground cover or residue cover helps to protect the soil.

> Texture, surface roughness, climate, field length, and vegetative cover all influence wind erosion.

FOCUS ON . . . WIND EROSION

1. Where in the United States is wind erosion a significant problem?

2. What is the critical wind velocity at which wind erosion occurs?

3. What is saltation?

4. What critical factors lead to wind erosion?

5. What size soil particles are most important in wind erosion?

Water Erosion

Water erosion accounts for approximately a billion metric tons (a metric ton is 1000 kg) of soil erosion yearly, or half to two-thirds of the annual erosive losses. As you might expect, water erosion is more significant in environments where there is abundant rainfall. Figure 12-10 shows water erosion from cropland in the United States; notice how most of these losses occur east of the Mississippi River.

Erosion by water occurs in three steps individually termed detachment, transport, and deposition:

> Up to two-thirds of soil erosion is caused by water.

❖ detachment—falling raindrops shatter surface aggregates and loosen soil particles.

❖ transport—detached soil particles move with flowing water.

❖ deposition—the soil is deposited when water velocity slows.

> Three stages in water erosion are detachment, transport, and deposition.

Falling rain strikes the soil at about 30 km/h, which imparts a tremendous amount of energy to the soil it strikes. Although the size of a raindrop is important, the impact energy of a raindrop increases as the square of its

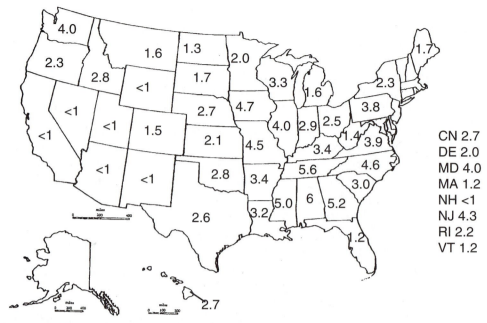

CN 2.7
DE 2.0
MD 4.0
MA 1.2
NH <1
NJ 4.3
RI 2.2
VT 1.2

FIGURE 12-10 Average annual sheet and rill erosion, by state, on nonfederal cultivated land in tons/acre/year. *(Adapted from the USDA National Resources Inventory, 2002)*

velocity (energy = mass x velocity^2), so a small change in velocity has a greater impact than a similar change in raindrop mass. The force of rain impacting bare soil can hurl soil particles 2 m away and essentially restructure the soil surface, clogging soil pores and potentially leading to soil sealing and preventing adequate water infiltration. The direct movement of soil by rainfall impact is called **splash erosion.** If the rain persists, water will begin to run off the soil surface—the greater the soil slope, the greater the velocity of the moving water and the greater its erosive force. This leads to several other types of erosion:

> Splash erosion, sheet erosion, rill erosion, and gully erosion are all types of water erosion.

❖ **sheet erosion**—a thin layer of topsoil is uniformly removed and sometimes imperceptibly removed as a sheet. This form of erosion usually occurs on gentle slopes or the summit of steep slopes and may pass unnoticed until the subsoil appears. Sheet erosion can often be observed as lighter-colored soils on knolls in rolling fields. The lighter color is due to the loss of SOM during erosion and may also be due to the color of the underlying subsoil that is now exposed.

❖ **rill erosion**—this erosion is visible as small channels on a slope. The channels represent the paths of least resistance for water flow on a slope. Rill erosion is modest enough that the channels can be filled in by subsequent tillage.

❖ **ephemeral gully erosion**—ephemeral gullies are large rills. They can be crossed by farm machinery but tillage no longer completely fills them in. Subsequent rainfall will cause water to collect here and initiate further erosion.

❖ **gully erosion**—Gullies are so large that they cannot be crossed by machinery and represent the most obvious evidence of erosion. Gullies form at the bottom of slopes, where water velocity is at its maximum, and gradually work their way up the slope as the sides of the gullies collapse with each new rainfall. Although gullies appear to have the most significant erosion, in fact most soil erosion is due to sheet and rill erosion.

Four factors have a direct effect on water erosion:

❖ *soil texture and structure*—soil texture is important because it influences the ease with which soil particles are detached by rain and running water. Silt-sized particles are among the most susceptible to erosion. Soil structure is important because the more water that infiltrates the soil, the less runoff will occur. Soils with strong, water-stable aggregates and granular structure will have higher infiltration rates than compacted or less-structured soils.

❖ *topography*—the length and grade (steepness) of a slope are important. As each increases, the velocity of running water also increases. Furthermore, longer slopes collect more water, so the mass of running water is also greater. Long, gentle slopes can have as much erosive potential as short steep slopes (Table 12-2)

❖ *soil cover and roughness*—bare soil is exposed to the full impact of rain. As with wind erosion, vegetative cover or residue can significantly decrease erosion. The vegetation or residue dissipates much of the energy of falling rain before it hits the soil. A complete vegetative cover almost completely eliminates soil loss (Table 12-3). Rough soil surfaces also impede water flow by increasing the tortuosity of water movement down the slope and thereby decreasing its velocity. Surface roughness is largely determined by tillage practice, with conservation tillage leaving the soil rough and less prone to runoff (Table 12-4).

❖ *climate*—the amount and intensity of rainfall on an annual or storm-event basis.

> Water erosion is affected by soil texture and structure, topography, soil cover and roughness, and climate.

Tillage influences the first three factors and is perhaps the single most important factor contributing to erosion in sloping cropland, precisely because tillage exposes soil to climatic activity (Rasnake, 1983). The amount of erosion also depends on:

❖ the kind of tillage and tillage implements

❖ the intensity of tillage

❖ the direction of tillage in relation to the slope of the landscape

❖ the timing of tillage and the period that elapses between the time of tillage and sufficient growth of the planted crop to provide soil cover

> The kind of tillage, its intensity, its direction, and its timing all influence soil erosion.

FOCUS ON . . .WATER EROSION

1. What region of the United States appears to be most susceptible to erosion by water?

2. What three steps occur during water erosion?

3. What is the difference between sheet erosion and rill erosion?

4. How does soil texture and structure affect erosion by water?

5. What tillage factors influence erosion?

Predicting Soil Erosion

Considerable effort has gone into predicting the extent of erosion that can occur in a landscape and the tolerance of various soil types to annual erosion. The **Universal Soil Loss Equation (USLE)** and its revised form (RUSLE) are used to predict site-specific erosion. "Universal" is something

> Soil loss is predicted by the Universal Soil Loss Equation (USLE).

TABLE 12-2 The interacting effect of slope grade and length. The erosive force in each combination is equivalent. *(Adapted from Plaster, 1997)*

Grade (%)	Slope Length, Feet (m)	
4	1,000	(305)
6	200	(61)
8	100	(30)
10	50	(15)
12	30	(9)
14	20	(6)

TABLE 12-3 Effect of cropping system and continuity of vegetative cover on runoff and erosion. *(From Miller and Krusekopf, 1932)*

Treatment	Runoff as % of Rainfall	Soil Loss (Mg/ha)
No crop, plowed 10 cm, regularly cultivated	31	94
Continuous corn	29	45
Continuous wheat	23	22
Corn-wheat-clover rotation	14	7
Bluegrass sod	12	0.7

TABLE 12-4 Effect of no-tillage versus plow tillage on soil loss in southern Illinois. *(From Blevins, 1981)*

Tillage System	Soil Loss (Mg/ha/yr)	
	5% slope	9% slope
Plow tillage, wheat and corn double-cropped	6.81	21.10
No-tillage, wheat and corn double-cropped	0.76	1.21
No-tillage continuous corn	0.56	0.81

of a misnomer because for some soils and cropping systems the equations have not been accurate. The USLE only predicts sheet and rill erosion and is based on annual losses rather than losses from single, extremely erosive events. Nevertheless, USLE and RUSLE are the foremost tools of soil conservation and the USDA's official erosion prediction and conservation planning tools (Renard et al., 1994).

The USLE assumes that soils can tolerate some erosion each year, generally 2240 to 11,200 kg/ha (1 to 5 tons/acre), which depends on the type and depth of soil. The tolerable soil loss is given as a "T" value for each soil on its soil survey report. The USLE allows you to predict whether the conditions on a given soil for a particular set of circumstances are likely to exceed its "T" value. The USLE is as follows:

$$A = R \cdot K \cdot LS \cdot C \cdot P$$

where　A = the predicted soil loss (annual tons of soil lost per acre)
　　　　R = the rainfall and runoff factor
　　　　K = the soil erodibility factor
　　　　LS = the slope factor
　　　　C = the cover and management factor
　　　　P = the supporting practices factor

The four factors most responsible for water erosion are represented in the USLE. The R factor assesses the potential erosive force of storms and is based on local weather conditions. The R values are obtained from published maps by the USDA. The K factors, which depend on texture, structure, and soil OM content, are typically given in soil survey reports for mapped soils. Length and slope are separate factors for each site but can be combined to form a single variable representing the effect of topography on erosion as published tables for determining the LS factor from measured slopes and lengths. Cover and management practices (the C factor) depend on the cropping system, amount of residue, tillage, and soil cover. These are site specific, so local Natural Resource Conservation Service (NRCS) offices have appropriate charts for their regions. Supporting practices such as strip cropping or contour plowing will reduce erosion, the extent of which is reflected in published tables from the USDA that take into account the land slope, length of slope, and width of strip crop. Because the P value is always < 1, it has the effect of reducing the predicted erosion loss when the supporting practice is used.

The RUSLE equation, first released in 1992, is a computer-based version of the USLE, which takes into account refinements in some of these factors and allows for different cropping patterns. However, it operates along basically the same principles. Another model used to predict potential erosion is **Water Erosion Prediction Project, WEPP.** To predict wind erosion a program called **Wind Erosion Equation, WEQ,** and its computer-based equivalent **Wind Erosion Prediction System, WEPS,** are used.

> Factors for use in the USLE are found in soil surveys.

Focus on . . . Predicting Soil Erosion

1. What is the underlying assumption of the "T" value for soil?
2. What do the USLE and RUSLE predict?
3. What is the best place to find information about the K factor of soils?
4. What does the C factor in the USLE assess?
5. How can you predict wind erosion?

Conservation Measures

The founding of the Natural Resource Conservation Service (originally the Soil Conservation Service) in the 1930s was a direct result of the dramatic soil erosion that occurred in the United States during the Dust Bowl era. As a result of NRCS activity, agricultural soils were mapped to categorize the nation's soils in terms of their potential productivity and susceptibility to erosion (see Chapter 15). In addition, considerable efforts were made by the NRCS and Land Grant Universities to develop conservation methods that would help to preserve and protect soil resources from erosion, in addition to predicting the extent of erosion that was possible on various soils.

> The Soil Conservation Service (now NRCS) was born out of the United States' experience with erosion in the "Dust Bowl" era.

Preventing Wind Erosion

The single most effective method of preventing wind erosion is to maintain vegetative cover and minimize the amount of bare soil exposed to the force

FIGURE 12-11 Deposition of windblown soil on the downwind (lee) side of windbreaks. Wind velocity on the downwind side of obstacles declines dramatically, causing soil particles to fall out of suspension.

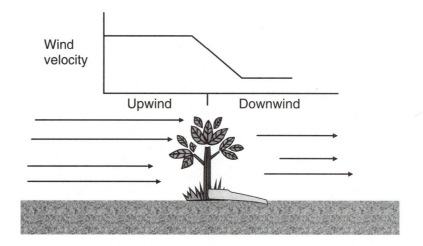

Maintaining surface residue is one of the best ways to prevent wind and water erosion.

of wind. Practices such as conservation tillage, which leaves abundant surface residue, and winter cover crops will accomplish this goal. A second strategy is to minimize wind velocity at the soil surface by tilling at right angles to the direction of wind and creating surface roughness. A third strategy is to plant wind breaks, rows of trees and large shrubs, to shorten long unobstructed fields, reduce wind velocity, and capture suspended soils, as demonstrated in Figure 12-11.

Preventing Water Erosion

There are a variety of methods to reduce or retard water erosion aimed at minimizing the start of runoff and subsequently minimizing the velocity of runoff water. Some of these methods are relatively easy to employ, such as leaving surface residue, employing cover crops to minimize periods when the soil is left uncovered, contour plowing (plowing parallel rather than perpendicular to a slope), and maintaining erodible soils or soils with steep slopes in permanent pasture through such practices as the Conservation Reserve Program (CRP). In addition, where water naturally gathers on a slope and would be prone to gully formation, a **grassed waterway** can be used, which is simply a shallow, grass-covered ditch that is designed to carry off excess water.

Some conservation methods involve more complicated crop management systems of restructuring of slopes. Strip cropping, for example, alternates strips of closely growing crops between row crops such as soybeans and corn. The intervening strip may be grass, for example. The idea is to slow runoff water before its velocity grows too great. Terracing restructures the slope by reducing the grade at intervals down the length of the slope to create flatter areas.

For water leaving cropland or construction sites other methods are used. Silt fences, for example, are fabric barriers that are placed around construction sites to retard soil movement off-site. Vegetative or grass filter strips are buffers between cropped land and surface waters or sinkholes that are designed to reduce runoff velocity, increase infiltration, and cause suspended sediment to fall out of solution (Figure 12-12). Riparian buffer strips are zones of shrubs and trees by waterways through which runoff water must pass. In riparian buffers suspended sediment also falls out of solution and dissolved nutrients are absorbed by the permanent growing vegetation.

Riparian buffers and grass filters help prevent sediment from running off soils into surface water.

FIGURE 12-12 Plots used to test the effectiveness of grass filter strips for preventing runoff.

FOCUS ON . . . CONSERVATION MEASURES

1. What are basic measures to prevent wind erosion?

2. Why does snow and silt accumulate on the downwind side of fences?

3. What features do conservation measures for water control have in common?

4. What is the CRP?

5. What purpose does a grass filter strip serve?

ORGANIC MATTER AND RESIDUE

To a great extent, managing the physical properties of soil is the same as managing the organic matter of soils through the tillage method, the types of plants grown, and the organic residues returned to soil. There is a direct relationship between the amount of soil OM and runoff (Figure 12-13). No-tillage is among the best methods for preserving soil OM in agricultural systems (Table 12-5). It is also among the best methods for preserving surface residues, which will reduce the effects of both wind and water erosion (Figure 12-14). The type of crop grown is equally important because different crops return different amounts of residue to soil (Table 12-6).

No-tillage is one of the best ways of controlling erosion because it maintains SOM and surface residue.

FOCUS ON . . . ORGANIC MATTER AND RESIDUE

1. What effect does increased organic matter content in soil have on runoff?

2. How does no-tillage compare to various forms of plow tillage in terms of percent residue cover?

3. What is the relationship between tillage practice and the amount of soil nutrients lost in sediment?

4. Which crops tend to provide the most surface residue?

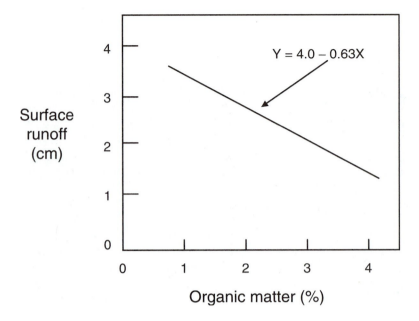

FIGURE 12-13 Effect of soil organic matter content on surface runoff. *(From Wischmeier and Mannering, 1965)*

Surface runoff (cm)

$Y = 4.0 - 0.63X$

Organic matter (%)

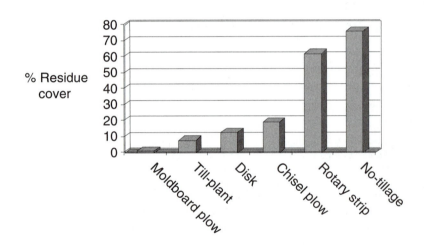

FIGURE 12-14 Effect of tillage system on residue cover in continuous corn. *(Adapted from Griffith et al., 1977)*

% Residue cover

Moldboard plow   Till-plant   Disk   Chisel plow   Rotary strip   No-tillage

TABLE 12-5 Soil erosion, nutrient, and C losses in eroded sediment from corn cultivation. *(Adapted from Follett et al., 1987)*

Tillage	Soil Erosion	Organic Carbon	Organic Nitrogen	Organic Phosphorus
	- 1000's of Mg -			
Conventional tillage	1,396,855	49,380	4,040	450
Conservation tillage	575,499	20,630	1,690	190
No-tillage	436,622	15,890	1,300	140

TABLE 12-6 Residue produced by several crops for typical yields. *(From Rasnake, 1983; Plaster, 1997)*

Crop	Residue Produced (kg/ha)
Flax	1,344
Soybeans	2,240–3,360
Rye	3,360–5,040
Sorghum	3,360
Barley	4,480
Hairy vetch	4,480
Wheat	4,480
Oats	4,816
Corn	6,720–7,840

SUMMARY

There are several ways in which soil can be degraded. Physical management of soil plays an important role in protecting or enhancing its degradation. Tillage is an important agricultural activity but it can have adverse effects on the soil environment. Consequently, several conservation tillage methods are used to minimize these adverse effects. Among the processes that lead to physical degradation of soil are desertification, compaction, and erosion. Desertification is a major problem in regions suffering from drought and overgrazing. Desertification is also related to salinization of soils when poor-quality irrigation water is used. Compaction is one of the consequences of vehicular traffic on wet soils. Compaction leads to poor root growth and increased runoff.

Runoff is one of the major causes of soil erosion. Of the four major types of soil erosion caused by runoff, gully erosion is the most obvious, but sheet erosion actually causes the greatest soil losses. In windy areas, such as the Great Plains, it is wind erosion that causes more runoff. In both cases, maximizing the organic matter content of soil and the amount of surface residue are two good practices to reduce soil erosion to tolerable levels.

END OF CHAPTER QUESTIONS

1. Which type of tillage practice leaves the most residue?
2. How does tillage contribute to soil aggregate destruction?
3. Why can tilled soils usually be farmed earlier in spring?
4. What are some of the off-site effects of soil erosion?
5. How has erosion been beneficial to human civilization?
6. Why will well-aggregated soils be less susceptible to wind erosion?
7. How extensive is wind erosion in the United States?
8. How do windbreaks work?
9. What are five basic types of water-induced erosion?

10. What evidence supports the use of cover crops to minimize soil erosion?

11. What types of conservation measures are used to prevent wind erosion?

12. What are some inexpensive means of preventing soil erosion in cropland?

13. What are the factors in the USLE?

14. In what ways is the USLE site specific?

15. What role does surface residue play in reducing wind and water erosion?

16. What role does residue management play in soil conservation?

FURTHER READING

There are several excellent Web sites with information on land use and soil conservation. To find out more about septic systems and soils, visit the National Small Flows Clearinghouse at http://www.nsfc.wvu.edu.

For recent maps cataloguing wind and water erosion, visit the NRCS Web site at http://www.nhq.nrcs.usda.gov/land/home.

For more information about erosion and its control, visit http://www.statlab.iastate.edu/survey.

Finally, for more information on using the USLE and RUSLE, visit the NRCS Web site at http://www.nrcs.usda.gov.

REFERENCES

Blevins, R. L. 1981. Cover crops and crop residues. *Soil Science News and Views*, Vol. 2, No. 9, Univ. KY. Coll. of Agriculture, Dept. of Agronomy.

Dubbin, W. 2001. *Soils.* London: The Natural History Museum.

Follett, R. F, S. C. Gupta, and P. G. Hunt. 1987. Conservation practices: Relation to the management of plant nutrients for crop production. In R. F. Follett, J. W. B. Stewart, and J. F. Power (eds.), *Soil fertility and organic matter as critical components of production systems.* Madison, WI: Soil Science Society of America, pp. 19–51.

Griffith, D. R., J. V. Mannering, and W. C. Moldenhauer. 1977. Conservation tillage in the eastern corn belt. *Journal of Soil and Water Conservation* 32: 20–28.

Harpstead, M. I., T. J. Sauer, and W. F. Bennett. 1997. *Soil science simplified*, 3rd ed. Ames, IA: Iowa State University Press.

Miller, M. F., and H. H. Krusekopf. 1932. *The influence of systems of cropping and methods of culture on surface runoff and soil erosion.* Columbia, MO: Res. Bull. 177. Missouri. Agric. Expt. Stn.

Nannipieri, P. 1994. Enzyme activity. In C. W. Fincke (ed.), *The encyclopedia of soil science and technology.* New York: Van Nostrand Reinhold.

Plaster, E. J. 1997. *Soil science and management,* 3rd ed. Clifton Park, NY: Thomson Delmar Learning.

Rasnake, M. 1983. *Tillage and crop residue management.* AGR-99. Lexington, KY: Cooperative Extension Service, University of Kentucky.

Renard, K. G., G. R. Foster, D. C. Yoder, and D. K. McCool. 1994. RUSLE revisited: Status questions, answers, and the future. *Journal of Soil and Water Conservation* 49: 213–220.

Wischmeier, W. H., and J. V. Mannering. 1965. Effect of organic matter content of the soil on infiltration. *Journal of Soil and Water Conservation* 20: 15–152.

Chapter 13

OUR WATER RESOURCES: THE HYDROLOGIC CYCLE AND WATERSHED MANAGEMENT

R. Kolka, USDA Forest Service North Central Research Station, Grand Rapids, MN

"We must come to understand our past, our history, in terms of the soil and water and forests and grasses that have made it what it is."

William Vogt, *Road to Survival*, 1948

OVERVIEW

> Water has many different uses.

Water is critical to the existence of Earth and our environment. We use water to drink, grow crops, produce energy, manufacture goods, and for recreation. Water is also important to wildlife because streams, lakes, and wetlands provide habitats for numerous plant and animal species. In this chapter you will examine the distribution of water on Earth, the most critical issues facing water resources in the twenty-first century, and be introduced to the hydrologic cycle and watershed management.

OBJECTIVES

After reading this chapter, you should be able to:

✔ Roughly describe the distribution of water on Earth.

✔ Discuss six important water resource issues.

✔ Identify the factors necessary for precipitation to occur.

✔ Indicate which factors influence evapotranspiration.

✔ Understand the effects of texture on water storage and movement.

✔ Know how to measure soil water potential in the field.

✔ Distinguish between different types of groundwater systems.

✔ Know the factors that go into making a water balance.

✔ Understand some of the consequences of mechanized agriculture for watersheds.

✔ Identify three different ways in which multiple uses can be distributed across a watershed.

✔ Name four steps involved in watershed management planning.

KEY TERMS

albedo	evapotranspiration	riparian areas
aquicludes	hydrophytic	stemflow
aquitard	methemoglobnemia	stream gauging
convective	orographic	tensiometers
precipitation	precipitation	throughfall
discharge	piezometers	watershed
eutrophication		

DISTRIBUTION OF WATER ON EARTH

Approximately two-thirds of the planet is covered with water (Figure 13-1). Unfortunately only about 0.7 percent is available to use, as over 97 percent is salt water and another 1.8 percent is frozen in ice caps. Of the 0.7 percent of freshwater available, almost 50 percent (48.7 percent) is relatively unavailable because it resides deep (> 0.5 miles deep) in the earth as subterranean groundwater (Figure 13-2). Another 48.7 percent is available

> Of all the water on Earth, only about 0.7 percent is available for use.

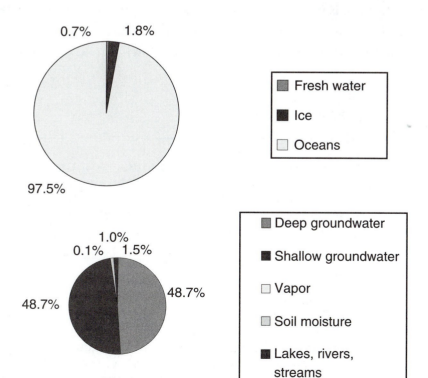

FIGURE 13-1 Distribution of water on Earth. *(From van der Leeden et al., 1990)*

FIGURE 13-2 Distribution of freshwater on Earth, not including water in our ice caps. *(From van der Leeden et al., 1990)*

TABLE 13-1 Availability of water on Earth. One m³ of water is equal to about 264 gallons. *(From van der Leeden et al., 1990)*

Continent	Water Available m³/Person/Year
South America	21,100
North America	7,640
Africa	5,500
Australia	2,750
Europe	2,100
Asia	1,960

as shallow groundwater: 1.5 percent is found in lakes and rivers, about 1 percent is found in soils, and about 0.1 percent is in the atmosphere as water vapor.

The freshwater available for use is not distributed equally across the globe (Table 13-1). South America has considerably more water available per person than any other continent. Not surprisingly, Asia, with large areas of desert and high populations, has the least amount of water available per person. Of course, even within each continent there is much variability. In North America, for example, there is considerably more water available in the east than in the west.

Focus on . . . Water Distribution

1. How can the Earth be water rich and water poor at the same time?
2. What is the largest source of freshwater on Earth?
3. Is water uniformly distributed per capita in various countries?

Water Resource Issues

Although there are numerous issues related to water, most revolve around the quantity of water available and the quality of that water. In some cases, entire chapters or even entire books have been devoted to these issues. A brief overview of six important issues related to water resources is presented in this section.

Water Scarcity

In many parts of the United States, people don't give a second thought about watering their lawns or washing their cars. However, there are many regions in the world where quality drinking water and sanitary facilities are nonexistent. After natural disasters such as earthquakes and tsunamis one of the most critical emergency needs is a supply of potable (drinkable) water. A recent United Nations report estimated that one-third of the world's population lives in areas experiencing moderate to high water stress (United Nations Division of Sustainable Development, 1998). By 2025 that percentage could grow to two-thirds, as populations increase and more demand is placed on the water resource. As water becomes scarcer, agricultural and economic development will be limited and public health will become more of a concern. These effects will have a dramatic impact on food supplies, the financial stability of developing countries, and the potential for outbreaks of disease resulting from contaminated water, and will likely lead to additional degradation in many ecosystems.

> By 2025 water scarcity may affect two-thirds of the Earth's people.

Water Pollution

Water can become polluted through numerous avenues. Both groundwater and surface water are susceptible to various pollutants. Most pollutants are the result of practices that occur on land. Agriculture, industry, and urban areas are all considerable sources of various pollutants. One of the largest concerns in the United States is sediment in surface waters (U.S. Environmental Protection Agency, 1992). Sediment results from soil erosion, stream or lake bank failures, or stream down-cutting, with soil erosion generally the largest contributor. Sediment covers fish and invertebrate habitat, limits light transmission through water thereby affecting the photosynthesis of submerged plants, and is the vehicle by which numerous other pollutants are transported to surface waters.

> One of the largest concerns in the United States is sediment in surface waters.

Phosphorus, which is typically a constituent of fertilizer, tends to bind tightly with sediment. Generally vegetation growth in freshwater bodies is limited by P. If too much P gets into surface water, plant growth explodes. Once the plants die, microbes consuming the dead plants use the O_2 in the water, leaving little for fish and other aquatic organisms. Many times the end result is a fish kill. This process is called **eutrophication**. Sediment is also responsible for transporting various heavy metals such as Hg, industrial organics such as TCE, pesticides, herbicides, and other nutrients.

> One symptom of eutrophication is excessive aquatic plant growth.

Dissolved chemicals are also important pollutants in surface waters and groundwater. Probably the two most important dissolved pollutants are nitrate (NO_3^-) and sulfate (SO_4^{2-}). Like P, NO_3^- or other forms of N are typically used in fertilizers. Unlike P, NO_3^- is not held in soils and leaches easily to groundwater or surface waters. Nitrate in drinking water is a health hazard because it can cause a blood disorder in children called **methemoglobnemia** (blue baby syndrome), in which the children are unable to absorb sufficient O_2 onto their hemoglobin. During the last decade or so, NO_3^- has also been implicated as the main cause of hypoxia (O_2 deficiency) in the gulf coast region of the southern United States. Unlike freshwaters that are P limited, oceans tend to be N limited. Areas along the coast of the Gulf of Mexico are turning into dead zones as a result of eutrophication.

> The Environmental Protection Agency (EPA) drinking water standard for NO_3^- −N is 10 ppm to protect against methemoglobnemia.

Sulfate originating from smokestacks is considered the main cause of acid rain. Sulfur is an important component of coal, and when coal is burned sulfur dioxide (SO_2) is released, which is transformed to SO_4^{2-} in the atmosphere. When SO_4^{2-} is dissolved in water, sulfuric acid results (H_2SO_4) and acid rain is produced. In areas with deep, rich soils much of the SO_4^{2-} is held in the soil and little surface water acidification occurs. However, in areas with shallow and/or nutrient-poor, sandy soils, SO_4^{2-} is readily transported to surface waters and can cause acidification and considerable damage to aquatic systems.

Although sediment, P, NO_3^-, and SO_4^{2-} are probably the most widespread pollutants, numerous others can be very important locally and regionally. Mercury and PCBs (polychlorinated biphenyls) are contaminants in lakes and rivers. Both can accumulate in the food chain, resulting in

Pesticides, heavy metals, bacteria, and viruses are some other important water contaminants.

high levels in fish. Pesticides and herbicides such as DDT and atrazine can be found in groundwater and surface waters and are important health concerns. Industrial solvents such as TCE can also be of concern in areas where they have been used in the past. Finally, outbreaks of bacteria and viruses found in poorly processed drinking waters kill hundreds if not thousands of people each year.

Climate Change

Water vapor contributes to global warming.

Water plays an essential role in climate. Local, regional, and global circulation patterns are the result of the cooling and heating of water in the atmosphere and on Earth's surface. There is still much to learn about climate, climate prediction, and whether climate change is occurring. Although the rise in carbon dioxide and other greenhouse gases in the atmosphere are considered the cause of climate change, water vapor in the atmosphere also strongly absorbs infrared radiation. Unfortunately, we understand very little about the movement of water vapor and clouds in the atmosphere, and even less about how potential temperature increases will affect water in the atmosphere and distribution on the planet.

Wetlands and Riparian Areas

Hydrophytic vegetation, hydric soils, and periodic flooding are three characteristics of wetlands.

Water is the life-blood of wetlands and riparian areas. Wetlands are areas that under normal circumstances have **hydrophytic** (water-loving) vegetation, hydric (wet) soils, and wetland hydrology. **Riparian areas** are the areas of land that are adjacent to streams, rivers, lakes, or wetlands and contain vegetation that, due to the presence of water, is distinctly different from the vegetation of adjacent upland areas. Around the world and in the United States, wetlands and riparian areas have been lost at an astonishing rate. Currently in the lower forty-eight states approximately 53 percent of wetlands have been destroyed (Dahl, 1990). Wetland losses in the United States continue at a rate of 117,000 acres per year, much of which is occurring on forested freshwater wetlands. As of the mid 1980s, states such as Kentucky, Ohio, Illinois, Indiana, and Iowa had lost greater than 80 percent of their original wetlands. Although few numbers exist for riparian areas, losses are probably comparable. Approximately 90 percent of wetland losses were originally caused by agriculture, although today urban development and road construction are also major contributors.

Wetlands are important because they are the last line of defense between upland activities and water resources.

Once wetlands are drained either through ditching or tiling, their organic-rich soils are very productive for agriculture. Losses of wetlands and riparian areas have considerable impact on wildlife, groundwater resources, pollution transport, and flooding. Approximately 46 percent of threatened and endangered species in the United States depend on wetlands for some part of their life cycle. Wetlands and riparian areas are important because they can be both groundwater recharge and **discharge** areas. They also can be very effective at filtering out pollutants, and they can lessen floods because they work much like a sponge in storing excess water. Wetlands and riparian areas are really the last line of defense when considering the effect of upland activities on the water resource.

Flooding

Flooding is a natural part of the cycle of rivers.

Flooding causes more damage, death, and monetary losses than any other natural disaster. The 1993 flooding of the Mississippi River alone led to an estimated $12 billion in damages and accounted for forty-eight deaths (National Oceanic and Atmospheric Administration, 1993). Flooding results when flows in rivers and streams are greater than the capacity of the channel. On average, rivers flood about once every 1.5 years. Flooding is a natural part of the cycle of rivers; however, land use changes in the **watershed** can affect the timing and magnitude of flooding.

In the past, dams have been erected in many places to provide power and flood protection. Reservoirs behind dams have the capacity to hold water back and release it at levels that are not above the downstream banks. In many areas, rivers have been straightened for flood protection. Straightening of a river provides a more efficient conduit for the water to flow, lessening the chances of flooding. Unfortunately these engineered solutions sometimes fail and catastrophic flooding occurs. One can imagine the effect of upstream straightening on downstream flooding if the downstream portion of the stream has not been straightened.

Lakes and wetlands provide natural flood attenuation. Attenuation stretches an event over a longer period so that less water travels through the channel at any particular moment. Lakes and wetlands can provide short-term storage for excess water if they are not at their water-holding capacity before the event. The 1993 flood of the Mississippi River would not have had nearly as great an impact if wetlands in the watershed were still functioning, but as noted earlier, many of the states in the Midwest have lost a major portion of their original wetlands.

Groundwater—Quantity and Quality

Two important issues facing society are the lack of potable drinking water and the need for water for irrigation to sustain agricultural production. Groundwater is the source of much drinking and irrigation water. Groundwater can easily be contaminated by the leaching of chemicals applied on the soil surface. As discussed, NO_3^- from N fertilizers, by-products of herbicides and pesticides, and industrial solvents are major threats to groundwater systems.

Equally important to groundwater quality is the quantity of water pumped from the groundwater resource for irrigation. The Ogallala aquifer in the Midwest is the most important source of water for U.S. agriculture. It stretches from Texas to South Dakota and is the primary source of water for irrigated agriculture in the Midwest. During the development of irrigated agriculture from 1940 to 1980, some areas of the aquifer dropped over 100 feet (United States Geological Survey, 1995). From 1980 to 1994 the aquifer has been declining at an average rate of 0.11 feet per year. As water levels drop it becomes more expensive to extract groundwater, and the economic viability of crop production decreases. This leads to higher food prices and in some cases the abandonment of irrigated crops if the costs become too high.

> The Ogallala aquifer in the Midwest, the most important source of irrigation water for U.S. agriculture, dropped 0.11 feet per year between 1980 and 1994.

As can be seen from these issues, it is important to understand how water is transported so that you can effectively manage for water quality and quantity. The hydrologic cycle helps illustrate this process.

FOCUS ON . . . WATER RESOURCE ISSUES

1. How many of the Earth's people are already experiencing some sort of water stress?

2. Where does sediment in surface waters come from?

3. What is eutrophication?

4. Why are high NO_3^- levels in water a problem?

5. What is one of the major concerns with mercury and PCBs in water?

6. How does water vapor contribute to climate change?

7. Historically, what has caused most wetland losses?

8. What are three functions of wetlands?

9. What is one consequence of stretching or straightening a river?

10. What are some threats to groundwater aquifers?

THE HYDROLOGIC CYCLE

Understanding the hydrologic cycle is critical to effectively managing water resources (Figure 13-3). In this section the hydrologic cycle will be broken down into its individual components and the water balance will then be examined.

Precipitation

Rain, snow, and dew are the dominant types of precipitation.

Precipitation is the major input of water to land and water surfaces. Water can exist in either liquid, solid, or vapor phases. In most regions of the world, rainfall is the dominant form of precipitation. In higher latitudes and at high elevations snow is also an important input to the hydrologic cycle. In some areas, such as the Pacific Northwest, condensation or dew

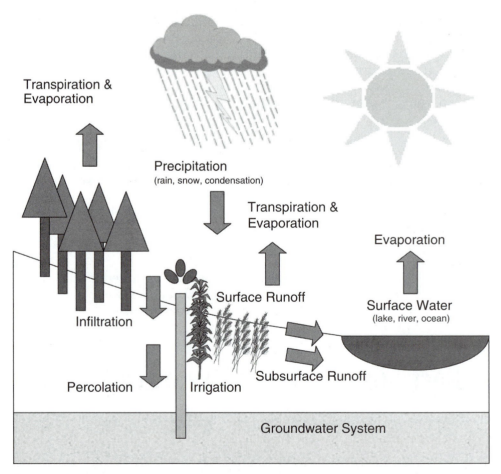

FIGURE 13-3   Schematic of the hydrologic cycle.

can also be a significant input. For precipitation to occur three conditions in the atmosphere must be met:

❖ The atmosphere must be saturated with water vapor.

❖ Small particles (e.g., dust) must be present in the atmosphere for condensation or sublimation.

❖ The water droplets or ice formed must grow large enough to reach the earth.

The relative humidity indicates how saturated the atmosphere is with water vapor. A relative humidity of 100 percent indicates that the atmosphere is saturated. Warmer air can hold more water vapor than cooler air, so it takes more water vapor to saturate the atmosphere at higher temperatures. The temperature at which an unsaturated parcel of air becomes saturated is its dew point temperature. Generally, saturation occurs when an unsaturated air mass is cooled to its dew point temperature through a lifting in the atmosphere. The lifting of air masses can occur as a result of fronts, orographic effects, or through convection.

Warm fronts and cold fronts occur as two air masses with dissimilar temperatures meet. Cold air will necessarily be warmed. If warmed to the dew point precipitation will occur. **Orographic precipitation** occurs when warm air is lifted and cooled as air masses go over mountains. The lifting and cooling results in precipitation commonly occurring in the upper elevations of the windward side of the mountain (where the air mass is coming from) and little to no precipitation on the leeward side of the mountain (the opposite side of the mountain). This is commonly called the *rain shadow effect.* The Atacama Desert in Chile is a good example of the rain shadow effect. **Convective precipitation** occurs when a moist layer of air near the Earth's surface is heated, rapidly lifted, and cooled. If the uplifting is high enough in the atmosphere ice crystals can result. As additional warm air is lifted into the atmosphere turbulence occurs, sometimes resulting in hail, high winds, and even tornados.

> Orographic precipitation occurs when warm, moist air is lifted over mountains.

Typically precipitation is measured with rain gauges that are put in openings to prevent interception (see next section). Although many households have rain gauges, the National Weather Service has a network of standardized rain gauges placed across the United States that are used to measure precipitation inputs to the hydrologic cycle.

Interception

Only a fraction of the precipitation that occurs actually hits the soil surface. Trees, shrubs, crops, and grass catch or intercept precipitation prior to its reaching the soil surface (Figure 13-4). Interception is that amount of precipitation that is caught by vegetation and is ultimately evaporated back to the atmosphere. Interception by forests tends to be greater than that of other vegetation communities because forests generally have more leaf surface area for storage. Also, as the canopy becomes denser, more interception occurs. In forest systems interception can account for less than 10 percent to greater than 40 percent of incoming precipitation, although typical values are 15 to 25 percent (Kolka et al., 1999). In addition to the vegetation in forests, the forest floor also intercepts incoming precipitation.

> Forests tend to intercept more precipitation than other types of vegetation because they have more leaf surface area.

Precipitation that passes through the forest canopy is termed **throughfall.** Precipitation that runs down the stems of plants is termed **stemflow** (Figure 13-4). Generally stemflow is a small component (1–4 percent) of the overall input of water to a forested system, but can be important chemically. The precipitation that is available to infiltrate the soil (throughfall + stemflow−litter interception) is termed *net precipitation* and is the amount

> Stemflow can be important because of the nutrients and contaminants it carries.

FIGURE 13-4 Schematic illustrating the components of interception.

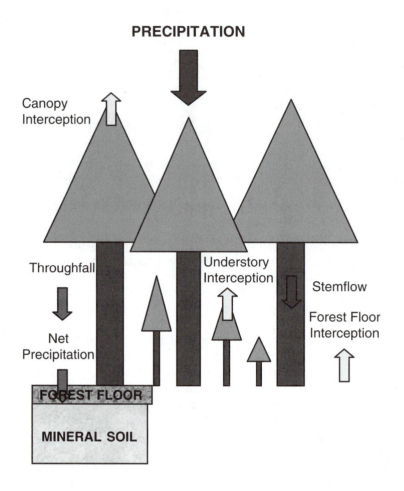

of precipitation that must be known to develop a water budget or to understand infiltration and runoff processes.

Throughfall can be measured with standard rain gauges or by placing troughs under the canopy. Because canopies have considerable variability in cover, many rain gauges are necessary to characterize throughfall inputs. Troughs, depending on their length, can integrate canopy variability by sampling a variety of canopy conditions with a single sampler.

Stemflow is typically measured by affixing a collar to a tree and collecting all the flow that runs down the stem. Stemflow is affected by the canopy shape, the attitude of the branches, and most importantly by the roughness of the bark. Trees with rough bark can retain and store more water than smooth-barked species that tend to promote stemflow.

Evapotranspiration

Evapotranspiration is composed of the combined effects of evaporation and transpiration.

Evapotranspiration (ET) is the combined water losses to the atmosphere through the two processes of evaporation and transpiration. Evaporation is the loss of water from a surface when water changes state from liquid to vapor. Transpiration is the loss of water from plant leaves by evaporation through the leaf stomata. For ET to occur from a surface: (1) energy, essentially heat, must be available at the evaporating or transpiring surface; (2) there must be a supply of liquid water at these surfaces; and (3) there must be a flow of vapor away from the surfaces so that a vapor gradient can be sustained.

In the United States, ET losses back to the atmosphere account for the bulk of precipitation (Figure 13-5). On average, about 67 percent of incoming precipitation is either evaporated or transpired back to the atmo-

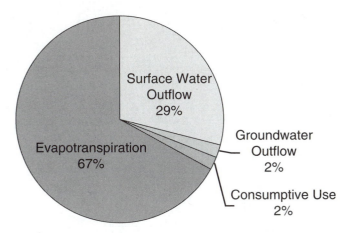

FIGURE 13-5 Distribution of losses of incoming precipitation. *(U.S. Geological Survey, 1990)*

sphere. Approximately 29 percent of incoming precipitation results in outflows to oceans, 2 percent discharges from groundwater, and 2 percent is consumed through humans, plants, and industrial processes.

Numerous factors influence the ET process, but solar energy is the ultimate driver. Other important factors include the availability of soil moisture, wind speed, type of vegetation cover, the color and associated solar radiation reflectivity of the landscape (termed **albedo**), latitude, and time of the year. For transpiration to occur, soil moisture must be available to be taken up by plants. As wind speed increases, the vapor gradient between the evaporating surface and the atmosphere becomes greater, promoting ET. The type of vegetation cover and the rooting depth influence the amount of uptake of water available for transpiration and the amount of interception and subsequent evaporation that can occur. Generally, forests transpire and intercept more water than do grassland or crop systems because of deeper rooting depths and more surface area available for interception. Dark-colored (low albedo) landscapes such as those dominated by conifer forests absorb more solar radiation; thus more energy is available for ET. As latitude increases, solar radiation decreases and less energy is available for ET processes. The time of year also influences the amount of solar radiation received by the Earth as more solar radiation is received in the summer than in the winter. When we consider all the factors affecting ET, the southeast United States has the highest ET because of the considerable availability of soil moisture, presence of conifer forests in many areas, and the relatively low latitude (Figure 13-6). As one goes west soil moisture becomes limiting and as one goes north latitude becomes higher and solar radiation decreases.

Measurement of ET is difficult. Evaporation is measured by the U.S. Weather Service with metal evaporation pans. Pans are 122 cm (4 ft) in diameter and 30 cm (1 ft) deep with water maintained at about 18 to 20 cm (7 to 8 in.). The amount of water evaporated is measured daily. Pan evaporation is somewhat comparable to lake evaporation, although pan evaporation tends to be greater because of the solar radiation absorbed by the pan itself. Coefficients multiplied by the pan evaporation (typically 0.70 to 0.75) are used to estimate lake evaporation. Although pan evaporation somewhat mimics lake evaporation, it is not very comparable to terrestrial evaporation or transpiration.

For grassland and crops, evaporation, transpiration, or the combination (ET) can be measured with lysimeters. Lysimeters are essentially boxes of soil with plants, and the input of precipitation is measured and the weight of the box is measured. As water is evaporated from the soil or plant surfaces the weight of the box decreases and ET can be calculated. To separate evaporation from transpiration, the surface of the soil can be covered

> The important factors driving evapotranspiration are solar energy, soil moisture, wind speed, vegetation, and albedo.

> Open pans and lysimeters are used to measure ET.

> In vegetated systems ET can be measured by changes in weight.

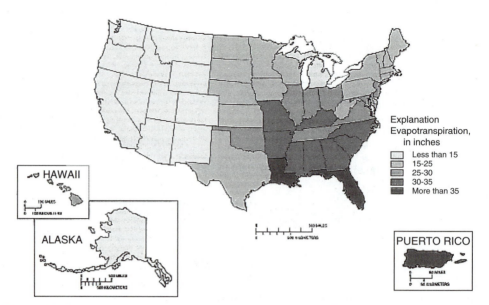

FIGURE 13-6 Annual ET distribution in the United States and Puerto Rico. *(Hanson, 1991)*

so that only transpiration is taking place. Although lysimeters work well for smaller plants, ET measurement on larger plants such as trees is difficult and few reliable methods have been developed. One method that is sometimes used is the sapflow method, in which a chemical or heat tracer is injected into the sapflow of the tree and sensors track the time it takes for the tracer to move up the stem. Both the lysimeter and sapflow methods are expensive and typically are only used for research purposes. To estimate actual evapotranspiration (AET) at the landscape or watershed scale numerous empirical methods have been developed, some based on the concept of PET, or potential evapotranspiration.

| Potential ET is frequently estimated by numerical models.

Potential evapotranspiration is the maximum ET that can occur given an adequate supply of soil moisture, and is really an index of actual evapotranspiration (AET). Based on a multitude of studies, numerous empirical equations have been developed to calculate PET. In their simplest form, they relate ET to temperature (such as the Thornthwaite equation); more complicated forms use wind speed, vapor pressure, and net solar radiation (e.g., Penman's equation). Once calculated, PET can be used with a water budget approach to estimate AET given precipitation inputs and soil water storage (see following box).

A second method used to estimate AET is the paired watershed approach. The paired watershed approach has been commonly used in forested landscapes where measurement of ET is difficult. With this method, two or more similar watersheds are chosen and predictive relationships for streamflow are developed among watersheds. One or more watersheds are then treated while one or more of the watersheds are left as untreated controls. A common treatment to estimate AET is a complete harvest. Additional streamflow resulting from the harvest is considered to be a consequence of the lack of ET provided by the forest canopy.

Soil Water Storage

| Soils vary in their ability to store water.

You previously used soil water storage as a term in the water budget approach for estimating evapotranspiration (box on page 287). Soils vary in their ability to store water. Remember that sandy soils store little water compared to silty or clayey soils (Figure 13-7). This is partly an effect of

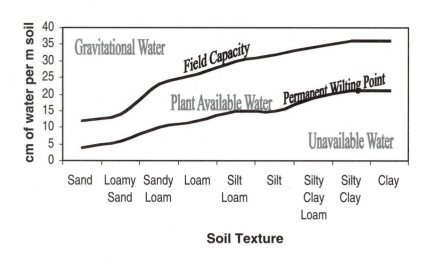

FIGURE 13-7 Water-holding capacity of different soil textures. *(Values from ASCE, 1990)*

USING A WATER BUDGET APPROACH TO ESTIMATE AET AND RUNOFF FOR LEXINGTON, KENTUCKY

In establishing a water budget one has to measure the precipitation going in as well as the runoff, evaporation, and transpiration going out. In the example that follows, all units are in cm of water.

	Jan.	Feb.	Mar.	Apr.	May	June	July	Aug.	Sept.	Oct.	Nov.	Dec.	Annual
Avg. Rainfall[1]	7.11	7.62	10.67	10.92	12.19	10.16	11.18	10.16	10.16	7.11	9.65	10.67	117.60
Initial SM[2]	12.10	12.10	12.10	12.10	12.10	12.10	9.07	4.10	0.00	0.00	1.67	8.95	
Total AM[3]	19.21	19.72	22.77	23.02	24.29	22.26	20.25	14.26	10.16	7.11	11.32	19.62	
PET[4]	0.10	0.51	2.78	5.80	9.76	13.19	16.15	15.21	10.75	5.44	2.37	0.61	82.67
AET[5]	0.10	0.51	2.78	5.80	9.76	13.19	16.15	14.26	10.16	5.44	2.37	0.61	81.13
Remaining SM[6]	19.11	19.21	19.99	17.22	14.53	9.07	4.10	0.00	0.00	1.67	8.95	19.01	
Final SM[7]	12.10	12.10	12.10	12.10	12.10	9.07	4.10	0.00	0.00	1.67	8.95	12.10	
Runoff[8]	7.01	7.11	7.89	5.12	2.43	0.00	0.00	0.00	0.00	0.00	0.00	6.91	36.47

[1]Average precipitation over the watershed for each month.

[2]Initial soil moisture. Moisture content at the beginning of the year (Jan.) is equal to the water-holding capacity of the soil.

[3]Total available moisture. Moisture available for ET and runoff. Equals the sum of average precipitation and initial soil moisture.

[4]Potential evapotranspiration. Calculated using an empirical formula. For this example the Thornthwaite equation was used.

[5]Actual evapotranspiration. Is equal to PET when total available moisture is sufficient. Equals the total available moisture when PET exceeds total available moisture.

[6]Remaining soil moisture. Soil moisture remaining after AET is satisfied. Equals the total available moisture minus AET.

[7]Final soil moisture. The final soil moisture at the end of the month. Equals the soil water-holding capacity if remaining soil moisture is equal to or greater than the soil water-holding capacity. Equals the remaining soil moisture when AET demands become greater.

[8]Water that will runoff as a result of excess soil water. Equals the difference between remaining soil moisture and final soil moisture.

In this example, the soil water-holding capacity is equal to 12.10 cm (soil moisture content in January). For most of the year AET is equal to PET because there is plenty of available water; however, in August and September when ET demands are high and soil moisture is not being replenished fast enough to compensate for ET demands, AET becomes less than PET.

pore size, which varies among soils with different textures. Because sand grains are relatively large soil particles (> 0.05 mm), pores between grains also tend to be large. Conversely, clay particles are relatively small (< 0.002 mm) and pores between clay particles are small. Water is more tightly held in small pores than in large pores, leading to greater water-holding capacities in finer-textured soils than in coarse-textured soils.

After water enters a soil, several forces are operating to either hold or transport that water. The force of gravity attempts to transport water downward through the soil profile. Within the soil other forces are attempting to hold the water within the soil profile. These forces are also called potentials. Although there are several potentials operating at any time within the soil, the two most significant are the gravitational and matric potentials in upland soils. You have seen these concepts before in Chapter 4, but it is worth repeating them here in the context of the larger soil environment.

> The movement of water in soil is controlled by forces called *potentials*.

Matric potential is the force by which water is held in soil by capillary and adsorptive processes. Matric potential becomes increasingly important as soil texture becomes finer and soil pores become smaller. Because of matric forces, soils can hold water very tightly. In some cases, matric forces hold water so tightly that plants cannot extract the water. Water held that tightly is unavailable to plants (Figure 13-7). The proportion of water that is unavailable becomes greater in fine-textured soils as pore size decreases and matric forces become stronger. When soil water content is low and only unavailable moisture exists, plants are moisture-stressed and wilt. The soil water content where this occurs is called the *permanent wilting point* (Figure 13-7).

Gravitational potential is the force by which water drains through soil as a result of gravity. Somewhat the reverse of matric potential, gravitational potential becomes increasingly important as soil texture becomes coarser and soil pores become larger. Following a precipitation event, soil moisture deficits are satisfied and excess water drains through the soil profile. The water that drains due to gravity is called gravitational water (Figure 13-7). Gravitational water is also unavailable to plants because it drains below the rooting depth before plants have an opportunity to use it. Once gravitational water drains through the soil profile, the soil moisture content is at field capacity, the maximum amount of water a soil can hold after gravitational forces have been satisfied (Figure 13-7).

> Gravitational water and very tightly held matric potential water are unavailable to plants.

The amount of water available for plant growth is that water held by matric forces between the field capacity and the permanent wilting point (Figure 13-7). Sandy soils have less available water than medium- and finer-textured soils because much of incoming precipitation drains as a result of gravity. Fine-textured soils such as clays generally have less plant-available water than medium-textured soils because most of the water is held too tightly by matric forces. Generally, medium-textured soils such as loams, silt loams, and silts have the greatest amount of plant-available water because of a relative balance between matric and gravitational potentials (Figure 13-7).

> Tensiometers, ceramic blocks, and TDRs are three ways of measuring water availability in soil.

Soil moisture is measured by various methods. Typically, soils are sampled, oven-dried, and soil water content is determined. In place or *in situ* methods include **tensiometers,** moisture blocks, and time domain reflectometry (TDR). Tensiometers measure the force or tension that results from a water-filled column in contact with the soil through a porous ceramic cup. Water enters the soil through the ceramic cup, which creates a suction in the tube. The drier a soil is, the more suction or tension created in the tube. The amount of tension is related to the soil water content. Soil moisture blocks work in much the same way. Dry porous blocks, usually constructed of gypsum or ceramic, are positioned in the soil, and water will either move in or out of the blocks depending on the soil water condition. The water in the blocks is determined indirectly by electrical resistance or heat capacity and is then related to the moisture content of the soil.

Time domain reflectometry is a relatively new method being used to measure soil water. In this method, two metallic probes are inserted into the soil and an electrical signal is sent down the probes. The electrical signal is reflected back from the end of the probe. Travel time and velocity of the signal is related to soil water content. Because electric current is transferred differentially among soil, water, and air, the detector can relate the signal to the soil water content.

Infiltration and Runoff

When net precipitation reaches the soil surface, water can be stored on the surface, run off the surface, be held in the soil, or be drained (by gravitational forces) vertically to groundwater or laterally to streams, lakes, and wetlands. Infiltration is the process by which water enters the soil and determines the proportion of water that is available for surface runoff. During intense rainfall or snowmelt events, water can be in excess of the capability of the soil to accept the water. When this occurs, water builds up on the soil surface and can run off to lower parts of the landscape. The water that runs off the soil surface is called surface runoff (Figure 13-8). The infiltration capacity is the maximum rate that water can enter a particular soil and decreases during an event as the soil pores fill with water. The infiltration capacity is determined by the soil porosity at and near the soil surface. Infiltration rate and capacity is typically measured by pressing a ring into the soil and measuring the time it takes for a known volume of water to enter the soil surface within the area of the ring.

> Infiltration refers to the process by which water enters soil.

Management that affects porosity also affects infiltration rate and capacity. Any management activity that tends to compact the soil such as grazing or plowing in agricultural settings, skidding of trees in forestry settings, and road development in urban settings, will also decrease infiltration capacity and enhance surface runoff. The generation of surface runoff is a problem because the force of water detaches soil particles and leads to erosion. Once detached, soil particles are transported downhill with the potential of entering surface water bodies such as streams and lakes.

> Grazing, plowing, and tree skidding are all activities that can reduce infiltration.

Water that infiltrates the soil surface can either be held in the soil by matric forces or transported either vertically or laterally. Water drained vertically will likely enter either local or regional groundwater. Water drained laterally, also called *interflow* or *subsurface flow*, flows downhill within the soil profile. Interflow is the result of horizons in the soil profile that are less permeable than the horizons above (for example, a Bt horizon below an E horizon). Interflow waters tend to flow on top of these horizons, ultimately entering low spots in the landscape, wetlands, rivers, or lakes.

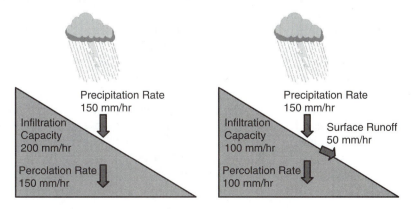

Precipitation Rate
150 mm/hr

Infiltration Capacity
200 mm/hr

Percolation Rate
150 mm/hr

Precipitation Rate
150 mm/hr

Infiltration Capacity
100 mm/hr

Surface Runoff
50 mm/hr

Percolation Rate
100 mm/hr

FIGURE 13-8 Schematic of the processes of infiltration, percolation, and surface runoff. In the left-hand example, the precipitation rate is not in excess of the infiltration capacity, resulting in no surface runoff. In the right-hand example the precipitation rate is greater than the infiltration capacity, and surface runoff occurs. The percolation rate is the rate at which infiltrated water moves through the soil.

FIGURE 13-9   Below-ground
water profile.

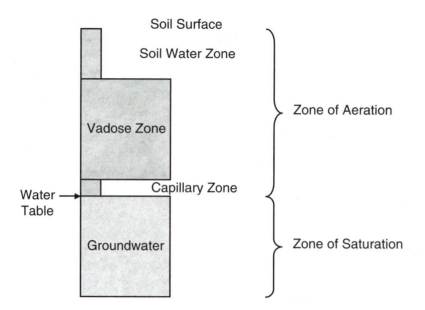

FIGURE 13-9   Below-ground
water profile.

Groundwater and Groundwater Consumption

As discussed earlier in the chapter, groundwater is an important resource for human consumption, industrial purposes, and irrigation. Beneath the soil surface, soil and geologic matrices can either be aerated or saturated. Groundwater is water that occurs in saturated zones (Figure 13-9). Above the groundwater table, soil and geologic materials are aerated. Directly above the groundwater table is a zone called the *capillary fringe* where matric and capillary forces wick water up into the soil or geologic matrix. The capillary fringe is typically a narrow zone with thickness increasing as particle size becomes smaller. Water between the capillary fringe and the soil water zone is considered to be in the vadose zone. The vadose zone is below the soil profile or rooting depth and is composed of geologic materials that have had minimum exposure to soil-forming processes. The vadose zone will vary in thickness and depends on the depth of geologic materials between the bottom of the soil profile and the water table. Soil water is that water found within the soil profile or rooting zone (Figure 13-9).

Groundwater can range in depth from at or near the surface of the Earth to deep within the Earth's crust. The presence of groundwater is the result of a relatively impermeable (low-porosity) soil or geologic layer beneath a relatively permeable (high-porosity) layer. When discussing groundwater systems we consider *perched, unconfined,* and *confined* groundwater (Figure 13-10). Perched groundwater occurs above a relatively impermeable layer in the soil or vadose zone. Perched groundwater is not continuous and can occur high up in the landscape. Where perched groundwater reaches the soil surface, springs are often present.

Unconfined groundwater is continuous under the landscape and is typically the type of groundwater system used for drinking and irrigation water. A relatively impermeable layer underlying an unconfined aquifer is the **aquitard.** Unconfined aquifers are considered unconfined because there is no impermeable layer above them and, as such, they are under normal atmospheric pressure and are able to expand towards the Earth's surface.

Groundwater present between impermeable layers is considered confined because it cannot expand toward the Earth's surface. In confined aquifers, the water is essentially trapped, and as a result is typically under pressure. Because of the pressure, water will come shooting out of a well placed in a confined aquifer, also called an artesian well. Completely impermeable lay-

> The vadose zone is the unsaturated water between the capillary fringe and the soil water zone.

> Groundwater systems include perched, unconfined, and confined groundwater.

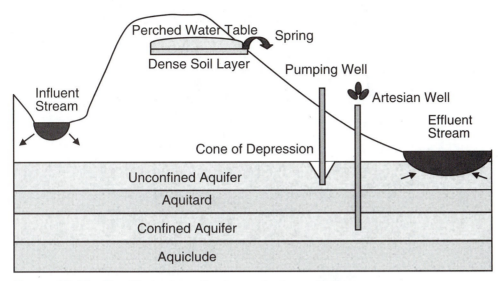

FIGURE 13-10 Simplified schematic of groundwater systems.

ers such as some types of bedrock are called **aquicludes.** Aquitards have very slow permeability whereas aquicludes are essentially impermeable.

Recharge is the process by which water that originally fell as precipitation moves through the soil profile and vadose zone and enters the groundwater. Most recharge occurs in upland landscapes that are comprised of porous, coarse-textured soils. Lower parts of the landscape can also be areas of recharge but typically streams, lakes, and wetlands are discharge zones. Approximately 30 percent of all streamflow in the United States is the result of groundwater discharge (Brooks et al., 1997). Streams that recharge groundwater are termed *influent streams* while those that discharge groundwater are *effluent streams* (Figure 13-10).

In 1995, the United States pumped 76,400,000,000 gallons of groundwater per day (Solley et al., 1998). Approximately 64 percent of the total groundwater pumped was for irrigation, 24 percent was used for public and domestic water supplies (drinking water), 5 percent for industrial uses, and 3 percent for livestock watering. Thermoelectric power generation, mining, and commercial uses comprise the remaining 4 percent. Of all the water used for agricultural production (mostly irrigation), 37 percent was groundwater. Approximately 51 percent of all drinking water was groundwater and 99 percent of rural drinking water was groundwater.

The pumping of groundwater for human consumption, industrial uses, or irrigation can have a dramatic effect on the resource. As discussed in the water resource issues section at the beginning of this chapter, groundwater is easily contaminated by the leaching of chemicals applied on the soil surface. Nitrate from nitrogen fertilizers, herbicides and pesticides, and industrial solvents contaminates groundwater systems. Pumping or overpumping of groundwater can also lead to land subsidence and saltwater intrusion near coasts. Land subsidence can occur because the pore pressure of groundwater in the soil or geologic matrix is actually holding the material in place. After the water is extracted the pressure lessens and sinkholes, for example, can result. In coastal areas, overpumping of groundwater can create areas of saltwater intrusion. If the cone of depression (Figure 13-10) is deep or wide enough that it intersects a saline aquifer, then the well becomes contaminated with saline water, which is not potable unless the salt is extracted from the solution—an expensive process.

Groundwater table elevations are typically measured with wells and **piezometers.** Groundwater wells are typically slotted their whole length to

> Uplands are typically groundwater recharge areas and lowlands are typically discharge areas.

> Groundwater pumping can lead to subsidence and saltwater intrusion.

FIGURE 13-11 Relationship between pathways of flow from a watershed and the resultant streamflow hydrograph: A = channel interception; B = surface runoff; C = subsurface runoff or interflow; D = groundwater or baseflow; Q = streamflow. The top curve of the hydrograph is the sum of A, B, C, and D. *(Brooks et al., 1997)*

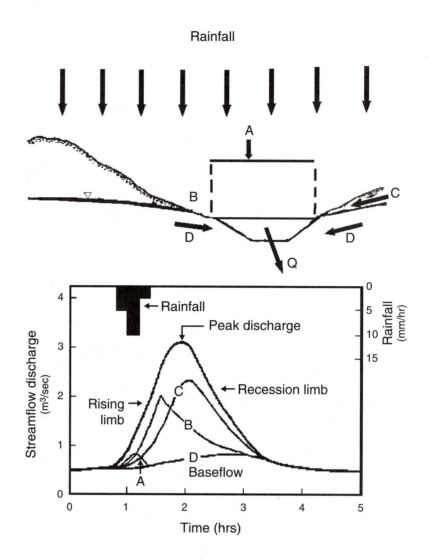

measure the elevation of water in unconfined aquifers. Piezometers are only slotted for a distinct distance and are designed to measure water table elevations at a point in a soil or geologic layer. Piezometers are typically used to understand water table elevation and pressure in confined aquifers.

Streamflow

Surface runoff, interflow, and in some cases groundwater are the water sources for our streams, lakes, and wetlands. At the initiation of a precipitation event, streams begin receiving water, first from direct deposition on the stream (channel interception), then from surface runoff and interflow as the event proceeds (Figure 13-11). Many streams and larger rivers also have a baseflow component that is the result of groundwater seepage into the channel. A graph relating time to streamflow is called a hydrograph. The magnitude (peak discharge) of the stormflow hydrograph is dependent on the size of the event and the level of saturation of soils in the watershed. For example, a watershed receiving a 10-cm (4-in.) rainfall that had little precipitation in the last week would have a lower peak flow than a similar one that had a 5-cm (2-in.) rainfall followed by a 10-cm (4-in.) rainfall two days later.

The magnitude of the peak discharge and the shape (that is, the shape of the rising and falling limbs) of the stormflow hydrograph are also de-

Baseflow is the result of groundwater seepage into a stream channel.

Land uses compacting soil will cause greater peak flow.

pendent on the land uses occurring in the watershed. In short, land uses that compact the soil such as agriculture, grazing, and urban environments will increase the proportion of surface runoff, leading to earlier and greater peak discharge. Watersheds that contain porous, uncompacted soils such as those covered by forests will have later and lower peak discharge. If wetlands occur in the watershed, peaks will also be later and lower as wetlands tend to retain water that runs off the uplands.

Although urban and agricultural land uses promote surface runoff, little surface runoff generally occurs in forested or wildland watersheds. The variable source area concept (VSAC) explains mechanisms of stormflow for watersheds that produce little surface runoff. The VSAC considers how water arrives to the stream channel during a storm event. Stated simply, the VSAC says that as a precipitation event proceeds, more area contributes flow to the channel. Areas low in the landscape, ephemeral and intermittent channels, and areas on shallow soils that are not typically sources of water to the stream become saturated from upland sources and begin to contribute water as interflow. As the event continues, more of these subsurface source areas contribute to flow in the stream. Essentially, the stream network expands as the rainfall event proceeds. After the precipitation event ceases, the network contracts and ultimately goes back to baseflow conditions if there is no additional precipitation.

Several methods are available to either measure or estimate streamflow discharge. Measurements of discharge are always preferred to estimation techniques. Discharge (Q) is the product of velocity (V) and cross-sectional area (A) of the stream ($Q = VA$). Discharge is commonly measured in cubic feet per second (cfs) or cubic meters per second (m^3/s). The cross-sectional area of the water-filled channel is relatively easy to measure. Typically a tape measure is strung across the water surface and the depth of water is measured at intervals across the channel. The cross-section of the stream is easily plotted and area between each interval is summed for the entire width of the stream.

> Discharge is the product of stream cross-sectional area times velocity.

Velocity is the more difficult measurement. The easiest method to measure velocity is to drop a floating object (for example, a bobber or ping pong ball) into the stream and measure the time it takes to cover a certain distance. The floating object method is not entirely accurate because water on the surface and the middle of the stream typically flows faster than most of the other water in the stream due to friction along the banks and bottom. As a result, the floating object method tends to overestimate true streamflow.

Another method is to use a flow meter. Flow meters come in several types but the end result is a measurement of water velocity at a point. Velocity is typically measured in the vertical sections between intervals during the measurement of the cross-sectional area. Established protocols call for one measurement at 60 percent depth for sections less than 0.5 m in depth and two measurements at 20 percent and 80 percent depth for sections greater than 0.5 m in depth. The velocity of a section greater than 0.5 m is the average of the two measurements. The velocities of each section are then multiplied by their respective areas and summed for the entire stream. Measuring streamflow at several different water levels (stage height) allows for a relationship to be developed relating stage height to flow. Once the relationship or rating curve is developed, only the stage height at one location needs to be measured to accurately predict flow. The process of developing a rating curve and ultimately measuring stage height to predict flow is called **stream gauging.**

> Once a stream is gauged, one can estimate the discharge simply based on stream height.

Precalibrated structures such as flumes and weirs are also used to accurately measure streamflow in small watersheds. A weir is essentially a dam with an opening across the top that is built across a stream. The opening is typically a notch where water is directed to flow through. Based on the water height in the opening and the geometry of the opening, hydraulic equations exist that can very accurately calculate the flow. Flumes are chutes or artificial channels that are put into the stream. Like the weir, the downstream opening of the flume has a certain geometry (typically notch-like) and the water level in the flume is related to flow with hydraulic equations.

One method commonly used to estimate velocity, and hence flow, is Manning's equation:

$$V = (1.49/n) \, (R_h^{2/3})(s^{1/2})$$

where V = velocity (ft/sec), R_h is the hydraulic radius, and s is the channel slope (ft/ft). The hydraulic radius $R_h = A/WP$, where A = cross-sectional area of flow (ft^2) and WP = the wetted perimeter (ft) or the distance from where the water intersects the bank on one side of the stream, across the stream bottom to where the water intersects the bank on the other side of the stream. The roughness coefficient (n) is a number associated with the roughness of the channel and is derived from previous research. Tables of roughness coefficients exist that describe the stream conditions associated with a certain roughness. Typically roughness coefficients range from 0.016 for little roughness to 0.200 for very rough channels (Brooks et al., 1997).

Water Balance

You've now seen the individual components of the hydrologic cycle, but how do you construct a water balance for a watershed? The water balance is: Inputs = Outputs, or

$$P = Q + ET + \Delta l + \Delta S$$

Where inputs = precipitation (P), Outputs = streamflow (Q), evapotranspiration (ET), change in groundwater storage (Δl), and change in surface storage (ΔS). The change in groundwater storage (Δl) is the difference between groundwater seepage out of the watershed (l_o) and groundwater seepage into the watershed (l_i) ($\Delta l = l_o - l_i$). Surface storage consists of changes in soil moisture, lakes, ponds, and wetlands (ΔS). The $\Delta S = S_2 - S_1$, where S_2 = storage at end of period and S_1 = storage at beginning of period.

Given the following information, could you calculate annual streamflow for a 180 ha watershed?

Throughfall = 100 cm

Stemflow = 4 cm

Groundwater seepage into the watershed = 30,000 m^3

Groundwater seepage out of the watershed = 15,000 m^3

Evapotranspiration = 70 cm

ΔS: S_1 = 150,000 m^3
      S_2 = 140,000 m^3

Watershed Area = 180 * 10,000 = 1,800,000 m^2

$\Delta S = 140,000 - 150,000 = -10,000 \text{ m}^3$

$P = (100 + 4) = 1.04 \text{ m}, 1.04 * 1,800,000 = 1,872,000 \text{ m}^3$

$ET = 0.7 \text{ m} = 0.7 * 1,800,000 = 1,260,000 \text{ m}^3$

$I = 0$

$1,872,000 = Q + 1,260,000 + 15,000 + (-10,000)$

$Q = 607,000 \text{ m}^3$

Now that you have seen the hydrologic cycle, it is time to apply this information to effectively managing watersheds.

Focus on . . . The Hydrologic Cycle

1. Which holds more water vapor, warm air or cool air?
2. What fraction of precipitation do forests typically intercept?
3. What accounts for the bulk of precipitation?
4. Why isn't pan evaporation quite the same as lake evaporation?
5. Which soil holds more water, coarse or fine?
6. What is gravitational water?
7. What does a tensiometer measure?
8. What forces control water movement in upland soils?
9. What determines the infiltration capacity at the near soil surface?
10. What kinds of activities reduce soil infiltration?
11. What is the "vadose zone"?
12. About how much of streamflow consists of groundwater discharge?
13. Why are piezometers useful?
14. How does compaction affect "peak" flow in a stream?
15. Why would you want to gauge a stream?

Watershed Management

To understand watershed management you first need to understand the concept of a watershed. A watershed is a topographically delineated area drained by a surface water body (that is, a stream, lake, or wetland). Stated more simply, a watershed is the total land area above some point that drains past that point (Brooks et al., 1997). The management of watersheds necessarily involves people and decisions. Watershed management is the process of organizing and guiding land and other resource use on a watershed to provide desired goods and services without adversely affecting soil and water resources. Embedded in the concept of watershed management is the recognition of the interrelationships among land use, soil, and water and the linkages between upland and downstream areas (Brooks et al., 1997). Watershed management is about understanding how

A watershed is the total land area draining a given discharge point.

the terrestrial system affects the aquatic resource, and more importantly, how changes in the terrestrial system affect aquatic resources.

Land Use Effects on Water Quantity

One goal of watershed management is to provide goods and services without adversely affecting soil and water resources.

As stated in the previous definition, the goal of watershed management is to produce desired goods and services *without adversely affecting soil and water resources*. All watersheds provide goods and services to our society; the key is understanding the effect on the soil and water resource. Land use and changes in land use over time in a watershed have the greatest impact on the soil and water resource. Originally, much of the United States was either forest or prairie, with considerable areas of wetlands and lakes. The onset of agriculture changed the landscape as forests, prairies, and wetlands were converted to cropland. Conversion of these areas altered the hydrologic cycle. Areas once in forest and prairie now had less interception. Less interception leads to greater soil moisture, increased potential for surface runoff, and ultimately greater streamflow. Wetlands that were drained had less ability to store excess upland-derived waters, also leading to greater streamflow.

Agriculture generally leads to greater streamflow.

As agriculture became more mechanized, large implements began to be used on the landscape. Large tractors, plows, and other heavy machinery tend to compact the soil, further enhancing the potential for surface runoff and greater streamflow. Generally, areas converted to agriculture now have greater streamflow than they once did. The effect of greater streamflow results in the downcutting and widening of streams and a much greater potential for flooding. The greater potential for flooding occurs not only because there is more water available for streamflow but also because water enters the stream much quicker because surface runoff becomes a greater contributor to the hydrograph (Figure 13-11).

As a result of much research we are beginning to understand ways to grow crops that lessen the negative impact on aquatic resources. Research has led to the development of best management practices (BMPs) for agriculture. BMPs such as contour plowing and conservation tillage are aimed at lessening the potential for surface runoff. In many agricultural areas, wetlands that were once used to grow crops are now being restored. Restoring wetlands is a science in itself because it is very difficult to understand how to restore wetlands so that they function as they once did. Nonetheless, wetland restoration is a move forward in our thinking and a tool that is available to watershed managers.

Urbanization reduces surfaces through which water can infiltrate, leading to greater surface runoff.

Another important land use change that is occurring at an accelerated rate today is urbanization. Forest, prairie, and, probably most notably, agricultural land is being converted to urban areas as cities sprawl across the landscape. Generally, as areas become urban, a percentage of soils in the watershed are covered with pavement (e.g., sidewalks, streets, and parking areas), and other soil areas become very compacted (e.g., lawns, parks, and golf courses). The overall result is an increase in surface runoff, leading to potentially even greater streamflow than areas that are in agriculture. Understanding how urban sprawl is affecting watersheds is a critical issue today and we are only beginning to understand how urbanization is affecting the hydrologic cycle.

Land Use Effects on Water Quality

The most widespread impact land use can have on the soil resource and ultimately on the aquatic resource is through erosion. Erosion is the process by which soil particles become detached because of raindrop impact or are transported some distance due to moving water or wind. Although wind erosion can be problematic in some areas such as the Central Plains of the United States, most surface erosion occurs as a result of surface runoff. Any land use that promotes surface runoff also promotes erosion. As discussed, the conversion of vast acreages from forests, prairies, and wetlands to agriculture and urban environments has dramatically increased the potential for surface runoff and erosion.

Erosion is problematic for two important reasons. The obvious problem is that once soil is detached and transported in surface runoff it may reach an aquatic system. The second, less obvious concern is related to the productivity of the area from which the soil is eroded. Surface soils are typically much richer in organic matter and nutrients and have greater water-holding capabilities than subsurface soils. Erosion of surface soils leads to less-productive soils because of lower organic matter and nutrients and less water-holding capabilities of soils left behind.

Erosion leads to increased sediment in streams and lakes, causing numerous problems to the aquatic habitat. Soil that is eroded also contains many chemicals, many of which are derived from the land use where erosion is occurring. Soils eroded from agricultural and urban areas tend to have high concentrations of nutrients, especially N and P. The nutrients present are mainly the result of fertilization of crops, lawns, golf courses, and so on. Various pesticides, both insecticides and herbicides, can also be transported with eroded soils. Through considerable research, BMPs have been developed to lessen the transport of eroded soil and the various chemicals eroded soil contains. The BMPs discussed earlier that lessen the generation of surface runoff necessarily lessen the transport of eroded soil. However, if surface runoff does occur, probably the most effective BMP to stop the eroded soil from getting to the aquatic resource is through the implementation of riparian areas (Figure 13-12).

Riparian areas, sometimes referred to as riparian zones or buffers, slow down surface runoff because of the vegetative roughness present. Once surface runoff is slowed, sediment in runoff drops out of suspension and is deposited in the riparian area. Dissolved constituents in surface runoff and interflow, such as NO_3^- from fertilizers, TCE from industrial processes, and atrazine from herbicides, can also be effectively taken up and/or broken down by riparian area vegetation and soils, thereby decreasing the chemical load to the aquatic resource. Effectively managing functioning riparian areas is critical to watershed management.

> Erosion is the most widespread impact that land use can have on water resources.

> One of the problems with erosion is not just soil loss, but loss of the chemicals attached to soil.

Multiple Use in Watersheds

Management at the watershed scale is a difficult endeavor because there are many issues related to land use and ownership. Unfortunately watershed boundaries do not often coincide with jurisdictional boundaries such as those between private individuals, towns, counties, states, and even countries. The management of watersheds becomes increasing complex as

> Watershed management is difficult because of mixed land uses and because watersheds cross different jurisdictions.

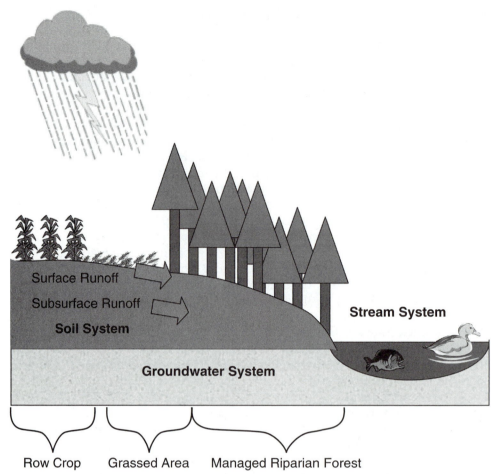

FIGURE 13-12   Schematic of a managed riparian area common in BMP guidelines for agriculture.

Multiple uses in a
watershed can be
separated by function,
space, and time.

jurisdiction becomes more fragmented within a watershed. A fundamental understanding of the hydrologic cycle is important, but in many cases watershed managers also need to understand social and economic factors that are important in the decision-making process.

Equally as complex as ownership issues is that most watersheds have multiple land uses occurring simultaneously. Multiple uses can be distributed in the watershed in three different ways (Brooks et al., 1997). First, an individual area of a watershed may have multiple uses. For example, a farm field may be used for agricultural production but also for wildlife management. A farmer may also hunt pheasant or deer in his crop fields. Second, some uses on a specific area may alternate over time. Commonly farmers alternate their crops on individual fields to prevent crop-specific disease organisms from building up in the soil. Finally, multiple uses may be separated in space. For example, in an individual watershed, forest production my be present in the steep upland portion of the watershed, corn production in flatter areas of the watershed, and urban housing and industry concentrated near the stream. It is entirely possible that all three of these types of multiple uses are occurring concurrently within an individual watershed. From a watershed perspective, as the number of uses increases, management becomes more difficult and complex.

The difficulty in management is twofold. First, it is very difficult to separate the individual effects of multiple land uses on soil and water resources. For example, suppose it has been determined that the stream draining a watershed has highly elevated levels of NO_3^- present. How do you determine the source so that you, as a watershed manager, can develop recommendations to fix the problem? Is the excessive NO_3^- a result of crop fertilization, home lawn and garden fertilization, golf course fertilization, natural geologic sources, atmospheric deposition, or some other source? In actuality, the NO_3^- probably results from a combination of sources.

Second, it is important to understand how individual uses present in a watershed affect other uses present. Individual uses in a watershed may be complementary, supplementary, or competitive with other uses present in the watershed (Brooks et al., 1997). Complementary uses occur when the product of one use increases the product of the second use. For example, a forest manager may conduct a harvest that not only provides goods (lumber) but also provides better habitat for deer. Supplementary uses have no influence on the products resulting from individual uses. For example, crops produced in one part of the watershed have no influence on forestry operations in another part of the watershed. Competitive uses occur when the product of one use decreases the product of a second use. For example, agriculture or urban development may have a deleterious effect on the aquatic resource.

One final concept related to multiple use is externalities. Externalities are off-site impacts that occur as a result of multiple uses present in watersheds both upstream and downstream from a watershed. Watersheds are imbedded in larger watersheds. A watershed can be very large such as the Mississippi River watershed or relatively small such as the area that drains the small creek behind your house. Activities conducted in Wisconsin, for example, affect both the water quantity and quality of the Mississippi River as it flows through Louisiana to the Gulf of Mexico. Understanding externalities means understanding how activities upstream affect uses in your watershed and how activities in your watershed affect downstream users. As is evident, multiple uses and externalities can be very complex and will influence how to plan for good watershed management.

> An externality is an off-site impact that occurs either upstream or downstream from your watershed.

Watershed Management Planning

The goal of watershed management may be as simple as providing clean water by managing forested watersheds that are known to be important contributors to stream water and/or groundwater. Most watershed goals are much more complex, with a variety of goods and services needing to be produced. Effective planning is critical if we are to manage watersheds to produce desired goods and services without adversely affecting soil and water resources. First we will discuss the planning process and then how we can use planning to effectively manage watersheds.

The planning process includes a number of steps, with each step being important to the development of an effective plan (Brooks et al., 1997). The first step is to monitor current or past activities to identify problems or possible opportunities. Unfortunately, in most cases watershed management is reactionary not proactive. Usually it is a problem that initiates the planning process. The second step is to characterize the problem (or opportunity) and to define the problem constraints so that strategies can be developed. An example might be that a stream has very high suspended sediment and turbidity. By investigating the problem it is determined that the main cause of the high turbidity is because off-road vehicles (ORVs) are driving up the stream bed and over the banks, causing bank failures and resuspension of stream bed material. Constraints to fixing the problem

> Problems usually stimulate watershed planning.

might be that these areas are difficult to limit access to and that the ORV organization is politically powerful in the area.

The third step in the planning process is to develop alternative actions given the constraints. Alternative actions for the ORV example may include posting these areas for no ORV use, building an ORV trail in another part of the watershed that will not impact the stream, or putting nails in the path of the vehicles so that their tires explode. The fourth step is to evaluate the alternatives from environmental, social, and economic points of view. After discussions with the ORV organization and law enforcement, suppose it is determined that posting signs is unlikely to change the current behavior and because of the remoteness, law enforcement will not be able to monitor the activity. Suppose that neither the ORV organization nor law enforcement are particularly excited about the alternative of putting nails in the path of vehicles; however, the ORV organization was excited at the potential for building a trail in the area.

The last step in the planning process is to prioritize the alternatives and recommend an action. In the ORV example, it appears clear that the recommended action should be to build an ORV trail in the uplands that will not significantly impact the stream. In the last step of the planning process, actions are implemented and monitored. In our ORV case, suppose it was discovered that although ORVs were using the designated trail they also continued to use the stream bed. Planning is an iterative process and we learn from previous efforts. After looking at our original alternatives we might decide to also post signs in the area where ORVs are entering the stream. Not only might we put signs in the area to keep ORVs out and direct them to the upland trail, but we might also develop a small educational kiosk to alert ORV users of the damage they are causing to the aquatic ecosystem. Suppose further monitoring confirms that the additional signs and education kiosk dramatically decreased use of the stream by ORVs. The ORV example is relatively simple in that we have a distinct group of users and relatively few constraints. When multiple use and multiple ownership is overlaid on a problem (or opportunity), the planning process becomes increasing complex.

> Multiple use and multiple ownership complicate the planning process.

Inevitably one will find in the planning process that critical information is of poor quality or entirely lacking. Gaps in our knowledge that will help us make better planning decisions should lead to the initiation of monitoring and/or research. Many watersheds now have monitoring programs that were implemented because data was needed. The typical monitoring program is a water quality program where waters are sampled periodically and analyzed for chemicals of interest. Wildlife monitoring programs are also common (for example, deer censuses and bird surveys). Research is conducted when simple monitoring is not providing the critical information needed. As in the example of the elevated NO_3^- in the stream discussed earlier, simple monitoring will not provide the information needed to discover sources. A research program should be initiated to track sources of nitrate from the terrestrial environment to the stream.

> Measurement is part of watershed planning.

Focus On . . . Watershed Management

1. What is a simple definition of a watershed?

2. How are flooding and streamflow related?

3. How does urbanization contribute to increased streamflow?

4. What are two ways in which erosion is important to water resources?

5. What is an externality?

6. What are some measurements that would be made as part of watershed planning?

SUMMARY

In this chapter we have looked at the distribution of water on Earth and discussed some of the important issues facing water resources today. Only a small percentage of the water on Earth is actually available for use. As a result, we need to conserve water resources. The important issues facing water resources can be summed up into two categories: water quantity and water quality. Understanding the hydrologic cycle is critical if we are to better manage water quantity and quality. Management of water quantity and quality is conducted through watershed management. Understanding the concepts of watershed management and multiple use allow for better planning and ultimately better decisions concerning our soil and water resources.

END OF CHAPTER QUESTIONS

1. What percent of the total water on Earth is not only fresh, but readily available?

2. What continent has the greatest per capita supply of fresh water?

3. What are some of the effects of global water scarcity?

4. What are some of the problems associated with sediment in surface water?

5. How do excessive nutrients in water contribute to fish kills?

6. What is regarded as one of the major causes of the "dead zone" in the Gulf of Mexico?

7. What is *bioaccumulation* and why is it a problem?

8. How does water vapor contribute to global warming?

9. What three characteristics are used to identify a wetland?

10. In addition to flood control, what other roles do wetlands have?

11. Are rivers likely or unlikely to flood?

12. What are some costs of declining aquifer levels?

13. Compare and contrast orographic and convective precipitation. Where is each most likely to occur?

14. What is the difference between *stemflow* and *throughfall* and how do you measure each?

15. What factors are most important in evaporation, and why?

16. How do you measure ET?

17. How does potential ET differ from measured ET?

18. What kind of water is affected by matric potential?

19. What are three methods of measuring soil water content?

20. What are the two most important "potentials" controlling water movement in upland soils?

21. How do you measure infiltration?

22. How does texture influence the thickness of the capillary fringe?

23. How dependent are rural populations on groundwater resources?

24. What are two adverse consequences of groundwater pumping?

25. How do you measure discharge of a stream?

26. How big can a watershed be?

27. Why is urban sprawl a critical water resource issue?

28. Why are retention basins necessary by parking lots?

29. What types of compounds are carried by erosion?

30. What are some of the consequences of land use decisions in the Midwest important for the Gulf of Mexico?

31. What can help explain increases in average streamflow in the last forty years?

32. Would the types of measurements made as part of watershed planning differ if the watershed was agricultural? What would you look for and why?

FURTHER READING

Black, P. E. 1996. *Watershed hydrology*, 2nd ed. Chelsea, MI: Ann Arbor Press, Inc.

Brooks, K. N., P. F. Folliett, H. M. Gregersen, and L. F. DeBano. 2003. *Hydrology and the management of watersheds*, 3rd ed. Ames: Iowa State Press.

Dunne, T., and L. B. Leopold. 1978. *Water in environmental planning.* New York: W. H. Freeman Publishers.

Heathcote, I. W. 1998. *Integrated watershed management: Principles and practice.* New York: John Wiley and Sons, Inc.

REFERENCES

ASCE, American Society of Civil Engineers. 1990. Evapotranspiration and irrigation water requirements. In M. E. Jensen, R. D. Burman, and R. G. Allen (eds.), *ASCE manuals and reports on engineering practice No. 70.*

Brooks, K. N., P. F. Folliett, H. M. Gregersen, and L. F. DeBano. 1997. *Hydrology and the management of watersheds*, 2nd ed. Ames: Iowa State Press.

Dahl, T. E. 1990. *Wetlands losses in the United States, 1780s to 1980s.* Washington, DC: Department of the Interior; U.S. Fish and Wildlife Service.

Hanson, R. L. 1991. Evapotranspiration and droughts. In R. W. Paulson, E. B. Chase, R. S. Roberts, and D. W. Moody (eds.), *National Water Summary 1988–89: Hydrologic events and floods and droughts.* U.S. Geological Survey Water-Supply Paper 2375, pp. 99–104.

Kolka, R. K., E. A. Nater, D. F. Grigal, and E. S. Verry. 1999. Atmospheric input of mercury and organic carbon into a forested upland/bog watershed. *Water, Air and Soil Pollution,* 113 (1/4): 273–294.

National Oceanic and Atmospheric Administration. 1993. *The summer of 1993: Flooding in the Midwest and drought in the Southeast.* TR 93-04.

Solley, W. B., R. P. Pierce, and H. A. Perlman. 1998. *Estimated use of water in the United States in 1995.* US Geological Survey Circular 1200. 71p.

United Nations Division of Sustainable Development. 1998. *Comprehensive assessment of the freshwater resources of the world.* Available online at http://www.un.org/esa/sustdev/sdissues/water/Documents/ Comprehensive_Assessement_Freshwater_resources.pdf.

United States Geological Survey. 1995. *Water-level changes in the High Plains aquifer, 1980–1994.* FS-215-95.

U.S. Environmental Protection Agency. 1992. *Managing nonpoint source pollution.* USEPA Rep. 506/9-90. Washington, DC: Office of Water, USEPA.

U.S. Geological Survey. 1990. *National water summary 1987: Hydrologic events and water supply and use.* U.S. Geological Survey Water-Supply Paper 2350, p. 553.

Van der Leeden, F., F. L. Troise, and D. K. Todd. 1990. *The water encyclopedia,* 2nd ed. Chelsea, MI: Lewis.

Chapter 14

SOIL FERTILITY AND NUTRIENT MANAGEMENT

"Whoever could make two ears of corn or two blades of grass to grow upon a spot of ground where only one grew before, would deserve better of mankind, and do more essential service to his country than the whole race of politicians put together."

Jonathan Swift, *Gulliver's Travels*, 1726

OVERVIEW

What is **soil fertility?** It is the ability of soil to hold plant nutrients and make them available for plant growth. There are essentially two kinds of fertility—native fertility and managed fertility. Native fertility is the natural capacity of a soil to hold and supply nutrients because of its parent material and organic matter content. Managed fertility refers to plant nutrients and other amendments that are added to soil to make it fertile.

Fertility is not identical to productivity. A fertile soil may not be productive for lack of water or proper crop management. An unfertile soil may be very fertile if irrigated and supplied with a timely amount of plant nutrients. The purpose of this chapter is to explore the basis of soil fertility. Why are some soils fertile and others not? Why can some soils deliver plant nutrients and others not? How can fertility be improved through amendment?

OBJECTIVES

After reading this chapter, you should be able to:

✔ Name the crucial nutrients required by plants and people and know their chemical symbols.

✔ Identify the forms in which plant nutrients occur in soil.

✔ Indicate where plant nutrients come from, how they are held in soil, and how they become available.

✔ Prescribe the forms and strategies for adding inorganic and organic fertilizers to enhance native soil fertility.

✔ Understand current issues surrounding nutrient management in agriculture.

✔ Calculate a P index.

KEY TERMS

anion exchange
cation exchange
  capacity
colloid
essential nutrients

isomorphous
  substitution
macronutrients
micronutrients

silicate clay
soil fertility
soil solution
trace element

THE ESSENTIAL NUTRIENTS FOR PLANTS AND PEOPLE

There are at least sixteen **essential nutrients** for plants, and there are additional nutrients (such as chromium [Cr]) required for human nutrition. You can divide these nutrients into three groups: basic nutrients, **macronutrients**, and **micronutrients** (Table 14-1). It is important to remember that simply because plants take up a nutrient doesn't mean that they require it. Some plants in the genus *Astragalus* (loco weeds) accumulate arsenic (As), which causes them to be toxic to animals. Indian mustard (*Brassica juncea*) takes up various heavy metals such as lead (Pb) and mercury (Hg), but these metals play no role in growth (although they do make indian mustard useful in phytoremediation). Tobacco (*Nicotinia tabaccum*) accumulates approximately 1.5 $\mu g/g$ cadmium (Cd), which contributes to tobacco's potential to cause increased lung cancer rates in smokers. Table 14-2 identifies some of the specific roles nutrients play in plant growth.

> The sixteen essential plant nutrients are divided into basic nutrients, macronutrients, and micronutrients.

Macronutrients

The basic nutrients carbon (C), hydrogen (H), and oxygen (O) are the building blocks of plant structure in compounds such as carbohydrates (starch, cellulose), hydrocarbons (fatty acids), and lignin. Because these elements come from water and carbon dioxide (CO_2), you generally don't concern yourself with their availability in soil. Water is a structural requirement (it provides plant cell turgor). In a strict biochemical sense it is also an

TABLE 14-1 Essential nutrients.

Basic Nutrients	Macronutrients	Micronutrients	Beneficial Nutrients[a]
Carbon (C)	Nitrogen (N)	Iron (Fe)	Silicon (Si)
Hydrogen (H)	Phosphorus (P)	Zinc (Zn)	Sodium (Na)
Oxygen (O)	Potassium (K)	Manganese (Mn)	Vanadium (V)
	Sulfur (S)	Copper (Cu)	Cobalt (Co)
	Calcium (Ca)	Boron (B)	Iodine (I)
	Magnesium (Mg)	Molybdenum (Mo)	Fluorine (F)
		Chlorine (Cl)	Strontium (Sr)
		Nickel (Ni)	

[a]Beneficial nutrients, with the possible exception of silicon (Si) and sodium (Na), don't appear to be an absolute requirement for plant growth. But some plants grow better if one or more of these beneficial elements are present, and some of the beneficial elements can substitute for required micronutrients.

TABLE 14-2 Functions of essential plant nutrients.

Plant Nutrient	Function in Plant
Macronutrients	
Nitrogen (N)	Part of amino acids, nucleic acids, chlorophyll
Phosphorus (P)	Part of lipids, energy transfer compounds (ATP)
Potassium (K)	Regulation of water use and osmosis
Calcium (Ca)	Calcium pectate in cell walls
Magnesium (Mg)	Central atom in chlorophyll
Sulfur (S)	Amino acids
Micronutrients	
Iron (Fe)	Chlorophyll, metal cofactor in enzymes
Zinc (Zn)	Plant hormones, metal cofactor in enzymes
Manganese (Mn)	Electron transport, metal cofactor in enzymes
Copper (Cu)	Electron transport, plant oxidases
Molybdenum (Mo)	Nitrate reduction, nitrogen fixation
Boron (B)	Growth and development of new meristematic cells
Chlorine (Cl)	Regulation of osmosis
Nickel (Ni)	Ureases and hydrogenases, grain filling
Cobalt (Co)	Nitrogen fixation

important substrate for metabolic reactions. But if a plant has enough water to grow, it has enough water for its nutritional and metabolic needs. About 90 percent of plant dry weight is made up of these basic elements.

> Macronutrients are those nutrients required in the greatest amount by plants.

The other macronutrients are so-named because they are needed in the greatest amounts by growing plants (Table 14-3). Nitrogen (N), potassium (K), calcium (Ca), phosphorus (P), magnesium (Mg), and sulfur (S) all have important roles in structural compounds such as proteins, lipids, or nucleic acids, or they play an important role in plant metabolic compounds involved in enzymatic activity and energy transfer.

Micronutrients

All of the beneficial nutrients listed in Table 14-1 are micronutrients with the exception of silicon (Si) and sodium (Na). Silicon may be required by some plants as an essential part of the plant structure. In some plants it is taken up and deposited naturally, which makes the stems feel gritty. Diatoms, single-celled algae in soil and aquatic systems, absolutely require Si because Si makes up part of their cell wall. Diatomaceous earth is almost entirely composed of these Si cell walls. Sodium is important in halophillic (salt-loving) plants. Sodium may play an essential role in survival in salt-stressed environments.

> Cobalt is an essential micronutrient required by nitrogen-fixing organisms.

Sometimes the beneficial nutrient is only an absolute requirement for plants undertaking certain functions. For example, nitrogen-fixing plants can grow perfectly well without nitrogen fixation as long as they are supplied with inorganic N. However, when these plants need to fix N they have to have trace quantities of the micronutrient cobalt (Co). Cobalt is an essential metal cofactor for the enzyme complex nitrogenase, which is actually responsible for nitrogen fixation. (You could argue that because nitrogen fixation is a strictly bacterial process, it really isn't the plants that need the Co, but this misses the point.)

TABLE 14-3 Elemental composition of various plants (dry weight basis). *(Adapted from Troeh and Thompson, 1993)*

Element	Maize Silage	Maize Grain	Wheat Grain	Soybean Seed	Alfalfa Hay	Bluegrass Hay
Basic Nutrients (g/kg)						
O	450					
C	440					
H	63					
Macronutrients (g/kg)						
N	13	14.4	21.1	60.6	24.5	13.1
Si*	12					
K	9	2.9	4.2	15.0	19.7	16.7
Ca	2.5	0.2	0.4	2.5	14.7	4.0
P	1.6	2.7	3.9	5.9	2.4	2.7
Mg	1.6	1.0	1.4	2.8	3.1	1.9
S	1.5	1.2	2.0	2.2	2.9	1.2
Micronutrients (mg/kg)						
Cl	1500	400	800	300	2800	5500
Al*	1100					
Na*	300	100	600	2200	1500	1000
Fe	90	30	60	80	270	150
Mn	60	5	40	30	60	80
Zn	30	10	5		50	
B	10					
Cu	5	3	8	15	18	9
Mo	1					

*Not essential for most plants.

Some beneficial nutrients can substitute for micronutrients. A good example of this is the substitution of vanadium (V) in molybdenum (Mo)-requiring proteins. Two important molybdoproteins in which this is observed are nitrate reductase and nitrogenase. Nitrate reductase is required by plants to convert nitrate (NO_3^-) to nitrite (NO_2^-) and eventually ammonium (NH_4^+) for growth. As you read before, nitrogenase is the essential enzyme complex in nitrogen fixation.

Humans need various micronutrients that are not required by plants. These nutrients include selenium (Se), fluorine (F), chromium (Cr), and iodine (I). Fluorine (F) plays a critical role in the formation of fluoroapatite. Fluoroapatite is a mineral that is part of the enamel on our teeth. The teeth become brittle and prone to cavities without sufficient F. Selenium, Cr, and I play important roles in enzyme function.

Micronutrients are a good example of the adage that "the dose is the poison." Although the micronutrients are essential for growth, too much can be harmful. The form of the micronutrient can also have drastic effects on its benefits. Fluorine is intentionally added to potable water supplies and iodine is added to some salts because they are essential human nutrients (Figure 14-1). But too much F can cause unsightly tooth mottling. Reduced Cr (Cr [III]) is an essential enzyme cofactor, but chromate (Cr [VI]) is toxic. The dividing line between toxic and beneficial levels of micronutrients can sometimes be very small (Figure 14-2).

"The dose is the poison" means that large amounts of required micronutrients can be toxic.

FIGURE 14-1   Iodine has traditionally been added to salt to prevent goiters (enlarged thyroids) in people without access to seafood, which is naturally high in this essential human nutrient.

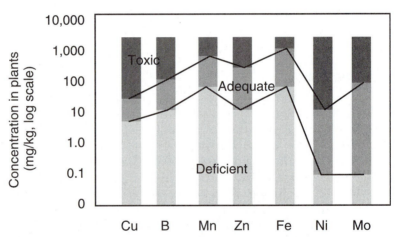

FIGURE 14-2   The distinction between beneficial and toxic levels of many micronutrients is quite narrow. The exact boundary between different levels is not discrete. *(Adapted from Brady and Weil, 2002)*

FOCUS ON . . . ESSENTIAL NUTRIENTS

1. What are examples of basic nutrients, macronutrients, and micronutrients?

2. Why is the chemical composition of a plant sometimes misleading as to the nutrient requirements for its growth?

3. What are beneficial nutrients?

4. What are some of the essential roles of the macro- and micronutrients?

5. What does it mean when you say "the dose is the poison"?

THE FORMS OF NUTRIENTS IN SOIL

Plant nutrients exist in several forms:

1. Minerals
2. Inorganic cations and anions bound to soil **colloids**
3. Nutrient complexes
4. Soluble or **soil solution** ions
5. Organic matter

Minerals are the building blocks of soil. Minerals such as feldspar contain elements like K and Ca. Cations (positively charged nutrients) and anions (negatively charged nutrients) are bound to the cation and **anion exchange** sites of soil that you read about in Chapter 8. Nutrient complexes are nutrients that have precipitated or formed chemical associations with other compounds in soil. An example is the interaction of P anions with the surface of calcium carbonate ($CaCO_3$). Soluble ions are the nutrients in solution that move with soil water and are most readily available for plant uptake. The complex molecules in organic matter have many immobilized nutrients that become available as soil microbes decompose this material.

Nutrient availability differs greatly depending on the form in which the nutrient exists. Phosphorus is a good example (Figure 14-3). Mineral

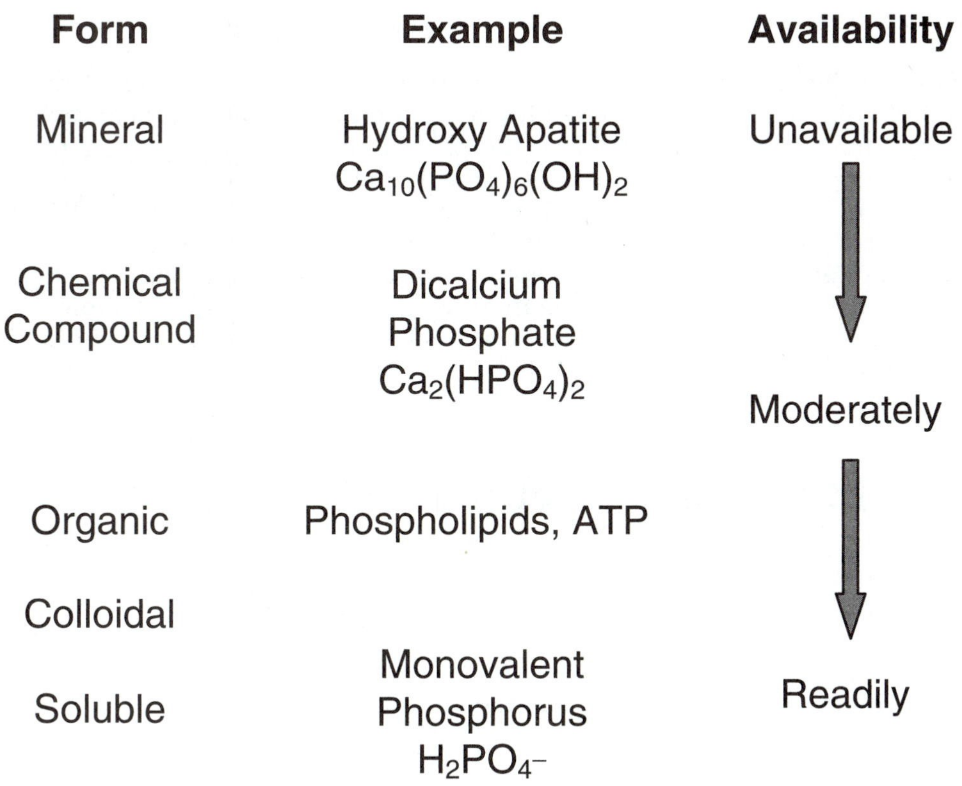

> A nutrient complex consists of nutrients that have precipitated or formed chemical associations with other compounds.

Form	Example	Availability
Mineral	Hydroxy Apatite $Ca_{10}(PO_4)_6(OH)_2$	Unavailable
Chemical Compound	Dicalcium Phosphate $Ca_2(HPO_4)_2$	
		Moderately
Organic	Phospholipids, ATP	
Colloidal		
Soluble	Monovalent Phosphorus $H_2PO_4^-$	Readily

FIGURE 14-3 Relative availability of different forms of phosphorus (P).

TABLE 14-4 Soluble forms of essential plant nutrients.

Plant Nutrient	Soluble Forms	
	Cations	Anions
Basic Nutrients		
Carbon (C)		HCO_3^-, CO_3^{2-}
Hydrogen (H)		H_3O^+
Macronutrients		
Nitrogen (N)	NH_4^+	NO_3^-, NO_2^-
Phosphorus (P)		$H_2PO_4^-$, HPO_4^{2-}
Potassium (K)	K^+	
Calcium (Ca)	Ca^{2+}	
Magnesium (Mg)	Mg^{2+}	
Sulfur (S)		SO_4^{2-}, S^{2-}
Micronutrients		
Iron (Fe)	Fe^{2+}, $Fe(OH)_2^+$ $Fe(OH)^{2+}$, Fe^{3+}	
Zinc (Zn)	Zn^{2+}	
Manganese (Mn)	Mn^{2+}, MnO_4^{2+}	
Copper (Cu)	Cu^{2+}, $Cu(OH)^+$	
Molybdenum (Mo)		MoO_4^{2-}, $HMoO_4^-$
Boron (B)		$H_2BO_3^-$
Sodium (Na)	Na^+	
Silicon (Si)		SiO_3^{2-}
Chlorine (Cl)		Cl^-
Nickel (Ni)	Ni^{2+}, Ni^{3+}	
Cobalt (Co)	Co^{2+}	

forms of P such as strengite are extremely insoluble and virtually unusable by plants. Soluble P is readily available at neutral pH, but can become unavailable very quickly if the pH changes to become either more acid or more basic. Nitrogen, however, is usually present in moderately or readily available forms in soil. The problem is that most of the Earth's N is present as gaseous N_2. Gaseous N_2 forms 78 percent of the atmosphere, and is unavailable to plants except for those few (such as legumes) capable of bacterially mediated N_2 fixation, as you read in Chapters 9 and 10.

Soluble nutrients and nutrients attached to soil colloids usually have the forms identified in Table 14-4.

FOCUS ON . . . FORMS OF NUTRIENTS

1. What are five forms in which nutrients exist in the soil?

2. How does the availability of nutrients differ?

3. Do nutrients in the same form have the same availability?

4. Why is the N in the atmosphere abundant, but not available?

5. Does P become more, or less, complex as it becomes more available?

SOURCES OF NATIVE FERTILITY

What are the sources of soil macro- and micronutrients? The sources of nutrients will be explored in this section.

Parent Material

Most of the plant nutrients come from the parent material from which the soils formed. Several types of minerals have naturally high levels of certain nutrients, and can even be mined commercially based on their nutrient content (Table 14-5). Most of the micronutrients occur as **trace elements** in primary and secondary minerals such as oxides, sulfides, silicates, and carbonates.

TABLE 14-5 Mineral sources of nutrients. *(Essential macro- and micronutrients are in bold)*

Feldspars	Silicates
Albite $Na(AlSi_3O_8)$	Andalusite Al_2SiO_3
Anorthite $\mathbf{Ca}(Al_2Si_2O_8)$	Fayalite $\mathbf{Fe_2}SiO_4$
Leucite $\mathbf{K}(AlSi_2O_6)$	Olivine $\mathbf{(Fe,Mg)_2}SiO_4$
Nepheline $(Na,\mathbf{K})(Al,Si)_2O_4$	Kaolinite $Al_4Si_4O_{10}(OH)_8$
Orthoclase $\mathbf{K}(AlSi_3O_8)$	Quartz SiO_2
	Augite $\mathbf{Ca(Mg,Fe}, Al)(Al,Si)_2O_6$
	Biotite (black mica) $\mathbf{K(Mg,Fe)_3}AlSi_3O_{10}(OH)_2$
	Muscovite (white mica) $\mathbf{K}Al_3Si_3O_{10}(OH)_2$
	Serpentine $\mathbf{Mg_3}Si_2O_5(OH)_4$
	Talc $\mathbf{Mg_3}Si_4O_{10}(OH)_2$
	Hornblende $\mathbf{Ca_2Na(Mg,Fe)_2}(Al,\mathbf{Fe},Ti)_3Si_6O_{22}(O,OH)_2$

Carbonates	Sulfates
Azurite $\mathbf{Cu_3}(CO_3)_2(OH)_2$	Gypsum $\mathbf{CaSO_4}\cdot 2H_2O$
Calcium carbonate (calcite) $CaCO_3$	
Dolomite $\mathbf{CaMg}(CO_3)_2$	
Siderite $\mathbf{Fe}CO_3$	

Oxides	Sulfides
Cassiterite SnO_2	Bornite (Cu sulfide) $\mathbf{Cu_5Fe}S_4$
Hematite $\mathbf{Fe_2}O_3$	Chalcocite $\mathbf{Cu_2S}$
Limonite (bog iron) $\mathbf{Fe}O(OH)\cdot H_2O$	Chalcopyrite $\mathbf{CuFeS_2}$
Magnetite $\mathbf{Fe_3}O_4$	(Fe sulfide) $\mathbf{Fe}S_2$
	Galena Pb$\mathbf{S}$
	Shalerite $\mathbf{ZnS}$

Phosphatic Minerals	Trace Elements	
Apatite $\mathbf{Ca_5(F,Cl)}(PO_4)_3$	Carnotite	$\mathbf{K_2}(UO_2)(VO_4)_2$
Strengite $\mathbf{Fe}PO_4\cdot 2H_2O$	Fluorite	$\mathbf{CaF_2}$
Variscite $AlPO_4\cdot 2H_2O$	Halite (Rock Salt)	Na$\mathbf{Cl}$

Organic Matter

Carbon, H, and O are supplied by air and water. Most of the N that is found in soil originally came from either N_2 fixation or from the precipitation of inorganic N from the atmosphere. Lightning discharge in the atmosphere, for example, will produce a small amount of NO_3^-. More recently, fossil fuel burning and internal combustion engines have produced considerable amounts of NOx (standing for **N**itrogen **Ox**ides) that are also ultimately transformed to NO_3^- and fall to earth. The N is rapidly taken up by plants and soil microbes and recycled in this readily available component of soil organic matter. Some of the N also becomes stably incorporated into humic and fulvic substances in soil, which have greatly diminished availability.

> Parent material and SOM are two important sources of plant nutrients.

One consequence of storing C and N in organic matter is that this pool fluctuates according to the decomposition processes that occur in soil. The amount grows when the environment is cool and wet, and organic matter accumulates. The amount shrinks as temperatures rise and organic matter decomposition accelerates. Soils with substantial amounts of soil organic matter, such as Mollisols, therefore have considerable amounts of nutrients in storage; the deeper the A horizons of these soils, the greater the potential organic matter storage.

FOCUS ON . . . SOURCES OF FERTILITY

1. What is the source of most of the essential plant nutrients in soil?
2. What do you call metals present in sufficient deposits that they can be mined?
3. What kinds of minerals are trace elements found in?
4. Why is organic matter a fluctuating pool of nutrients?
5. How does N differ from its sources compared to other nutrients?

RETENTION AND RELEASE OF SOIL NUTRIENTS

There are various factors that affect plant nutrient availability. Two of the most important are texture and organic matter content. Texture, of course, is determined by the percentage of sand, silt, and clay. The nutrient storage capacity of sand and silt are modest compared to clay. Clay, along with organic matter, is the major "warehouse" of available nutrients in soil. Like a warehouse, the availability of nutrients in clay and organic matter depends a lot on how big the warehouse doors are; the bigger the door, the more accessible the nutrients.

> Clay and SOM are important "warehouses" for plant nutrients.

> The type of clay in soil can affect fertility.

The type of clay is important because the dominant types of clay in the United States (kaolinite, illite, and smectite/montmorillonite) have low (2–10 $cMol_c$/kg), moderate (20–30 $cMol_c$/kg), and high (80–100 $cMol_c$/kg) **cation exchange capacity**, respectively. Think of them as increasingly large nutrient warehouses. Soils can have lots of clay and be relatively infertile if that clay is kaolinite, as might be the case with many Ultisols. Some soils could have relatively modest amounts of clay and still be fertile because that clay is montmorillonite, as is the case in many Mollisols.

Organic matter can have very high nutrient capacity (50–200 $cMol_c$/kg) depending on the pH. As the pH rises, the nutrient-holding capacity for cations tends to increase. Organic matter can hold and release nutrients,

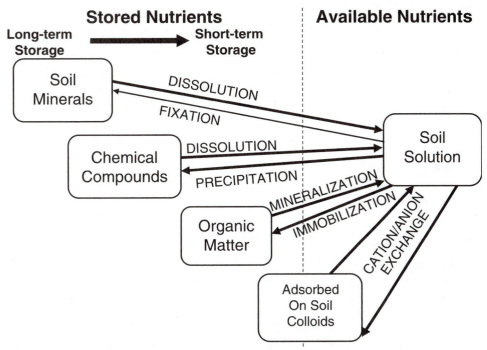

FIGURE 14-4 Sources of plant nutrients in soil. *(Adapted from Plaster, 1997)*

but when organic matter decomposes it releases the immobilized nutrients (Figure 14-4).

The factors that regulate nutrient release from clays and organic matter include decomposition, moisture, temperature, and pH. Without soil moisture these attached nutrients can't go into solution, which is a big problem in Aridisols. Too much rain causes leaching, which is a big problem in Oxisols. High temperatures promote rapid organic matter decomposition, which is also a problem in Oxisols. Low pH can dissolve minerals and make nutrients available for attachment to soil colloids. But low pH can also promote the availability of undesirable cations such as aluminum (Al) in the soil solution. High pH can immobilize many nutrients. Thus, a neutral pH is generally regarded as optimal for most plant systems.

Repulsion or Exchange

A review of the principles of cation and anion exchange will help to illustrate how soluble nutrients move between the soil solution, which holds the available nutrients, and the soil colloids, which act as short- and long-term nutrient storage sites. First, remember that soil colloids are very small (usually < 1 μm in diameter) but have a very large surface area (5 to 1000 m^2/g), which means that 10 g of colloids could have as much surface area as one hectare, 10,000 m^2 (Table 14-6). Second, remember that most colloids have a net negative charge, but some can have a net positive charge depending on factors such as pH. How does this charge develop? For a colloid like soil humus, positive charge at a low pH occurs because some of the organic molecules have a protonated N group ($-NH_4^+$). As the pH rises, these N groups become unprotonated and lose their charge. In contrast, carboxyl groups ($-COOH$) lose a proton and develop a negative charge (Figure 14-5).

Colloids such as the silicate clay minerals have a permanent negative charge due to **isomorphous substitution**—the substitution of a positively charged cation in the **silicate clay** structure by a similarly charged cation

> To appreciate fertility you have to understand cation and anion exchange in soil.

TABLE 14-6 Properties of important soil colloids.

Colloid	Type	Surface Area (m²/g)	Net Charge, cmol$_o$/kg
Humus	Organic	Variable	−500 to −100
Vermiculite	2:1 silicate	70–120	−200 to −100
Smectite (Montmorillonite)	2:1 silicate	80–150	−150 to −80
Mica	2:1 silicate	70–175	−40 to −10
Chlorite	2:1 silicate	70–100	−40 to −10
Kaolinite	1:1 silicate	5–30	−15 to −1
Allophane/ Imogolite	Silicate (Noncrystalline)	100–1000	−150 to +20
Gibbsite	Al-oxide	80–200	−5 to +10
Goethite	Fe-oxide	100–300	−5 to +20

of lesser charge (Figure 14-6). Cations and anions in the soil solution will swarm around the surface of colloids in soil, but at varying distances. Cations swarm closer than anions because the overall negative charge on most soil colloids repulses negatively charged anions (Figure 14-7). The distance that cations maintain partly depends on the charge of the cation; trivalent cations swarm closer than monovalent cations (Figure 14-8). Unfortunately, in terms of N fertility this means that NH_4^+ ions will be retained by soil colloids while NO_3^- ions (most of the inorganic N in soil solution) is at greater risk of leaching because it is more loosely held.

Sorption or Precipitation

The solubility product constant (K_{sp}) of a compound determines how readily it will precipitate.

Chemicals are loosely classified as soluble or insoluble depending on how much will go into solution. Chemicals have a maximum solubility in water (the solubility product constant or K_{sp}) depending on the chemical characteristics, temperature, and pH. If the K_{sp} is exceeded, then the material will

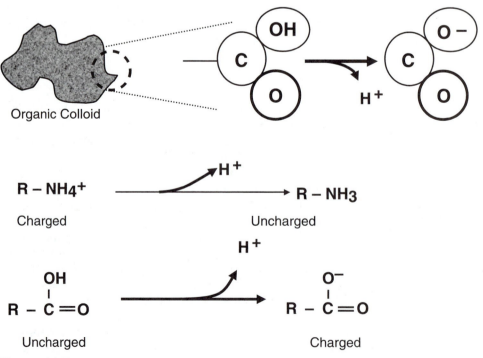

FIGURE 14-5   Development of pH-dependent charge in soil humus.

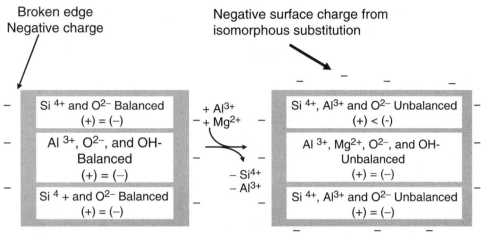

FIGURE 14-6 Development of negative charge in a layered silicate clay by isomorphic substitution.

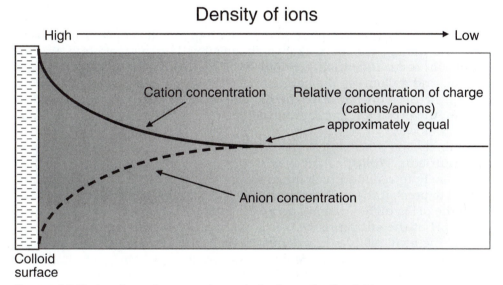

FIGURE 14-7 Ion dispersion around negatively charged soil colloids.

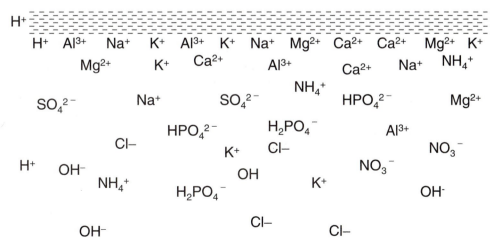

FIGURE 14-8 Distribution of ions in soil solution around a negatively charged soil colloid.

TABLE 14-7 Solubility product constants (Ksp) for relatively insoluble materials.

Compound	Ions	K_{sp}
Aluminum hydroxide	$(Al^{3+})(OH^-)^3$	1.6×10^{-34}
Calcium carbonate	$(Ca^{2+})(CO_3^{2-})$	4.96×10^{-9} @ 25 °C
Calcium sulfate	$(Ca^{2+})(SO_4^{2-})$	7.10×10^{-5} @ 25 °C
Ferric hydroxide	$(Fe^{3+})(OH^-)^3$	2.64×10^{-39} @ 25 °C
Hydroxy apatite	$(Ca^{2+})^5(PO_4^{3-})^3(OH^-)$	3.7×10^{-58}
Variscite	$(Al^{3+})(OH^-)^2(H_2PO_4^{2-})$	3.0×10^{-31}

The K_{sp} is calculated by multiplying the molar concentrations of the ions given using the formulas for the ions. For example, if the molar concentration of Ca^{2+} in solution was 1.0×10^{-6} M (one micromolar), and the concentration of soluble PO_4^{3-} was 1×10^{-5} M, and the concentration of OH^- was 1×10^{-7} M (i.e., the pH = pOH = 7), then the solubility product is:

$$(1 \times 10^{-6})^5 (1 \times 10^{-5})^3 (1 \times 10^{-7}) = 1 \times 10^{-55}$$

Because $1 \times 10^{-55} > 3.7 \times 10^{-58}$ (the K_{sp} of hydroxy apatite) then phosphorus (P) should start to precipitate in this environment.

precipitate, although precipitation does not happen all at once (Table 14-7). The lower the K_{sp}, the more likely a chemical will be to precipitate. Precipitation can occur indefinitely as long as there is a place for the precipitate to form, and as long as the solution concentration of a chemical is above the K_{sp}. If the concentration of the chemical in solution becomes low enough, the precipitated material will begin to dissolve, but this may happen very slowly.

Solubility and precipitation have an important effect on the availability of plant nutrients. While N and K are unlikely to precipitate, and if they do they are unlikely to remain precipitated for long, other essential nutrients are subject to precipitation. Phosphorus will precipitate as calcium phosphates when the pH is high, or as Al and Fe phosphates when the pH is very low.

The pH of the environment is one of the most important features controlling the availability of nutrients. At high pH, for example, Mn precipitates as manganese oxide (MnO_2) while at low pH Mo precipitates as MoO_4^{2-} (molybdate) compounds. Figure 14-9 illustrates the effect of pH on various plant nutrients.

Buffering

In soil chemistry, buffering refers to resistance to a change in pH. With respect to soil fertility it also refers to the capacity of soils to lose their base cations (Ca^{2+}, Mg^{2+}, K^+, Na^+) and have them replaced by H^+ and Al^{3+} in order to moderate changes in soil solution chemistry. The CEC of soils gives them their buffering capacity; the greater the CEC, the more H^+ it will take to change the pH. When basic ions (OH^- for example) are dissolved in the soil solution of a soil with a high CEC, H^+ can be released from the surface of the colloids to neutralize the charge. Likewise, when acidic ions such as H^+ and Al^{3+} are added to soil they can replace basic cations on the colloids, which replace them in solution. In either case, because of the high CEC and base saturation it takes more lime or acid to change the pH of the soil environment, resulting in better conditions for plant growth.

The CEC is critical to soil buffering capacity.

**Relationship of Plant Nutrient
Availability to Soil pH**

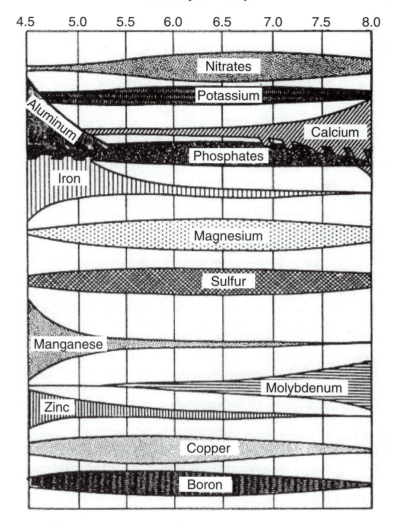

FIGURE 14-9 Relationship between pH and nutrient availability in soil.

FOCUS ON . . . MOBILITY OF NUTRIENTS

1. Why are soil colloids like a warehouse?

2. How do temperature and pH interact to affect nutrient retention and removal in soil?

3. How does the type of clay in soil affect nutrient-holding capacity?

4. What is the distribution of anions and cations around most soil colloids?

5. Why does isomorphous substitution help to repulse anions?

6. How does the concentration of various nutrients in soil affect their solubility?

7. What does buffering refer to?

FERTILIZERS AND FERTILIZATION

Native soil fertility is sometimes not enough to supply all the nutrients that plants could use for optimal growth, assuming temperature and water are not limiting. Fertilization is used to supplement the nutrients that native soil fertility provides to the soil solution from the slowly and readily available nutrient pools (Figure 14-10).

Types of Fertilizers

The four categories of fertilizers are: mineral, inorganic, organic, and synthetic organic.

Fertilizers can be grouped into four categories: mineral, inorganic, organic, and synthetic organic.

Mineral fertilizers are natural rocks and minerals that are rich in essential nutrients. Dolomite ([Ca,Mg]CO_3) is a rich source of Ca and Mg, but is primarily used to raise the pH of acid soils. Langbeinite ([K,Mg]SO_4) is a source of three essential nutrients. Apatite, or rock phosphate, is a rich source of P. The biggest problem with mineral fertilizers is that the nutrients within them are only slowly available, unless the material is finely ground. And even then it may be difficult for the materials to dissolve.

Inorganic fertilizers are mined or manufactured. Saltpeter (KNO_3), for example, is mined from desert deposits in countries like Chile. Anhydrous ammonia (NH_3) is manufactured via the Haber-Bosch method in which atmospheric N_2 is chemically reduced at high temperatures and pressures. The combination fertilizers such as ammonium nitrate (NH_4NO_3), ammonium sulfate ([NH_4]$_2SO_4$), and ammonium phosphate ([NH_4]H_2PO_4, ([NH_4]$_2HPO_4$) are produced by mixing anhydrous ammonium with nitric, sulfuric, and phosphoric acids, respectively.

Organic fertilizers have some type of organic C associated with them. This includes urea ([NH_2]$_2CO$), which is excreted in the wastes of mammals, including people. When people mention organic fertilizers, they usu-

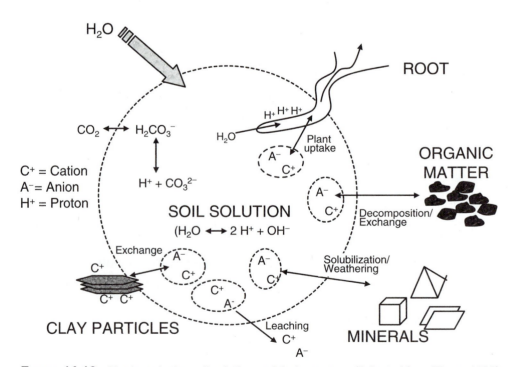

FIGURE 14-10   Nutrients in the soil solution and their sources. *(Adapted from Thom, 1995)*

ally mean manures, but composts, food by-products, crop residues, and cover crops can all serve as organic fertilizers by returning essential elements to soil.

Synthetic organic fertilizers are manufactured industrially, but are chemically organic because they contain C and H. The best example of this is the chemical synthesis of urea.

Source and Form

There are four groups of fertilizers that can be categorized on the basis of their form or method of delivery: pressurized liquids, fluids, dry fertilizers, and slow-release fertilizers.

> Fertilizers are usually delivered as pressurized liquids, fluids, dry fertilizers, or slow-release fertilizers.

Pressurized Liquids

Pressurized liquids are best typified by anhydrous ammonia (NH_3). Ammonia would normally be a gas at room temperature and pressure, but it can be stored as a liquid if the temperature falls below $-33°C$ ($-28°F$). The liquid is held in pressurized tanks until it can be injected into the soil where the ammonia rapidly converts to ammonium (NH_4^+). If the injection technique leaves cracks in the soil the NH_3 can be readily lost through volatilization.

Fluid Fertilizers

Fluid fertilizers are liquids that are either dissolved chemicals or suspensions of chemicals attached to very fine clays. Liquid animal wastes from lagoon storages, for example, are a combination of dissolved and suspended materials that are either sprayed onto the soil or injected beneath the soil surface (for odor control). Foliar fertilizers, for example, are liquid fertilizers. Golf courses that are managed by fertigation (mixing the fertilizer with the irrigation water) also employ the equivalent of liquid fertilizers. Many of the commercial solutions for houseplants are concentrated solutions of nutrients that are diluted to appropriate strength before use.

Dry Fertilizers

Dry fertilizers are generally broadcast on the surface of the soil and rely on a combination of tillage and rainfall to ensure that they mix with the soil. Some dry fertilizers are banded next to the growing crops. Dry fertilizers are typically pulverized, granulated, or prilled. Pulverized fertilizer is made by crushing the fertilizer to a fine powder. The advantage is greater reactivity, but the disadvantage is greater dust and caking.

Granulated fertilizer is easier to handle, but doesn't react as quickly. Handling can still produce fine material that cakes, and because the size of granules is not perfectly uniform, settling during transport of the granules to the field and during application can cause an uneven distribution of particles, which leads to uneven fertilizer application rates. Prills are smooth, round, and dust-free. Some prills are coated to prevent caking. In terms of application they spread easier than granules and are also free of fine material. But the extra processing makes them more expensive fertilizers to use.

Bulk-blend fertilizers are mixes of solid fertilizers added as individual components to the specifications of the producer prior to field application. Although this is a flexible way to produce custom blends for a specific field, it suffers from the disadvantage that even with premixing for uniformity, the different sizes of the starting materials will cause them to segregate en-route.

Slow-Release Fertilizers

Slow-release fertilizers are very popular horticultural applications. Slow-release fertilizers, as the name implies, release the nutrients they contain at a slower rate than other fertilizers. These are also dry fertilizers. Typically, the feature that makes slow-release fertilizers work is a coating of some relatively insoluble material around the fertilizer. Such is the case with S-coated urea. Urea is very soluble in water, but elemental S is not. Thus, by coating the urea with a thin film of elemental S, it is more difficult for the urea to come into contact with water and go into solution.

Plastics or polymers are also used to coat urea and other soluble nutrients. Osmocote is an example of this material. Sulfur and an additional coating, such as wax, plastic, or a polymer, have also been combined to increase the longevity of the fertilizer. The second coat surrounds the S coating. These products have been marketed by Scotts Co. and Purcell Industries.

Nitrification and urease inhibitors have been added to some fertilizers to slow the transformation of N. Urease inhibitors such as NBPT (N-[n-butyl]thiophosphoric triamide, trade name Agrotain®) inhibit urease activity in soil. Nitrification will be inhibited by DCD (dicyandiamide). The combination of the two with urea fertilizers will slow down the rate at which urea is hydrolyzed and thus NH_4^+ subsequently nitrified. These kinds of products are often referred to as "stabilized" N fertilizers.

Fertilizer Placement

The form of a fertilizer influences how and where it is applied. Fertilizers can be broadcast, broadcast and disked or plowed in, injected, banded beside the seed (starter fertilizer), banded with the seed (pop-up fertilizer), and top-dressed or side-dressed after plant emergence (Figure 14-11). Broadcast fertilizer is simply broadcast over the surface of the soil in a thin layer that is allowed to gradually infiltrate the soil surface. When a chisel-disc is used, the broadcast fertilizer can be injected within the first few centimeters of soil. Plowed-down fertilizer incorporates the fertilizer at a slightly greater depth.

Injection of liquid fertilizer or manure is much like banding in that a thin row of the fertilizer material is placed in discrete rows beneath the soil surface. When the fertilizer is solid and applied at the time of planting it is referred to as starter fertilizer. The fertilizer can also be applied directly in a band with the seed, or incorporated on the seed as a coating, in which case

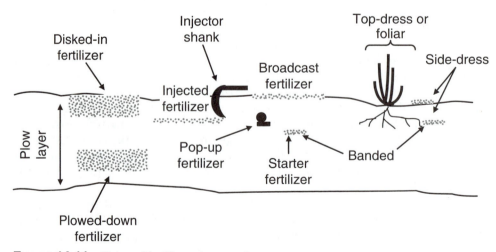

FIGURE 14-11   Types of fertilizer placement.

it is referred to as "pop-up" fertilizer. The greatest concern with both starter and pop-up fertilizer is that small amounts be used so that the germinating seed is not harmed by excessive salt in the soil solution.

> Care must be taken that fertilizers do not damage germinating seeds.

When row plants have started growing, another type of fertilizer placement is the side-dress, which can occur either on the side of the plant or in a band within the soil to the side of the plant. Fertilizer for corn and tobacco, for example, is frequently applied in two separate applications—half at planting and half as a side-dress several weeks later. Pastures and forages are frequently top-dressed to avoid disturbing the soil environment. The top-dress fertilizer is applied directly over the surface of the crop as either a solid or a liquid. Foliar fertilizer applications are also top-dressed, but since the nutrients added by this method are frequently more expensive than other types of fertilizer, the spray zone is typically smaller.

No-Tillage

No-tillage and conservation tillage present unique problems for fertilizer placement because they are based on the concept of minimally disturbing the soil environment for the purpose of residue preservation and minimizing erosion. However, this limits the potential placement of fertilizer by plowing and tillage. As a result, in no-tillage much of the fertilizer is broadcast, which influences the choice of fertilizers away from those with the potential for high volatility, such as urea, unless the material is likely to be incorporated into the soil by rainfall. Conservation tillage, which permits some residue burial by chisel-disc, allows somewhat more incorporation of the applied fertilizer.

> Because no-tillage doesn't involve incorporation by tillage, this affects fertilizer choice.

Fertilizer Management

Fertilizer management is a multifaceted process that involves:

- ❖ determining what the fertility status of the soil is
- ❖ deciding what plants will be grown
- ❖ evaluating how much fertilizer they need
- ❖ calculating the economically optimum fertilizer rate
- ❖ determining when and where to apply the fertilizer
- ❖ making any subsequent fertilizer amendments that can be used to address nutrient deficiency problems

Soil Tests

Most recommendations for fertility requirements are based on a soil test. Thus the most important part of ensuring a good fertility recommendation is collecting a representative soil sample. As a general rule, most soil samples are taken from the first 15 to 20 cm (6 to 8 in.), which represents a common depth to which many fertilizers are incorporated by tillage. In pastures or minimally tilled fields, the sample depth is shallower, only 8 to 10 cm (3 to 4 in.), because fertilizer and nutrient movement is not as deep.

> The key to accurate soil testing is representative soil sampling.

Each sample typically represents a composite of 10 to 20 soil cores per 8 ha (about 20 acres) taken by auger, soil probe, or shovel and mixed to be as homogeneous as possible. If there is known variability in the field, or there will be known variability in the management of the field, then separate soil samples should be taken for each management area, as Figure 14-12 illustrates.

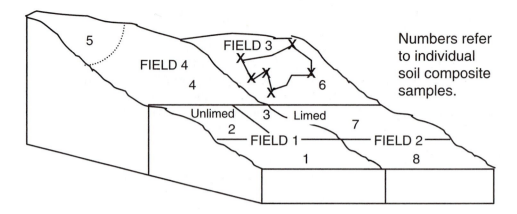

The four fields are divided into eight sampling areas based on field size, topography, and management. A random sampling scheme is illustrated in Field 3. One or more samples would be taken at the vertices of the polygon indicated by the 'x.'

FIGURE 14-12. Strategy for preparing composite soil samples for fields requiring different management. *(Adapted from Thom et al., 2003)*

Requirements

Different plants have different nutrient requirements. For any given state, this information is usually available from the state's Cooperative Extension Service, which publishes recommendations for fertilization based on soil test levels. Sample recommendations from the Kentucky Cooperative Extension Service are shown in Figure 14-13.

The Kentucky Cooperative Extension Service uses a conservative approach to fertilizer recommendations that is based on assumed climatic conditions, management conditions, and average yields. The recommendations are not designed to increase soil test values for nutrients such as P and K. Other Cooperative Extension Services use different approaches to fertilizer recommendations appropriate to their states.

Nitrogen fertilization recommendations are complicated by several factors: the influence of drainage on the potential for N loss through leaching and denitrification, the contribution of N from mineralization of soil organic matter and legume N_2 fixation, the variability of inorganic N soil test values from sample to sample, and the effect of tillage. It is often recommended to perform a soil test in fall in preparation for spring fertilization, but the soil test values for NO_3^- and NH_4^+ in this period will bear little resemblance to what they will be in spring. An example of N recommendations for corn are shown in Table 14-8.

| Leaching, immobilization, and denitrification are three reasons why testing for inorganic soil N gives extremely variable results. |

Residual Fertility

Native fertility is the level of available plant nutrients a soil can provide without additional fertilization. Residual fertility is this inherent native fertility (as measured by a soil test) plus some additional amount that becomes available during the year as the organic and mineral forms of nutrients decompose and weather. Residual fertility decreases with prolonged cropping and weathering, and it increases as a result of prolonged application of inorganic and organic nutrients. Beyond a certain level residual fertility increases to such an extent that no benefit to plant growth is obtained, and for some micronutrients toxicity is observed (Figure 14-14).

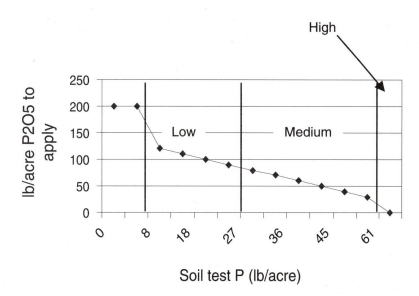

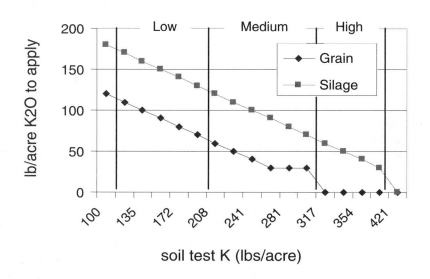

FIGURE **14-13** Sample phosphorus and potassium fertilizer recommendations for corn fertilization in Kentucky. *(Adapted from AGR-1, 2003)*

TABLE **14-8** Nitrogen recommendations for corn. *(Adapted from AGR-1, 2003)*

lb N/acre to Apply

Previous Crop	Conventional Tillage			No-Tillage	
	Well-Drained	**Moderately Well-Drained**	**Poorly Drained**	**Well-Drained**	**Moderately Well-Drained**
Large- and small-seeded grain Soybean Set-aside Fallow	100–125	150–175	175–200	125–150	175–200
Grass or grass-legume sod (< 4 yr old) Winter annual legume cover	75–100	125–150	150–175	100–125	150–175
Grass or grass-legume sod (> 5 yr old)	50–75	100–125	125–150	75–100	125–150

FIGURE 14-14 Residual soil fertility based on soil test levels and relative yield. *(Adapted from Thom, 1995)*

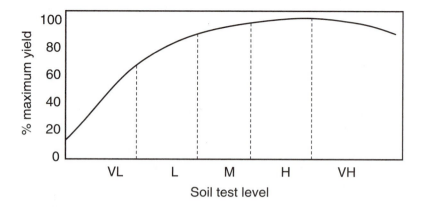

TABLE 14-9 Excess fertilizer beyond plant removal required to change soil test P and K values. *(Adapted from Thom, 1995)*[a]

Soil Test P (lb/acre)	P$_2$O$_5$ Required (lb P$_2$O$_5$/lb soil test P)	Soil Test K (lb/acre)	K$_2$O Required (lb K$_2$O/lb soil test K)
10	14.5	100	6.4
20	10.3	150	5.4
30	8.4	200	4.7
40	7.3	250	4.2
50	6.5	300	3.8

[a] The study was performed on a Belknap silt loam. Different soils would give a different response. Note that as the soil test P or K increases in each case, it requires progressively less fertilizer to cause a measurable increase in soil test values, which means the nutrients are increasingly more extractable.

Soil test levels for the major plant nutrients do not change on a kilogram for kilogram basis as fertilizer is added because of factors such as soil type, clay type, and the original soil test level. When soil test nutrients are low, some of the nutrients of added fertilizer become tightly held and less extractable during the soil test procedures. Thus it takes correspondingly more fertilizer to change the soil test value when fertility is low compared to when fertility is high (Table 14-9). It is also important to note that if fertilizer applications are halted because of high soil test values, future soil test values will decline more rapidly at high fertility levels than at low fertility levels.

Residual fertility also influences the placement of additional fertilizer in soil. If the residual fertility levels are low, banding is more effective than broadcasting fertilizer in terms of affecting yield. At low residual fertility levels more of the applied nutrients are fixed into less-available forms. By concentrating the fertilizer in a specific band, the effect is to saturate the capacity of the soil to immobilize the nutrients in a limited zone, and more nutrients will be available for plant uptake. At high residual soil fertility levels, it generally does not matter whether the fertilizer is banded or broadcast.

Rates

Crops will respond differently to fertilizers depending on whether the initial fertility levels are high or low.

Fertilizer application rates depend on the soil test values, desired yield, and cost of fertilizer. When fertility levels are low, plants are most likely to respond to additional fertilizer. Although, as you have read, when fertility levels are very low more fertilizer may have to be added than anticipated to see the desired result because of fertilizer immobilization in the soil environment. At very high fertility levels, there is generally little benefit from adding fertilizer (Figure 14-15).

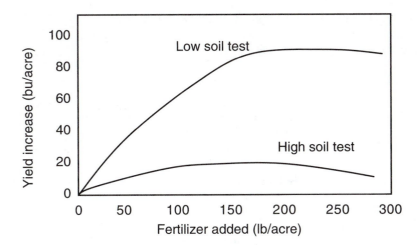

FIGURE 14-15 Crop response to fertilizer addition at high and low fertility levels. *(Adapted from Troeh and Thompson, 1993)*

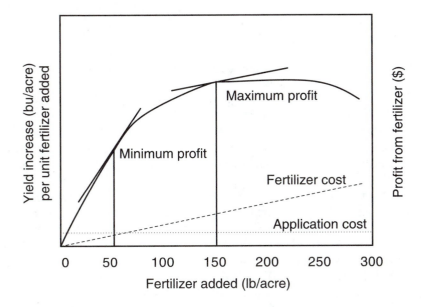

FIGURE 14-16 Profit response curve to a theoretical crop yield. For a given yield response curve the maximum profit is given as the tangent to the yield curve that is parallel to the fertilizer cost. The minimum economic profit is given as the tangent to the yield response curve originating at the cost of fertilizer application. *(Adapted from Troeh and Thompson, 1993)*

The economically optimum rate of fertilization can be determined from graphs like the one shown in Figure 14-16. Given a particular crop response to fertilization, application costs, and fertilizer costs, the maximum and minimum rates of economic return can be assessed by drawing tangents to the yield response curve.

Timing

Timing of fertilization depends on the type of fertilizer that is added. Fertilizer can be added before, during, or after crop planting. It can also be added to existing pastures. Some fertilizer nutrients, such as P, are relatively immobile after addition to soil, and so they can be added at any time without great fear of loss. Other nutrients, such as N, have great potential to be lost through volatilization, denitrification, and leaching. Fall-applied N, therefore, would be expected to show considerable losses before planting in spring. If the mobility of nutrients is expected to be great, then application closest to planting is recommended.

> Mobile plant nutrients should be applied close to the period of plant growth.

TABLE 14-10 Deficiency symptoms for essential nutrients.

Plant Nutrient	Deficiency Symptom
Macronutrients	
Nitrogen (N)	Yellowing of midrib, browning (firing) of lower leaves
Phosphorus (P)	Reddish-purple color of young plant leaves
Potassium (K)	Browning of outer edges of lower leaves
Calcium (Ca)	Suppressed growth of terminal buds and apical root tips
Magnesium (Mg)	Between-vein yellowing of middle and lower leaves
Sulfur (S)	Uniformly yellowed upper leaves and slow growth
Micronutrients	
Iron (Fe)	Between-vein yellowing in young leaves and in extreme cases complete leaf yellowing
Zinc (Zn)	Lack of terminal growth, spotted white or yellow areas between veins in the upper third of the plant
Manganese (Mn)	Between-vein yellowing in younger leaves
Copper (Cu)	Yellowing and stunting of young leaves
Molybdenum (Mo)	Yellowing of the midrib of lower leaves
Boron (B)	Pale green young leaves and suppressed terminal growth
Chlorine (Cl)	Partial wilting and loss of leaf turgor even when water is present

Quick Tests.   Several tests exist to determine whether nutrient levels in soil and plant tissue are sufficient for optimal growth. The preside-dressed NO_3^- test (PSNT), for example, tests the soil for NO_3^- at early stages of plant growth and makes recommendations for side-dressing N applications accordingly. Nitrate in plant tissue can also be sampled by a simple test using diphenylamine. The diphenylamine reacts with NO_3^- in plant tissue to form a blue color, the intensity of which varies with the amount of NO_3^- present. The more NO_3^- in plant tissue, the less likely the plant will be to respond to additional N.

A relatively new approach for monitoring N fertilization requirements is to use a chlorophyll meter. Nitrogen is an essential constituent of chlorophyll, so leaf greenness will be positively related to chlorophyll content and also positively related to nitrogen (N) content in the plant tissue (Murdock, 1997). Leaf greenness can be monitored by a chlorophyll meter. If the readings fall below a certain level additional N is suggested. One advantage of this technique is that it can be done without injury to the growing plant.

The appearance of plant tissue itself is diagnostic of certain nutrient deficiencies. These deficiencies are outlined in Table 14-10. Some deficiencies, such as N deficiency, show up in the older leaves because the nutrients are moved from older to younger tissue. Other nutrients, such as copper (Cu) and Fe, are less mobile, so the deficiency symptoms occur in the younger leaves.

Although micronutrient deficiencies can sometimes be remedied by a foliar application during growth, in general, by the time a nutrient deficiency is observable in plant tissue yield has been suppressed. Observation of nutrient deficiencies is also merely diagnostic. An analysis of plant tissue and soil test for macro- and micronutrient content would be required to correctly assess what nutrients are limiting and how much is required. Nutrients, particularly micronutrients, may be present but unavailable because of soil factors such as drainage and pH. These also have to be taken into consideration in any additional testing.

Nitrogen status of plants can be determined remotely by use of chlorophyll meters.

FOCUS ON . . . FERTILIZATION AND FERTILIZERS

1. What are different types of fertilizer?

2. What are some different forms of fertilizer?

3. What do fertilizers supplement?

4. What are different types of fertilizer placement?

5. How should fields with different management be treated for soil testing?

6. What are some factors that determine fertilizer recommendations for different crops?

7. What is the importance of residual fertility?

8. Why is the rate of fertilization for maximum crop yield and maximum economic return different?

9. Why does the timing of fertilization differ for some nutrients?

10. What is a chlorophyll test, and what nutrient is it used to monitor?

PROBLEM SOILS AND SOLUTIONS

There are certain soils for which amendments have to be made as part of a fertility program in order to address specific problems such as acidity or alkalinity. There are also some soils that have such an abundance of salts that they become inferior for crop production. Some soils suffer from drainage problems that can affect nutrient availability. Each of these problems will be addressed in turn in this section.

Acid and Alkaline Soils

Plants have very different tolerance for the pH of their environment (Table 14-11). One of the goals of nutrient management is to ensure that the pH of the soil is appropriate for the type of plant growing in it. Most soils in humid climates where sufficient rainfall is present to support plants without irrigation will ultimately show decreased pH with time. This is a function of leaching of bases (Ca^{2+}, Mg^{2+}, K^+, and Na^+) by percolating water and their replacement by Al^{3+} and H^+. It is also a function of base cations being taken up during plant growth and the subsequent excretion of H^+ by the plant roots to maintain charge neutrality. Other processes that will acidify soil are:

❖ formation of carbonic acid (H_2CO_3) from CO_2

❖ dissociation of carboxyl groups in organic matter (R-COOH → R-COO$^-$ + H^+)

❖ Oxidation of reduced N, S, Fe, and Mn compounds

❖ atmospheric deposition of combustion residues (H_2SO_4 and HNO_3)

Liming is a soil amendment process in which the pH is raised. Frequently, additional plant nutrients such as calcium (Ca) and magnesium (Mg) are added as a by-product of liming, but the main reason for the process is to decrease soil acidity. Liming and fertilization can therefore be

> Liming is a generic term for raising the soil pH.

TABLE 14-11 Optimum pH ranges of selected plants. *(Adapted from Spurway, 1941)*

Forages/Pastures		Flowers		Oak, white	5.0–6.5
				Pine, jack	4.5–5.0
Alfalfa	6.2–7.8	African violet	6.0–7.0	Pine, loblolly	5.0–6.0
Bluegrass, KY	5.5–7.5	Alyssum	6.0–7.5	Pine, red	5.0–6.0
Clover, red	6.0–7.5	Begonia	5.5–7.0	Pine, white	4.5–6.0
Clover, sweet	6.5–7.5	Calendula	5.5–7.0	Spruce, black	4.0–5.0
Clover, white	5.6–7.0	Carnation	6.0–7.5	Spruce, Colorado	6.0–7.0
		Chrysanthemum	6.0–7.5	Spruce, white	5.0–6.0
Grains		Gardenia	5.0–6.0	Sycamore	6.0–7.5
		Geranium	6.0–8.0	Tamarack	5.0–6.5
Barley	6.5–7.8	Lily, Easter	6.0–7.0	Walnut, black	6.0–8.0
Corn	5.5–7.5	Orchid	4.0–5.0	Yew, Japanese	6.0–7.0
Oats	5.0–7.5	Phlox	5.0–6.0		
Rice	5.0–6.5	Poinsettia	6.0–7.0	**Forest Plants**	
Rye	5.0–7.0	Snapdragon	6.0–7.5		
Sorghum	5.5–7.5	Snowball	6.5–7.5	Heather	4.5–6.0
Wheat	5.5–7.5	Sweet William	6.0–7.5	Moss, club	4.5–5.0
		Zinnia	5.5–7.5	Moss, sphagnum	3.5–5.0
Field Crops					
		Shrubs		**Weeds**	
Bean, field	6.0–7.5				
Flax	5.0–7.0	Azalea	4.5–5.0	Dandelion	5.5–7.0
Pea, field	6.0–7.5	Barberry	6.0–7.5	Dodder	5.5–7.0
Peanut	5.3–6.6	Burning Bush	5.5–7.5	Foxtail	6.0–7.5
Soybean	6.0–7.0	Holly	6.0–8.0	Goldenrod	6.0–7.5
Sugar cane	6.0–8.0	Ivy	6.0–8.0	Grass, crab	6.0–7.0
Tobacco	5.5–7.5	Lilac	6.0–7.5	Grass, quack	5.5–6.5
		Magnolia	5.0–6.0	Horse tail	4.5–6.0
Vegetable Crop		Quince	6.0–7.0	Milkweed	4.0–5.0
		Rhododendron	4.5–6.0	Mustard, wild	6.0–8.0
Asparagus	6.0–8.0	Rose, tea	5.5–7.0	Thistle, Canada	5.0–7.5
Beets	6.0–7.5				
Broccoli	6.0–7.0	**Trees**		**Fruits**	
Cabbage	6.0–7.5				
Carrot	5.5–7.0	Ash, white	6.0–7.5	Apple	5.0–6.5
Cauliflower	5.5–7.5	Aspen, American	3.8–5.5	Apricot	6.0–7.0
Celery	5.8–7.5	Beech	5.0–6.7	Blueberry	4.0–5.0
Cucumber	5.5–7.0	Birch, white	4.5–6.0	Cherry, sour	6.0–7.0
Lettuce	6.0–7.0	Cedar, white	4.5–5.0	Cherry, sweet	6.0–7.5
Muskmelon	6.0–7.0	Fir, balsam	5.0–6.0	Crab apple	6.0–7.5
Onion	5.8–7.0	Fir, Douglas	6.0–7.0	Cranberry	4.2–5.0
Potato	4.8–6.5	Larch	5.0–6.5	Peach	6.0–7.5
Rhubarb	5.5–7.0	Maple, sugar	6.0–7.5	Pineapple	5.0–6.0
Spinach	6.0–7.5	Oak, black	6.0–7.0	Raspberry	5.5–7.0
Tomato	5.5–7.5	Oak, pin	5.0–6.5	Strawberry	5.0–6.5

regarded as complementary practices, and liming is an essential part of most nutrient management plans whether the nutrients derive strictly from inorganic fertilizers or whether the nutrients are supplied by organic materials.

The most common liming material is limestone rock, either various purities of calcium carbonate ($CaCO_3$) or dolomite ($[Ca,Mg]CO_3$). Many people mistakenly assume that simply because a material has Ca or Mg it will be a good liming material. This is not true. Adding compounds such as $CaCl_2$ (calcium chloride) or $CaSO_4 \cdot 2H_2O$ (gypsum) will add base cations to soil, but will not affect the pH. The easiest way to see why this is true is to look at an example of a liming reaction. The increase in pH occurs because the liming agent adds bicarbonate (HCO_3^-) to the soil solution, which eventually consumes H^+ in the production of H_2O and CO_2.

$$(Ca,Mg)CO_3 + 2H_2O + 2\ CO_2 \leftrightarrow Ca^{2+} + 4HCO_3^- + Mg^{2+}$$
(Dissolution Phase)

$$\boxed{\text{Clay or Humus}}\ H^+ + 2\ Mg^{2+}/Ca^{2+} + 2HCO_3^- \leftrightarrow \boxed{\text{Clay or Humus}}\ Ca^{2+}\ + Al(OH)_3(s)$$

$$Al^{3+} \qquad\qquad\qquad\qquad\qquad\qquad\qquad Mg^{2+}\ + H_2O$$
$$+ CO_2\ (g)$$
(exchange phase)

The same would be true if calcium oxide (CaO) were used as the liming agent. It is the OH^- released by the dissolution of CaO that neutralizes the soil acidity, not the Ca.

$$CaO + H_2O \rightarrow Ca(OH)_2 \rightarrow Ca^{2+} + 2\ OH^-$$

$$2OH^- + 2\ H^+ \rightarrow 2H_2O$$

Liming materials are not equivalent in terms of neutralizing capacity. Calcium carbonate ($CaCO_3$) is used as the standard and given a value of 100 percent. All other liming materials are rated on the basis of their $CaCO_3$ equivalent (CCE); that is, how much pure $CaCO_3$ would they be equal to. Pure magnesium carbonate ($MgCO_3$), for example, has a lower molecular weight than $CaCO_3$ (84 vs. 100); therefore, the $CaCO_3$ equivalent of $MgCO_3$ is $100/84 = 1.2$ or 120 percent. That's to say that pure $MgCO_3$ would be 20 percent more effective in neutralizing acidity as an equal mass of pure $CaCO_3$. Calcium oxide has a molecular weight of 56. It has a CCE of $100/56 = 1.8$ or 180 percent.

In addition to the CCE, the other consideration with liming materials is the rate at which they dissolve, which is an important consideration. Agricultural limestone must be finely ground for it to be effective. If the limestone is added in particles > 5 mm in diameter it may take so long for the material to dissolve that it is of little use in neutralizing acidity. Liming material that passes through a 60-mesh screen is assumed to be 100 percent available within three years of application, while that held back by an 8-mesh screen (considerably coarser material) is assumed to be only 10 percent effective (Table 14-12).

The relative neutralizing value (RNV) is determined from the percent of material retained by each mesh size times the assumed availability, as specified in Table 14-12. The sum of these values gives the total amount of material that will be effective over a three-year period. The RNV of the material times the CCE gives the effective $CaCO_3$ equivalent (ECCE). This is the amount of a material that would have to be added to have the same neutralizing value as 100 percent effective $CaCO_3$.

Not all liming materials are equally effective.

TABLE 14-12 Effect of mesh size on the effectiveness of liming materials.

Mesh Size[a]	% Effectiveness
< 4	0
4–8	10
8–60	40
> 60	100

[a]These values represent the range of mesh through which the material may pass. For example, in mesh size 4–8 the material will pass through a 4-mesh screen but be retained by an 8-mesh screen.

DETERMINING EFFECTIVE CALCIUM CARBONATE EQUIVALENT (ECCE)

The ECCE is determined by sieving a representative portion of the liming material through sieves of different sizes, and calculating the percent mass retained by each mesh. The percent mass is multiplied by the assumed effectiveness of each sized material in terms of dissolution over a three-year period. The sum of these values is the relative neutralizing value (RNV) for the material. RNV times the CCE of the liming material gives the effective calcium carbonate equivalent (ECCE).

Assume you had 120 g of liming material with a CCE of 0.75 (75 percent). After passing it through sieves you obtained the following results:

Sieve Material Size	g of Material Retained	% of Retained
4 mesh	10	8
8 mesh	20	17
60 mesh	30	25
> 60 mesh	60	50

These values are multiplied by assumed effectiveness:

% of Material	Mesh Size	$\times$	Availability Coefficient	= % Effectiveness
8%	< 4 mesh	$\times$	0 availability	= 0% effectiveness
17%	4< $\times$ <8	$\times$	0.1 availability	= 1.7% effectiveness
25%	8< $\times$ <60	$\times$	0.4 availability	= 10% effectiveness
50%	60<	$\times$	1.0 availability	= 60% effectiveness
			Σ	= 71.7% effectiveness

Over a three-year period 71.7 percent of this material will be available to neutralize soil acidity. This is the relative neutralizing value or RNV. The RNV × CCE = ECCE, so:

(71.7% RNV) × (75% CCE) = 53.8% ECCE

When using this material as a liming agent, you will have to add 100/53.8 = 1.86 times as much material as an equivalent amount of calcium carbonate.

Excessively acid or alkaline soils can limit nutrient availability.

It is sometimes, but less frequently, necessary to lower the pH because the soils are too alkaline. This most often occurs when the soils are calcareous—have free lime accumulations—that raise the pH above 8. In alkaline soils the availability of Fe, P, Mn, B, Zn, and S are all reduced. Perversely, the availability of Mo increases in alkaline soils, and it becomes phytotoxic. Soybeans, for example, suffer from an Fe deficiency symptom called *iron chlorosis* in calcareous soils, in which the plants are stunted and the leaves are yellow or white as though they had been bleached. Some horticultural and fruit crops such as blueberries grow better in acid than alkaline soils for these reasons. Applying elemental sulfur (S^0) or sulfide (S^{2-}) will help to acidify soils because acidity is generated by microbial oxidation of reduced S compounds in soil. Sulfuric acid (3 percent) can also be added directly to acidify soils. Ferrous sulfate ($FeSO_4$) and aluminum sulfate ($Al_2[SO_4]_3$) may also be added to acidify soil. The oxidation of ferrous iron (Fe^{2+}) and the hydrolysis of Al both generate acidity.

Saline Soils

Saline soils are a particular problem in the western part of the United States where there is insufficient rainfall to leach bases from soils. Other reasons for the development of these soils are the evaporation of groundwater in low topographic positions and the weathering of parent materials that have exceptionally high base content.

Salinity causes physical and chemical problems in soils. Chemically, if the soils are alkaline because of their base content it reduces the availability of nutrients. The concentration of salts in solution may be so high that plants are unable to effectively absorb water. Physically, saline soils may suffer from dispersed colloids that clog soil pores and reduce water infiltration. This is primarily a concern with high levels of Na.

There are four distinct types of alkaline soils (Harpstead et al., 1997). High lime soils can occur in humid regions and develop in low topographic areas in which the parent material is high in calcium. The calcium carbonate content of these soils may approach 10 percent and they typically have a pH between 7.5 and 8.0. The most serious issue with these soils is the potential for nutrient deficiency.

Saline soils have a high content of soluble salt as measured by electrical conductivity of the soil solution. Typically, if the electrical conductivity exceeds 4 millimhos/cm, the soil is classified as saline. Another way of looking at it is that the soils contain more than 0.2 percent soluble salt. The soluble salt causes osmotic stress in growing plants because it makes water uptake more difficult. Saline soils, often called *white alkali* soils, are noticeable because of the salt crusts that form on the surface of soils where soluble salts carried by evaporating water have been deposited. The pH of these soils will typically range between 7.3 and 8.5.

Sodic soils have more than 15 percent of their cation exchange sites occupied by Na ions. This causes colloids to disperse because the Na^+ becomes surrounded by a large sphere of water, and the soil permeability decreases. The dispersed colloids can form a dark soil crust due to the organic matter, which is one reason why these soils are often called *black alkali* soils. The pH of these soils exceeds 8.5 because chemical reactions in soil generate OH^- ions. Sodic soils are the most alkaline of all soils and some of the hardest to reclaim because of their low permeability.

Saline sodic soils combine high salt content with high Na content. They usually have a pH between 8.0 and 8.5. The structure of saline sodic soils is seemingly better than sodic soils because of the influence of the soluble salts that minimize colloid dispersion. However, if the soils are leached to remove the soluble salts, these soils can be turned into sodic soils with all the associated permeability problems.

> Four types of alkaline soils are high-lime soils, saline soils, sodic soils, and saline-sodic soils.

Treating Alkaline Soils

Drainage is critical for alkaline soils because excess soluble salts must be removed before the soils become usable. In high-lime soils excess Ca^{2+} can compete with K^+ adsorption by plants and cause nutrient imbalances. Excess Ca^{2+} can also force K^+ into nonexchangeable positions in the interlayers of clay micelles. However, the treatment of high-lime soil is not very difficult. Additional K fertilizer can be added to make up for deficiencies. Immobilized micronutrients can be added as foliar sprays as necessary. Organic matter can be added as a source of K^+ and mineralizable N that will ultimately help to stabilize the pH. Because the pH is inherently high in lime, alkaline-tolerant legumes such as alfalfa can be grown preferentially on these soils.

Saline soils that are nonsodic can be readily reclaimed if there is enough water to leach out the excess soluble salts. Without adequate drainage the effect is temporary because additional irrigation will raise the water table and lead to salt evaporation on the soil surface. Some irrigation waters are already saline, in which case much larger volumes of irrigation water must be used to remove the soluble salts beyond the root zone. A variety of plants can tolerate saline soils (Table 14-13) and it may be preferable to use these plants rather than those that require more soil manipulation.

> Saline soils can be renovated by leaching soluble salts.

TABLE 14-13 Salt tolerance of various plants. *(USDA Salinity Laboratory Staff, 1954)*

High	Medium	Low
Date palm	Olive	Pear
Garden beets	Grape	Apple
Kale	Cantaloupe	Orange
Asparagus	Tomato	Grapefruit
Spinach	Cabbage	Prune
Saltgrass	Cauliflower	Plum
Bermudagrass	Lettuce	Almond
Rhodes grass	Potato	Apricot
Canada wild rye	Carrot	Peach
Western wheatgrass	Peas	Strawberry
Barley	Squash	Lemon
Bird's-foot trefoil	Sweet clover	Avocado
Sugar beet	Mountain brome	Radish
Rape (canola)	Strawberry clover	Celery
Cotton	Dallis grass	Green bean
	Sudangrass	White dutch clover
	Alfalfa	Meadow foxtail
	Tall fescue	Alsike clover
	Rye	Red clover
	Wheat	Landino clover
	Oats	Field beans
	Orchard grass	
	Meadow fescue	
	Reed canary grass	
	Smoothbrome grass	
	Rice	
	Sorghum	
	Corn	
	Sunflower	

Saline-sodic soils must be amended before the soluble salts can be leached, otherwise dispersion will occur. The most common amendment is gypsum ($CaSO_4 \cdot 2H_2O$). The purpose of the gypsum is to replace the Na on the soil colloids. The soluble sodium sulfate (Na_2SO_4) can then be leached.

$$Na^+—Colloid—Na^+ + CaSO_4 \leftrightarrow Ca^{2+}—Colloid + Na_2SO_4$$

If enough gypsum is added most of the CO_3^{2-} can be removed from the soil as well.

$$2\,Na^+ + CO_3^{2-} + CaSO_4 \rightarrow CaCO_3 \text{ (s)} + Na_2SO_4$$

Sulfuric acid can also be added as an amendment because as it is neutralized by some of the CO_3^{2-} in soil, it is lost as CO_2. Gypsum is also formed.

$$H_2SO_4 + 2Na^+ + CO_3^{2-} \rightarrow 2Na^+ + SO_4^{2-} + H_2O + CO_2 \text{ (g)}$$

$$H_2SO_4 + CaCO_3 \rightarrow CaSO_4 + H_2O + CO_2 \text{ (g)}$$

The $CaSO_4$ that forms acts as the reactions for gypsum have previously illustrated.

Sodic soils are most difficult to treat. The physical condition of the soil is poor, which makes it difficult to evenly apply nutrients. Leaching is slow because Na attached to the colloids causes dispersion. Gypsum is once

Sodic soils are difficult to treat because of poor structure.

again the amendment of choice to replace Na in solution and on the cation exchange complex. Organic matter is another useful amendment as it improves soil structure by binding clay particles together and mitigates some of the dispersive effects of Na when the soil is wet.

Drainage and Fertility

Drainage is an issue in soil fertility principally because of the effect it has on N retention. Poor drainage makes fertilization with N less efficient, particularly fertilization with $NO_3^- - N$. In waterlogged conditions that develop in poorly drained soils, any NO_3^- that forms can be quickly lost via denitrification. In addition, poor drainage can also make some micronutrients such as Mn more available, because these micronutrients become more soluble in the reduced and waterlogged conditions. Poorly drained soils have less developed root structures and are colder in spring, which inhibits the timely uptake of nutrients.

The traditional solution to poorly drained soils was to install tile drains, which artificially lowered the water table. Due to increased emphasis on wetland protection, drainage of many of these soils is no longer possible.

> Greater emphasis on protecting wetlands has led to restrictions on drainage.

FOCUS ON . . . PROBLEM SOILS

1. What are the characteristics of saline soils?

2. Why do high-lime soils remain relatively productive?

3. How does a sodic soil differ from a saline soil?

4. Why must saline-sodic soils be amended first before they can be leached to remove soluble salts?

5. What nutrient often becomes limiting in poorly drained soils?

NUTRIENT MANAGEMENT PLANS

To reduce agricultural nonpoint source pollution, one of the most important current issues with respect to fertilizers and fertilization is the development and use of nutrient management plans. A nutrient management plan is a method for accounting for all of the nutrients that are present in an agricultural system with respect to inputs (fertilizer, manure) and outputs (leaching, volatilization, erosion, crop removal). Ideally, the nutrient management plan balances the amount of nutrients applied for crop growth with those removed by harvest. In practice, developing a nutrient management plan can be difficult, particularly when organic sources of nutrients are used.

> Nutrient management plans balance nutrient inputs with crop and livestock nutrient removal.

What's in a Nutrient Management Plan?

A nutrient management plan consists of seven key components (IP-71, 2001):

1. Soil maps with field designations (where things will be grown, on what soil)

2. Crop plan (what will be grown, what nutrients it requires, and how much)

3. Conservation practices (how erosion and runoff will be minimized)

4. Manure nutrient content (what quantity of nutrients will be supplied by manure and how quickly)

5. Manure use plan (where manure will be used)

6. Records: soil tests, fertilizer recommendations, previous manure applications, yield estimates or yield history

> Balancing nutrient inputs is one of the most difficult parts of a nutrient management plan.

Perhaps the most important component, particularly when manures are used to supply nutrients, is balancing the input of nutrients with that removed by the crops. Different crops have different requirements for N, P, and K, which are the most important nutrients (that is, the most limiting to optimal growth) for plant production. Unfortunately, manures used for fertilization usually do not supply nutrients in the same proportion that they will be removed by crops; hence, potential nutrient imbalances will develop in the soil with long-term manure use.

The Mismatch between Supply and Demand in Manures

Look at Table 14-14. If oats were being grown for grain they would typically remove 0.62 lb/bu N (0.28 kg/bu N), 0.25 lb/bu P (0.11 kg/bu P) as P_2O_5, and 0.19 lb/bu K (0.09 lb/bu K) as K_2O. The ratio of N:P:K removal in grain

TABLE 14-14 Nutrient removal values for selected crops. *(Adapted from the University of Kentucky Soil Testing Laboratory Manure Calculator, http://soils.rs.uky.edu/soildata.htm)*

Crop	Yield Unit	lbs per Yield Unit	Nutrients Removed (lbs per yield unit)		
			N	P_2O_5	K_2O
Hay					
Alfalfa	ton	2000	50	14	55
Warm season native grass	ton	2000	20	6.8	25
Bermuda grass	ton	2000	37.6	8.7	33.6
Reed canary grass	ton	2000	27	8.2	25
Eastern gamagrass	ton	2000	35	16.1	31.2
All other hay	ton	2000	35	12	53
Corn silage	ton	2000	7.5	3.6	8.0
Forage from pasture	ton	2000	10.5	3.6	15.9
Soybean					
Edible bean	bu	56	0.95	0.41	0.3
Dry bean	bu	60	3.0	0.7	1.1
Tobacco					
Burley	lb	1	0.07	0.011	0.075
Dark, air-cured	lb	1	0.07	0.06	0.06
Dark, fire-cured	lb	1	0.07	0.06	0.06
Grain					
Barley	bu	48	0.9	0.41	0.30
Corn	bu	56	0.7	0.4	0.35
Oats	bu	32	0.62	0.25	0.19
Rye	bu	56	1.16	0.33	0.32
Winter wheat	bu	60	1.2	0.5	0.3

would be 1.0:0.4:0.3. Corn removed for grain, however, would give a ratio of 1.0:0.6:0.5 for N:P:K, respectively. So corn would require relatively more P and K fertilizer than would oats.

What if these crops were fertilized with beef manure (Table 14-15)? The relative content of N, P, and K in beef manure is 11:7:10 lbs per ton, or a ratio of 1.0:0.6:0.9. For either crop, if you fertilized with beef manure to meet the nutrient removal of each crop for N you would overapply P and K. If you fertilized at rates to supply P and K, you would underfertilize with respect to N. Because most producers tend to fertilize with animal manures to meet the N requirements of the crop, this means that with time fields tend to accumulate unutilized P and K. In the long run this can lead to nonpoint source pollution. The situation is exacerbated because while 100 percent of the K and up to 80 percent of the P are available from manure (Table 14-16), no more than 60 percent of the N in manure is assumed to be available in the year of application. So, to meet the N needs of the growing crop for that year, the manure rate is typically increased by 40 percent or more, which only serves to add more excess P and K.

> Manure use for fertilizer typically results in the overapplication of P.

Excessive fertilization with manures can lead to a situation in which P, normally regarded as an immobile nutrient because of its adsorption to soil colloids, becomes increasingly mobile as the sites to which it can adsorb are saturated. A value of 0.2 ppm P (0.2 mg P per L) in the soil solution is generally regarded as adequate for most crop growth. Consequently, when the soil becomes saturated with P beyond the limits of crop uptake, particularly below the level of root penetration, P gets into surface waters and aquifers where it can cause significant problems with eutrophication at extremely low concentrations (Table 14-17).

> Excessive P addition can saturate soil and lead to P mobilization.

The P sorption by soils is closely related to the amount of Fe and Al oxides in acid soils and soluble Ca and $CaCO_3$ in alkaline soils. One way to determine how close a soil is to saturation is to calculate the DPS (degree

TABLE 14-15 Nutrient content of common manures. *(Adapted from the University of Kentucky Soil Testing Laboratory Manure Calculator, http://soils.rs.uky.edu/soildata.htm)*

	N	P_2O_5	K_2O	% Moisture
Solid manures (lbs/ton)				
Beef	11	7	10	80
Dairy	11	9	12	80
Swine	9	9	8	82
Broiler				
Fresh	55	55	45	20
Stacked	60	70	40	30
Cake	60	70	40	30
Pullet	40	68	40	25
Breeder	35	55	30	40
Layer	30	40	30	40
Liquid manures (lbs/1000 gallon)				
Holding pit				
Swine	36	27	22	96
Dairy	31	15	19	94
Lagoon				
Swine	4	2	4	99
Dairy	4	2	3	98

of P saturation),which describes the relationship between the extractable P in soil and the soil P sorption capacity (van der Zee and van Riemsdijk, 1988):

$$\text{DPS (\%)} = \frac{\text{Extractable P}}{\text{P sorption capacity}} \times 100$$

The soil is considered P-saturated when the DPS exceeds 25 percent.

TABLE 14-16 Coefficients of nitrogen availability from various manures. *(Adapted from the University of Kentucky Soil Testing Laboratory Manure Calculator, http://soils.rs.uky.edu/soildata.htm)*

Crop in Year of Application				Comparison to Fertilizer Nutrients	
				Availability Coefficient	
Nutrient	**Crop**	**Season Applied**	**Incorporation**	**Poultry or Liquid Manure**	**Other Manures**
Nitrogen					
	Corn	Spring	≤ 2 days	0.60	0.50
	Tobacco		3–4 days	0.55	0.45
	Sorghum		5–6 days	0.50	0.40
	Annual		≥ 7 days	0.45	0.35
	Grasses	Fall			
		w/o Cover crop		0.15	0.20
		w/ Cover crop		0.50	0.40
	Small grains (preplant)			0.50	0.40
	Pasture	Early spring		0.80	0.60
		Fall		0.80	0.60
Phosphorus (P_2O_5)				0.80	0.80
Potassium (K_2O)				1.0	1.0

TABLE 14-17 Critical phosphorus concentrations reported for surface waters. *(Adapted from Sims, 1998)*

Phosphorus Concentration (mg/L)	Comment
0.01	Dissolved P
	Critical concentration for lakes
	Target concentration allowed to enter Florida Everglades by 2000
0.05	Total P
	Critical concentration for lakes
	Dissolved P
	Critical concentration allowed to enter Florida Everglades
0.10	Total P
	Critical concentration for streams
1.0	Dissolved P
	Proposed flow-weighted annual allowable limit for agriculture

One consequence of the concern with overapplying P from animal manures is the development of P indexes. Phosphorus indexes integrate soil test P (STP) levels with other factors relating to the potential for soil P loss. The typical index uses eight characteristics of a site that are given numerical rankings corresponding to low, medium, high, or very high in terms of potential for P loss. Each characteristic is given a weighting factor to reflect its importance to P loss (erosion is a more significant factor than P application method, for example). The individual weights are site specific and currently based on professional judgment for a soil and its environment. Thus, the P index is site specific. If the P index is too high, it indicates that steps must be taken immediately to protect the environment from nonpoint source pollution by P. An example of a P index from Delaware is given in Table 14-18.

> P indexes are used to estimate the potential for P loss from soil.

TABLE **14-18** The phosphorus index. *(Adapted from Sims, 1996)*

Site Characteristics (weighting factors)	Phosphorus Loss Rating (value)				
	None (0)	**Low (1)**	**Medium (2)**	**High (4)**	**Very High (8)**
Soil erosion (1.5)	N/A	< 5 ton/acre	5–10 ton/acre	10–15 ton/acre	> 15 ton/acre
Irrigation erosion (1.5)	N/A	Infrequent irrigation Well-drained	Moderate irrigation slopes < 5%	Frequent irrigation slopes 2–5%	Frequent irrigation slopes > 5%
Soil runoff class (0.5)	N/A	Very low Low	Medium	High	Very high
Soil test P (1.0)	N/A	Low	Medium	High	Excessive
P fertilizer application method (0.5)	None	At planting > 5 cm deep	Incorporate before crop	> 3 mo before incorporation	Surface applied 3 mo before cropping
P fertilization rate (kg P/ha) (0.5)	None	< 15	15–45	46–75	> 75
Organic P application method (1.0)	None	At planting > 5 cm deep	Incorporate before crop	> 3 mo before incorporation	Surface applied 3 mo before cropping
Organic P application rate (kg P/ha) (1.0)	None	< 15	15–45	46–75	> 75

(continued)

TABLE 14-18 The phosphorus index. *(continued)*

Phosphorus Index for Site[a] Generalized Interpretation

<8	Low potential for P movement
	Current farming practices have minimal adverse impact to surface waters
8–14	Medium potential for P movement
	Some chance of adverse impact
	Some remedial action should be taken
15–32	High potential for P movement
	Likely adverse impacts unless remedial action is taken
	Soil and water conservation practices and P management practices necessary
> 32	Very high potential for P movement
	Definite adverse impacts
	Remedial action necessary to reduce risk of P loss
	All necessary soil and water conservation practices should be used
	A P management plan should be put into place

[a] Determined by the Σ (Weighting factor) $\times$ (P loss rating).

FOCUS ON . . . NUTRIENT MANAGEMENT PLANS

1. What are the components of a nutrient management plan?

2. Why does manure complicate a nutrient management plan?

3. Which nutrients tend to be overapplied when manure is used to meet nitrogen demands?

4. What determines a soil's P sorption capacity?

5. What characteristics are used to create a P index?

CALCULATING A PHOSPHORUS INDEX FOR SOIL

Phosphorus indexes are calculated to help assess the potential P loss from a site. The index is calculated by summing the product of site characteristics and weighting factors times P loss ratings:

$$\text{P index} = \Sigma \text{ (Characteristic weighting factor)} \times \text{(P loss rating)}$$

The higher the P index, the more likely the P loss from a site.

Here is an example of how the Delaware P index works. For each site characteristic, a relevant status is circled:

Site Characteristics (weighting factors)	Phosphorus Loss Rating (value)				
	None (0)	**Low (1)**	**Medium (2)**	**High (4)**	**Very High (8)**
Soil erosion (1.5)	N/A	⟨< 5 ton/acre⟩	5–10 ton/acre	10–15 ton/acre	> 15 ton/acre
Irrigation erosion (1.5)	⟨N/A⟩	Infrequent irrigation Well-drained	Moderate irrigation slopes <5%	Frequent irrigation slopes 2–5%	Frequent irrigation slopes > 5%

Site Characteristics (weighting factors)	Phosphorus Loss Rating (value)				
	None (0)	**Low** (1)	**Medium** (2)	**High** (4)	**Very High** (8)
Soil runoff class (0.5)	N/A	Very low Low	(Medium)	High	Very high
Soil test P (1.0)	N/A	Low	Medium	(High)	Excessive
P fertilizer application method (0.5)	(None)	At planting > 5 cm deep	Incorporate before crop	> 3 mo before incorporation	Surface applied 3 mo before cropping
P fertilization rate (kg P/ha) (0.5)	(None)	< 15	15–45	46–75	> 75
Organic P application method (1.0)	None	At planting > 5 cm deep	Incorporate before crop	> 3 mo before incorporation	(Surface applied 3 mo before cropping)
Organic P application rate (kg P/ha) (1.0)	None	< 15	(15–45)	46–75	> 75

If we take each of those characteristics and multiply the weighting factor by the P loss rating, we can construct the following table:

Characteristic	Weighting Factor		P Loss Rating		Value
Soil erosion	1.5	×	2	=	3.0
Irrigation erosion	1.5	×	0	=	0.0
Soil runoff class	0.5	×	2	=	1.0
Soil test P	1.0	×	4	=	4.0
P fertilization method	0.5	×	0	=	0.0
P fertilization rate	0.5	×	0	=	0.0
Organic P application method	1.0	×	8	=	8.0
Organic P application rate	1.0	×	2	=	2.0
				Σ =	18.0

The P index gives a value of 18, which for Delaware's index means a high potential for P movement (see Table 14-18) and likely adverse impacts unless remedial action is taken. This remedial action could consist of such things as incorporating the organic P rather than letting it sit on the soil surface, reducing the P application rate because the soil already has a high soil test P, or using inorganic P rather than organic P.

SUMMARY

In this chapter you examined the basis of soil fertility. You identified the major nutrients required by plants and people and identified their sources. You examined how the soil environment can hold and deliver nutrients for plants in terms of the sources of fertility and the mechanisms by which nutrients are delivered. You examined different types of fertilizers and the methods by which they are applied, and how misapplication of fertilizers can lead to either excessive or deficient nutrient levels in soil. You also examined soil amendments and procedures that are used to improve the fertility of soils that are either too acid or too saline. Finally, you explored some environmental issues associated with using manures as a nutrient source, and how P is one of the nutrients controlling nutrient management planning. In the next chapter you will read about how soil information is collected and classified to rate soils in terms of different land use.

END OF CHAPTER QUESTIONS

1. What are the essential plant macronutrients? Micronutrients?
2. What nutrients are needed by people, but not by plants?
3. What's the difference between a beneficial nutrient and a required nutrient?
4. What are the five forms in which nutrients exist in soil?
5. Are all the forms of nutrients in soil equally available?
6. What are examples of minerals in soil that contain essential nutrients?
7. Why does one refer to the soil organic matter as a "fluctuating" source of plant nutrients?
8. How does the charge on soil colloids develop?
9. How does clay type make a difference in the capacity to deliver nutrients for plant growth?
10. What causes some essential nutrients to precipitate and become unavailable?
11. What are four categories of fertilizer?
12. How does the method of delivery of fertilizers differ?
13. What are the different ways in which fertilizer is placed?
14. Does fertilizer placement affect nutrient availability? Explain.
15. How much land should soil sampling reflect?
16. Why isn't N one of the nutrients looked for in soil tests?
17. What is residual fertility?
18. Which changes faster, soil test results at low nutrient values or soil test results at high nutrient values? What causes this difference?
19. What are examples of quick tests for soil nutrients?
20. What is a chlorophyll meter actually reading and how is that information used?
21. Why does nutrient deficiency sometimes show up in older leaves and sometimes in younger tissue?
22. Why do soils become acid?
23. Why isn't the liming potential of all liming agents the same?
24. How does the size of a liming particle affect its effectiveness?
25. What's a saline soil?
26. Why do you have to amend sodic soils first before leaching them?
27. How can drainage influence the type of N fertilizer you add?
28. Why can continued application of manures to a soil lead to accumulation of one or more nutrients?

29. What does P saturation refer to?

30. Why are the characteristics used to make P indexes weighted differently?

FURTHER READING

Balancing manure nutrient supply and crop nutrient uptake is tricky. Several aids are available to assist the producer with this task. An example manure calculator can be found at the University of Kentucky Soil Test Lab Web site (http://soils.rs.uky.edu/soildata.htm).

REFERENCES

AGR-1. 2002–2003 lime and fertilizer recommendations. 2003. Lexington: University of Kentucky Cooperative Extension Service.

Brady, N. C., and R. R. Weil. 2002. *The nature and properties of soil,* 13th ed. Upper Saddle River, NJ: Prentice Hall.

Harpstead, M. I., T. J. Sauer, and W. F. Bennett. 1997. *Soil science simplified,* 3rd ed. Ames: Iowa State University Press.

IP-71. Nutrient management in Kentucky. 2001. Lexington: University of Kentucky Cooperative Extension Service.

Murdock, L. W. et al. 1997. *AGR-170. Using a chlorophyll meter to make nitrogen recommendations on wheat.* Lexington: University of Kentucky Cooperative Extension Service.

Plaster, E. J. 1997. *Soil science and management,* 3rd ed. Clifton Park, NY: Thomson Delmar Learning.

Sims, J. T. 1996. The phosphorus index: A phosphorus management strategy for Delaware's agricultural soils. *Soil Testing Fact Sheet ST-05.* Newark: University of Delaware, College of Agricultural Science.

Sims, J. T. (ed.). 1998. *Soil testing for phosphorus: Environmental uses and implications.* Southern Cooperative Series Bulletin No. 389. SERA-IEG 17. USDA-CSREES.

Spurway, C. H. 1941. *Soil reaction (pH) preferences of plants.* Michigan Agr. Exp. Stn. Special Bulletin 306. East Lansing: Michigan State University.

Thom, W. O. 1995. *AGR-144. The nature and value of residual soil fertility.* Lexington: University of Kentucky Cooperative Extension Service.

Thom, W. O., G. J. Schwab, L. W. Murdock, and F. J. Sikora. 2003. AGR-16. *Taking soil test samples.* Lexington: University of Kentucky Cooperative Extension Service.

Troeh, F. R., and L. M. Thompson. 1993. *Soils and soil fertility,* 5th ed. New York: Oxford University Press.

U.S. Salinity Laboratory Staff. 1954. *Diagnosis and improvement of saline and alkali soils.* USDA handbook 60. Washington, DC: U.S. Government Printing Office.

van der Zee, S., and W. H. van Riemsdijk. 1988. Model for long-term phosphate reactions in soil. *Journal of Environmental Quality* 17: 35–41.

SECTION 6

INTEGRATING SOIL WITH OTHER RESOURCES

SOIL INVENTORIES AND MAPPING

*"Soil mapping is possible only because men can examine a profile at
one point and successfully predict its occurrence at another point
where surface indications are similar."*

Author unknown

"Maps are a way of organizing wonder."

Peter Steinhart, 1986

OVERVIEW

Soil classification is a useful endeavor because without it you would not
have a full understanding of how the soil in a particular location is constructed, and therefore you would not have a good idea about how it can
be managed. In previous chapters the management practices for erosion,
organic matter, water, and fertility were all based on fundamental properties of the soils: what they were made from, how the soil particles were held
together and influenced water flow, and the potential nutrient-holding and
nutrient-releasing capacity.

Surveying and mapping build on that classification knowledge to provide information that will identify the appropriate uses for soils. Once the
identity of different soils is determined by survey, maps can be generated
to show their distribution throughout landscapes, and then the properties
of the soils within these landscapes can be used to make decisions about
sustainable land use.

The objectives of this chapter are to briefly describe how soil surveys are
conducted and the products of those surveys, and then to discuss the
types of information gathered from soils within a mapping unit that are significant in determining its use.

OBJECTIVES

After reading this chapter, you should be able to:

✔ Understand how soil surveys are conducted.

✔ Appreciate the difference between individual soil units and
associations.

✔ Describe how the presentation of soil survey data has changed with time.

✔ Interpret some basic information from a soil survey such as the soil series name and slope.

✔ List some basic soil characteristics that will determine land use.

KEY TERMS

catena   land capability class   soil survey
complex   mapping unit   toposequence
generalized soil maps   permeability   undifferentiated soils
inclusions   soil association

SOIL SURVEYS

The NRCS, Forest Service, and BLM all carry out soil mapping.

The Natural Resources Conservation Service (NRCS) has the overall responsibility for making **soil surveys** in the United States and developing the inventory of the nation's soil resources. In addition, soil surveys are conducted by the U.S. Forest Service and the Bureau of Land Management (BLM) for the soils over which they have jurisdiction. Soil surveys are published on a county-by-county basis, and the most recent surveys have been prepared in digital versions and can be obtained electronically. The soil survey will contain a map of soil locations, descriptions of the soils, and interpretations for their agronomic and engineering use.

Development of Survey Maps

Soil surveys have become more detailed and rely more on aerial photography than surveys of the past.

Early soil survey maps were usually published as a single sheet at a scale of 1:63,650 (1 inch per mile). Soils were color-coded and contained letter symbols for identification. Obvious types of drainageways and relief were noted, as well as the location of such cultural features as roads, cities, bridges, schools, churches, and cemeteries. Since 1935 most surveys have been based on aerial photography, and are much more detailed with scales of 1:15,840 (1 inch per quarter mile). These maps are published on separate pages and contain much more information about the location of individual soils.

Soil mapping in the United States is not yet complete.

The essential tools of the soil surveyor have been the base map (either a geologic map or an aerial map), a spade, an auger or probe, a clinometer to determine slope, a soil color chart or set of color vials, and experience (Figure 15-1). Experience is required to accurately determine texture-by-feel in the field, predict where changes in soil type occur in the landscape so that the number of soil test probes can be minimized, and identify the soil profile with that of previously described soil series. Because mappers are expected to survey large areas on a daily basis (approximately 300 acres or 121 ha per day), speed and efficiency are essential. Soil mapping in the United States continues today because not every county has a published soil survey, and changing information, due to land use and other factors, requires updating existing surveys.

(a)

(b)

(c)

FIGURE 15-1 The essential tools of the soil mapper have not changed greatly with time. They still include the base map (either a geologic map or an aerial map), spade, an auger or probe, clinometer to determine slope, a soil color chart or set of color vials, and experience, as these soil mappers illustrate while constructing a soil profile on Alaska native lands near Homer, AK (a). The soil mapper's tool kit has been extended by portable hydraulic soil probes (used here on soil in Sarpy County, NB) (b) and GPS technology, which allows precison location of mapped sites as demonstrated at this site in Washington County, VA (c). *(All photographs courtesy of the USDA-NRCS)*

Mapping Units

The areas delineated on a soil map are called **mapping units.** Most soil-mapping units represent *phases* of soil series. A phase of a soil series is a further division of that series to a level that has practical management applications. For example, a soil series may occur across a range of slopes that have implications for the way a soil is managed for erosion control. Or a soil series may have been eroded through past use. The erosion does not change the soil series, but it does change how it can be used and how it should be managed.

Map units represent phases of a soil series.

Soil-mapping units may have **inclusions.** Inclusions are small areas of one type of soil within a greater expanse of a different soil series. As long as the inclusions don't exceed 15 percent of a mapping unit they are generally not identified. A soil **complex** is used to distinguish a location where two or more soil series are so intermixed that they cannot be reasonably separated for mapping purposes. In contrast, **undifferentiated** map units are used when two soil series could be distinguished but are not because their differences are not important.

Soil associations represent groups of soils that occur in repeating patterns in landscapes.

Soil associations are groups of soils that occur in repeating patterns in a landscape. A typical pattern is soil that forms on a ridge, another on the hillside, and a third in the valley. In the McAfee-Maury-Braxton association, for example, the Maury series occurs on ridgetops, the McAfee in moderately sloping hillsides, the Braxton on steeper sloping hillsides, and the Huntington series in the toe slope positions. As you might expect, Huntington soils are colluvial and alluvial soils that have a tendency to flood.

The varied soil series in a soil association can be quite different, but the important thing is that they occur together regularly in the same positions in a landscape. Because soil associations tend to consistently appear, they are used to prepare **generalized soil maps.** Generalized soil maps are used to characterize large areas, such as a county, and provide a quick overview of what the soils look like in a location.

Catenas are toposequences forming from the same parent material.

A **toposequence** is a soil association that has differences related to the topography of a site. In a toposequence the soils developed in the same climate and vegetation, but did not necessarily develop with the same parent material. In contrast, a **catena** is a toposequence that formed entirely from the same parent material.

Interpreting Soil Survey Mapping Units

The soil survey map contains important information in the legend and also in the map units themselves (Figure 15-2). Once the code is understood, it makes reading the survey map much easier, and one can start developing an internal image of what the landscape looks like based on clues about each soil's position and slope. The example from Figure 15-2 is actually

FIGURE 15-2 Interpreting mapping unit symbols.

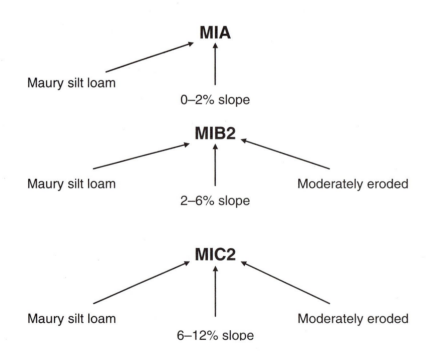

very easy to interpret. The letter code for each soil is followed by a letter code for slope and a numerical code for erosion if those are significant characteristics. In some cases the letter code for the soil is replaced by a numerical code for the soil. The same soil series can have a different letter code. For example, MnB and MpC2 both refer to the McAfee soil series. But the former (Mn) has a silt loam texture and the latter (Mp) has a silty clay loam texture. The MpC2 soil also has a steeper slope (C, 6–12 percent vs. B, 2–6 percent) (Table 15-1), and shows evidence that erosion has occurred (the "2" designation), which is not surprising considering the slope (Table 15-2). Knowing how to interpret soil survey maps is a powerful tool in being able to manage land resources.

> Letter codes indicate soil type, slope, and extent of erosion.

Information in the Soil Survey

There is a wealth of information in a soil survey beyond the distribution of different soil series in the landscape (Figure 15-3). Most soil surveys have four basic components:

> Most soil surveys have four basic components in addition to the maps of soil location.

1. General soil map unit descriptions.
2. Detailed but nontechnical descriptions of each map unit (i.e., soil series).
3. Use and management descriptions for agriculture and engineering purposes.
4. Technical descriptions of the soil series and its morphology.

TABLE 15-1 Designation of slope characteristics for soil-mapping units.

Map Unit Code	%	Slope Description
A	0–2	Nearly level
B	2–6	Gently sloping
C	6–12	Sloping
D	12–18	Strongly sloping
E	18–30	Severely sloping
F	30–60	Steep
	> 60	Very steep

TABLE 15-2 Designation of erosion characteristics for soil-mapping units.

Map Unit Code	Description
0 or none	No erosion
1 or P	Slight, 0 to 1/3 of the topsoil gone
2 or R	Moderate, 1/3 to 2/3 of the topsoil gone
3 or S	Severe, 2/3 or more of topsoil gone, up to 1/3 of subsoil gone
4	Heavy subsoil erosion and deposition of eroded soil

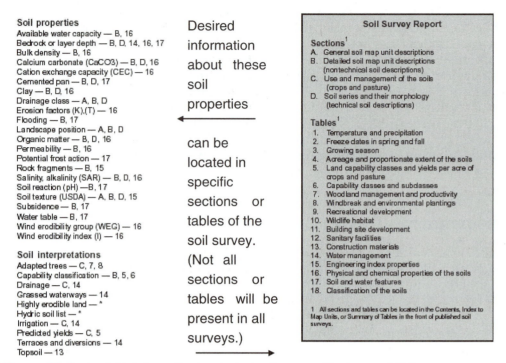

Soil properties
Available water capacity — B, 16
Bedrock or layer depth — B, D, 14, 16, 17
Bulk density — B, 16
Calcium carbonate (CaCO3) — B, D, 16
Cation exchange capacity (CEC) — 16
Cemented pan — B, D, 17
Clay — B, D, 16
Drainage class — A, B, D
Erosion factors (K),(T) — 16
Flooding — B, 17
Landscape position — A, B, D
Organic matter — B, D, 16
Permeability — B, 16
Potential frost action — 17
Rock fragments — B, 15
Salinity, alkalinity (SAR) — B, D, 16
Soil reaction (pH) —B, 17
Soil texture (USDA) — A, B, D, 15
Subsidence — B, 17
Water table — B, 17
Wind erodibility group (WEG) — 16
Wind erodibility index (I) — 16

Soil interpretations
Adapted trees — C, 7, 8
Capability classification — B, 5, 6
Drainage — C, 14
Grassed waterways — 14
Highly erodible land — *
Hydric soil list — *
Irrigation — C, 14
Predicted yields — C, 5
Terraces and diversions — 14
Topsoil — 13

Desired information about these soil properties

← can be located in specific sections or tables of the soil survey. (Not all sections or tables will be present in all surveys.) →

Soil Survey Report

Sections[1]
A. General soil map unit descriptions
B. Detailed soil map unit descriptions (nontechnical soil descriptions)
C. Use and management of the soils (crops and pasture)
D. Soil series and their morphology (technical soil descriptions)

Tables[1]
1. Temperature and precipitation
2. Freeze dates in spring and fall
3. Growing season
4. Acreage and proportionate extent of the soils
5. Land capability classes and yields per acre of crops and pasture
6. Capability classes and subclasses
7. Woodland management and productivity
8. Windbreak and environmental plantings
9. Recreational development
10. Wildlife habitat
11. Building site development
12. Sanitary facilities
13. Construction materials
14. Water management
15. Engineering index properties
16. Physical and chemical properties of the soils
17. Soil and water features
18. Classification of the soils

1 All sections and tables can be located in the Contents, Index to Map Units, or Summary of Tables in the front of published soil surveys.

FIGURE 15-3   Available information in a typical soil survey. *(Broderson, 2000)*

In addition, you can find useful information about specific items such as:

❖ Suitability ratings for engineering projects.

❖ Suitability ratings for water management projects such as building reservoirs or installing drainage.

❖ Suitability ratings for recreational development.

❖ Potential for cropping and typical yields.

❖ Woodland suitability and potential trees adapted to the soils.

❖ Potential for wildlife habitat.

FOCUS ON . . . SOIL SURVEYS AND MAPS

1. Who has authority for collecting and organizing soil survey data in the United States?

2. What was the scale of early soil survey maps? How were the different soil series represented?

3. What is one of the biggest differences between soil survey maps prior to 1935 and those of today?

4. What is a mapping unit?

5. What feature of soil associations makes them useful for developing generalized soil maps?

6. What are the basic components of a soil survey report?

7. What kinds of information are found in a soil survey report?

8. What key information do you typically find in a mapping unit code?

SOIL USE CLASSIFICATION

The lowest level of classification used by the NRCS is the soil series. However, the NRCS also classifies soils at a higher level of classification, distinct from a taxonomic classification, called a **land capability class**. The land capability class will group soils on the basis of similar hazards and limitations for use such as their erosion hazard. This type of functional classification may group soils of different taxonomies together. It is more subjective than a soil series classification, but more practical from the perspective of actual land use.

> The NRCS also classifies soils in terms of land capability class.

Types of Land Use Classification

There are eight land use classifications:

> Land use classes range from I to VIII, with Class VIII having the most restrictions.

* ❖ Class I: These lands can be cultivated safely with long-term productivity and good yields for adaptable crops without the need for special practices or treatments.

* ❖ Class II: These lands cannot be cultivated with long-term productivity to produce moderate to good yields unless some simple practices or treatments are made.

* ❖ Class III: The lands require extensive practices or treatments such as contour cultivation, strip cropping, terracing, tile drainage, fertilization, or systematic rotation. The practices listed indicate the sort of limitations these lands have—steeper slopes and potential erosiveness, drainage problems, and low native fertility and/or CEC.

* ❖ Class IV: These lands cannot be cultivated safely under any plan for continuous use, but they can be used safely for hay and pasture, which retain a continuous soil cover.

* ❖ Class V: These lands cannot be cultivated safely at any time and are only suitable for permanent cover.

* ❖ Class VI: These lands have extreme limitations that restrict their use to pasture, range, woodland, or wildlife.

* ❖ Class VII: These lands have even more severe limitations than Class VI soils and are primarily restricted to grazing, woodland, and wildlife.

* ❖ Class VIII: No plant production should occur on these lands. They should be reserved for recreation, wildlife, plant supply (seed banks), and aesthetic purposes (greenspace).

Figure 15-4 illustrates how these lands may be positioned within a landscape. You can see how plant use varies (cultivated row crops in the foreground shading into pasture and forest in back) and you can also see the major limitation that controls land use classification at this site: slope.

There are also four land capability subclasses that reflect specific limitations to each soil group (Table 15-3). Land use capability units (as opposed to soil-mapping units) have their own distinct code, which is made by adding the subclass code to the capability class designation. For example, a Class II capability soil that is most limited by the potential for erosion would be designated IIe; one that was most limited by moisture would be designated IIw. Table 15-4 illustrates how the soil-mapping units for Mason County, Kentucky are further classified by land use capability.

> There are four land use capability subclasses that reflect specific restrictions, such as erosion potential.

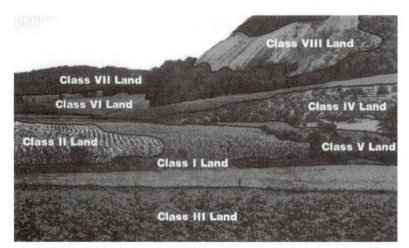

FIGURE 15-4 A representation of the land use classifications. *(Courtesy of the USDA-NRCS)*

TABLE 15-3 Land capability subclasses.

Code	Description
e	Eroded Existing or potential erosion and runoff Lands with slopes > 2% Some form of runoff control is needed
w	Wetness Poorly drained or occasionally flooded Some soils may be drained, some have been prior converted Some soils are statutory wetlands and must be maintained as such
s	Shallow Shallow root zone and tillage problems Soils tend to be stony, droughty, infertile, or saline Some potential for wind and water erosion
c	Climate Rainfall or temperature extremes make farming difficult

The NRCS has a color-coding system for land use capability that it uses during the preparation of conservation plans:

❖ Class I, Light Green
❖ Class II, Yellow
❖ Class III, Red
❖ Class IV, Blue
❖ Class V, Dark Green or White
❖ Class VI, Orange
❖ Class VII, Brown
❖ Class VIII, Purple

TABLE 15-4 Land use capability designations of soil-mapping units from Mason Co., Kentucky.

Soil Name	Map Symbol	Land Use Capability
Beasley	BaB	IIe
	BaC2	IIIe
	BeE3	VIe
Boonesboro	Bo	IIIw
Chavies	ChB	IIe
	ChC	IIIe
Dumps	Du	VIIIs
Eden	EdD2	IVe
	EfE2	VIIe
Elk	EkB	IIe
	EkC	IIIe
Fairmount	FrF	VIIe
Faywood	FwB	IIe
Lowell	LoB	IIe
	LoC	IIe
	LoD	IVe
Nicholson	NcB	IIe
Nolin	No	IIw
Otwell	OtB	IIe
Pits	Pt	VIIs
Wheeling	WhA	I
	WhC	IIIe

If soil-mapping units are coded according to their land capability class, a very different picture emerges of potential land use, in which much of the land has only minor restrictions on use, and problem areas for crop growth are clearly delineated (Figure 15-5).

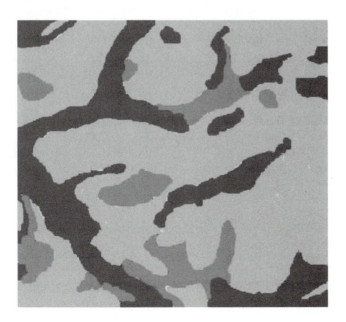

FIGURE 15-5 Mapping of soil in terms of land use capability. The darkest shades are Class III soils. The lightest shades are Class II soils. Intermediate shades are Class I and Class IV soils.

Soil Properties

Agronomic Use	Organic Matter	Flooding	Texture	Bedrock or Pan	pH	Subsidence	CEC	CaCO₃	Bulk Density	Permeability	Frost Potential	Available Water	Salinit/Alkalinity	Water Table	Wind Erosion	Erosion Factors K, T	Slope
Tillage	✓	✓	✓	✓					✓	✓				✓			✓
Erodibility Wind Water	✓		✓					✓	✓	✓				✓	✓	✓	✓
Irrigation		✓	✓	✓	✓	✓		✓	✓	✓		✓	✓	✓	✓	✓	✓
Drainage		✓	✓	✓		✓			✓	✓				✓			✓
Productivity	✓	✓	✓	✓	✓		✓	✓	✓		✓	✓	✓	✓			✓
Conservation Practices	✓	✓	✓	✓	✓	✓	✓	✓	✓	✓		✓	✓	✓	✓	✓	✓
Land Use Capability		✓	✓	✓			✓		✓			✓	✓	✓	✓	✓	✓
Plant Suitability	✓	✓	✓	✓	✓		✓	✓	✓	✓	✓	✓	✓	✓			✓

✓   Indicates that the soil properties listed in the soil interpretations data base affect the selected agronomic concerns.

FIGURE 15-6   Soil survey information that will influence agronomic use. *(Diagram adapted from Broderson, 2000)*

Soil Properties Affecting Land Use

Part of constructing the soil survey involves evaluating the physical and chemical properties of the soil, particularly with respect to their effect on agricultural activity. An outline of some of these measured soil properties is given in Figure 15-6.

Available water-holding capacity is important because it influences the amount of water plants can acquire during the growing season. Water-holding capacity is influenced by texture, organic matter content, compaction, and the depth to restrictive layers. There are at least eighteen types of restrictive layers recognized in soil surveys. Soils are considered restricted because they physically impede root growth or because they inhibit root growth. An example of the latter are saline or alkaline layers in soil.

Examples of soil layers that physically impede root growth are cemented pans, permafrost, fragipans, clay pans, and plowpans. Changes in soil texture and bulk density can also affect root growth. Table 15-5 provides some guidance for the maximum bulk density that should occur in different soil textures before root penetration is affected.

The soil survey recognizes five different categories of soil depth to bedrock from very shallow (< 10 inches, 25 cm) to very deep (> 60 inches, 152 cm; Table 15-6). However, the effective rooting depth can be signifi-

Some of the properties measured for land use are water-holding capacity, restrictive layers, and depth to bedrock.

TABLE 15-5 Root restriction guide for soil classification based on texture and bulk density characteristics.

Applicable Textures	Average Bulk Density (g/cm³)
Coarse sand, loamy coarse sand, loamy sand, fine sand, loamy fine sand	> 1.85
Very fine sand, loamy very fine sand, fine sandy loam, coarse sandy loam, very fine sandy loam, sandy loam, loam with $< 18\%$ clay	> 1.80
Loam, sandy clay loam, clay loam that has 18–35% clay	> 1.70
Silt, silt loam, silty clay loam that has $< 35\%$ clay	> 1.60
Clay loam, sandy clay, clay, silty clay loam, silty clay with 35–39% clay ($> 30\%$ in ertisols)	> 1.50
Clay soil with $> 60\%$ clay (except Vertisols)	> 1.36

TABLE 15-6 Designation of depth classes for land use classifications.

Depth (in.)	Designation
< 10	Very shallow
10–20	Shallow
20–40	Moderately deep
40–60	Deep
>60	Very deep

cantly affected by any of these other restrictive layers, which can play a critical role in such things as the siting of on-site waste disposal.

The calcium carbonate ($CaCO_3$) content, as you saw in Chapter 14, influences pH, and sensitive plants can be affected by micronutrient deficiencies (or toxicity in the case of Mo) when there is as little as 0.5 to 2.0 percent $CaCO_3$ in soil.

There are seven natural drainage classes used in soil surveys that range from very poorly drained to excessively well-drained (Table 15-7). These drainage classes are closely related to **permeability** rates in soils, which range from impermeable to very rapid (Table 15-8). Drainage class is an important feature for determining whether soils are suitable for on-site waste disposal from septic systems (Figure 15-7).

Another classification having to do with water is the frequency and duration of flooding that occurs in particular soil groups (Table 15-9). The flooding can be rare or frequent and the duration short- or long-term.

TABLE 15-7 Drainage classes used for land use characterization.

Class	Description
Very poorly drained	Water table at or near the soil surface most of the year Histic epipedons Too wet to support most crops Obvious gleying
Poorly drained	Usually wet Water table close to the soil surface much of the year Obvious gleying
Somewhat poorly drained	Wet for significant periods Mottles in lower A, B, and C horizons A horizon can be thick Crop growth possible, but improves with drainage
Moderately well-drained	Wet for a small but significant part of the year Influences crop choice, timing of management Mottles restricted to B horizon
Well-drained	No water table in the soil profile No mottles in the soil solum Optimal for plant growth
Somewhat excessively well-drained or excessively well-drained	Sandy and/or gravely Poor water-holding capacity No mottling, but plant productivity limited by available water

TABLE 15-8 Permeability classes for land use characterization.

Class	Permeability Rate	
	in./hr	cm/min
Impermeable	< 0.0015	6.35×10^{-5}
Very slow	0.0015–0.06	6.35×10^{-5}–0.003
Slow	0.06–0.20	0.003–0.008
Moderately slow	0.20–0.60	0.008–0.025
Moderate	0.60–2.0	0.025–0.085
Moderately rapid	2.0–6.0	0.085–0.254
Rapid	6.0–20	0.254–0.85
Very rapid	> 20	> 0.85

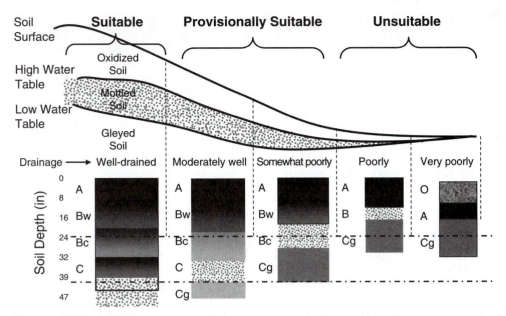

FIGURE 15-7 A hydrosequence showing the change in depth of mottling layers as water table depth changes. Suitability refers to suitability for an engineering function such as placement of an on-site septic system.

TABLE 15-9 Designation of flooding frequency and duration classes for land use classification.

Classification	Description
Flooding frequency (per 100 years)	
None	Near 0
Rare	0–5
Occasional	5–50
Frequent	> 50
Flooding duration	
Very brief	< 2 days
Brief	2–7 days
Long	7 days to 1 month
Very long	> 1 month

FOCUS ON . . . SOIL USE CLASSIFICATION

1. In what ways does soil use classification differ from taxonomic classification?

2. How many different land use classifications are there? Which are the most restrictive?

3. What do land use subclassifications indicate?

4. Does mapping on the basis of land use classification create more, or less, heterogeneity of mapping units?

5. What sorts of information contained in the soil survey are useful for agronomic purposes?

SUMMARY

What is the value of soil mapping and classification? As you have seen, mapping provides an inventory of where soils are in relationship to one another. Land use classification provides a better sense of a landscape's capability, which is vital for appropriate management and determining whether a particular location is suitable for an intended purpose. Land use classification is useful for revenue because potentially more productive lands can be taxed at a higher rate.

In this chapter you examined how the soil survey evolved and some of the very basics of creating a survey. You examined the different types of mapping units that will be found in a soil survey and how to interpret the codes associated with mapping units. In addition to the information on mapping units, you looked at other types of information that are found in a typical soil survey, which range from a taxonomic description of each soil series in the survey to an evaluation of potential agronomic and engineering uses of each soil.

The second part of this chapter was devoted to land use classification, which is a parallel system for evaluating soils that the USDA-NRCS employs. Land use classification may collect various taxonomically different soils together because it bases associations on criteria such as susceptibility to erosion, drainage, and soil depth. Of the eight land use classifications employed, the first three classes (I–III) are suitable for row crop production with increasing restrictions. The next two classes (IV and V) are suitable for pasture and grazing. The last three classes (VI–VIII) have such severe restrictions on use that they should only be used recreationally or for forest and wildlife.

Much of the soil survey information is devoted to soil characteristics that will affect agronomic use. This information for a particular location is extremely useful in terms of determining management schemes for a particular soil. The data is collected in various tables found in the soil survey that allow different soils to be compared.

In the next chapter you will examine the perception of soil as a commodity, and how social and environmental changes have influenced the way people deal with soils.

END OF CHAPTER QUESTIONS

1. What are several reasons for conducting soil surveys?

2. Published soil surveys after 1935 are very different from earlier surveys for two reasons. Can you name them?

3. How has the scale of published soil surveys changed with time?

4. Is a soil surveyor expected to map every different soil in a landscape?

5. How do toposequences differ from catenas?

6. What is useful to know about soil associations?

7. How much of a landscape can consist of inclusions before they have to be mapped?

8. What are the basic tools of a soil mapper?

9. What knowledge about soil formation helps a soil mapper reduce the number of soil probes that are made?

10. What's the smallest unit described in a soil survey?

11. In the code LyD3, what does each component of the code indicate?

12. Which soil, LoB, LoC, or LoC2, has the steepest slope? Which is most eroded?

13. Does erosion change soil classification?

14. If the soil survey indicates a soil is alluvial, what should you predict about the slopes of soil series on either side of it?

15. Should you grow corn on Class IV soil? Why or why not?

16. What do the subclassification terms for land use classification tell you?

(The following questions refer to Table 15-4.)

17. According to land use classification, what is the best soil in Mason County, Kentucky?

18. Which soil series are suitable for row crop production?

19. Which soil series are only suitable for pasture or grazing?

20. What feature of land use capability in these soils probably most limits its use?

21. What are the most likely major problems with the Pits soil series in terms of land use?

(The following questions refer to Tables 15-5 to 15-9.)

22. If the bulk density exceeds 1.7 in a loam soil, should you be concerned? Why or why not? What other soil properties in this soil might be affected?

23. Is a soil with a depth of 40 inches to bedrock considered very deep? If not, what should be its classification?

24. If you observe mottles in the lower A, B, and C horizons, what would you predict is the drainage class of this soil?

25. Explain how a soil can be too well-drained.

26. How does the position of the soil in a landscape affect its drainage? (See Figure 15-12 for clues.)

27. If the permeability of a soil is moderately slow, will 2 inches of rain likely cause runoff? Why or why not?

28. Soils that have occasional or frequent flooding are likely to be what type of soils?

29. If the duration of flooding in a soil exceeds one month, what might be its land use classification, and to what use might you put it?

30. For soils that have a flooding duration greater than seven days, what color might you expect to see in the subsoil?

FURTHER READING

A very short and succinct introduction to the topics in this chapter can be found in the USDA-NRCS publication *From the Surface Down: An Introduction to Soil Surveys for Agronomic Use* by W. Broderson (2000). Many of the figures from this chapter were drawn from this source.

REFERENCES

Broderson, W. 2000. *From the surface down. An introduction to soil surveys for agronomic use.* Washington, DC: USDA-NRCS.

Harpstead, M. I., T. J. Sauer, and W. F. Bennett. 1997. *Soil science simplified,* 3rd ed. Ames: Iowa State University Press.

Troeh, F. R., and L. M. Thompson. 1993. *Soils and soil fertility,* 5th ed. New York: Oxford University Press.

USDA-NRCS. 1968. *Soil survey: Fayette Co. Kentucky.* Washington, DC.

USDA-NRCS. 1988. *Soil survey of Mason County Kentucky.* Washington, DC.

USDA-NRCS. 2004. *National soil survey handbook.* Washington, DC.

Chapter 16

SOIL AS A NATURAL RESOURCE

The Ancient Rule—
"Whatever is affixed to the soil belongs to the soil."
Anglo-Saxon law

". . . the soil of any one place makes its own peculiar and inevitable
sense. It is impossible to contemplate the life of the soil for very long
without seeing it as analogous to the life of the spirit."
Wendell Berry, *The Unsettling of America*, 1977

OVERVIEW

In the preceding fifteen chapters you examined soil from the perspective of its formation and classification, structure and physical properties, chemical and biological activity, management, and evaluation. In this final chapter you will try to generate a holistic framework for the soil ecosystem with respect to its conservation and management as a true natural resource. You will focus on the roles that soil has played in society and resource conservation, and how changing perceptions of the use and value of soil have affected its management. You will end by examining how public legislation is critical in the evaluation and management of soil.

Perceptions of the use and management of soil have changed with time.

OBJECTIVES

After reading this chapter, you should be able to:

✔ Identify six ecosystem roles for soils.

✔ Draw a timeline of events important with respect to soil use and management.

✔ Discuss how public perception of the soil resource has changed with time.

✔ Provide examples of specific legislative acts that have influenced soil resource management.

✔ Indicate how soil resource conservation is being implemented at state and local levels.

KEY TERMS

C sequestration
Crop Reserve Program
  (CRP)
ecosystem

Federal Agriculture
  Improvement and
  Reform Act (FAIRA)

Food, Agriculture,
  Conservation, and
  Trade Act (FACTA)
Food Security Act
  (FSA)
takings

SOIL AS A NATURAL RESOURCE

What is soil? You have looked at many definitions of soil, but one that puts soil into an ecosystem perspective is as follows:

> *Soil is a complex dynamic ecosystem, sustaining physical processes and chemical transformations vital to terrestrial life.*

| Soils can have many different roles.

You have already seen that soil can be used for various purposes apart from its role in growing things. And even from that perspective, the soil is less a medium for plant growth than it is a resource from which humans use plants to extract nutrients and water. Soils have roles in road construction, buildings, dams and levees, bricks, ponds, and canals (as sealants). Soil is used to construct trails, to insulate, to decompose waste and to store waste, and as a source of raw materials. Soil has been used in painting, sculpture, and personal decoration.

| With soil, short-term abuse can mean long-term recovery.

You know that soil is a renewable resource based on Hans Jenny's model that soil forms as a consequence of (and in response to) parent material, time, climate, vegetation, relief, and human influences. Soil formation is a slow process. It may take 100 to 200 years to form productive topsoil and 200 to 1000 years to regenerate 2.5 cm of lost topsoil. So, to the extent that vegetated land surface has degraded, it may take many generations to recover the soil and the functions of soil without outside assistance.

Ecosystem Functions

| Soils perform six critical ecosystem functions.

The purposes for which soil is typically used pale compared to the role that soils play in global, regional, and local **ecosystems**. There are six critical ecosystem functions that soils carry out, and which, from the perspective of human populations, are vastly more important than the uses of soil as a raw material (Daily et al., 1997). Some of them you have already recognized.

1. Soils buffer and moderate the hydrological cycle.
2. Soils provide the physical support of plants.
3. Soils retain and deliver plant nutrients.
4. Soils dispose of (recycle) wastes and organic material.
5. Soils renew fertility.
6. Soils regulate major element cycles.

Keep these ecosystem functions in mind throughout the rest of the chapter. You typically think of soil from the perspective of its role in agriculture. However, when society loses its perspective of other roles that soils play, it leads to environmental degradation.

Focus on . . . The Different Roles of Soils

1. How do you define soil?

2. In what ways is soil a renewable or nonrenewable resource?

3. Which ecosystem functions of soil are directly related to agriculture, and which to environmental protection?

Historical Reflections on Soil and Soil Degradation

How old is the recognition of soil as a resource and recognition of the means to protect it?

Colonial America to the Dust Bowl Era

The mid 1700s to mid 1800s was a period of early American conservationists who recognized:

❖ soil-building crops

❖ effects of liming

❖ contour plowing

❖ healing of gullies

❖ wind erosion

For example, in 1813 Thomas Jefferson, former president of the United States, recommended horizontal (i.e., contour) plowing as a mechanism to reduce soil erosion. Edmund Ruffin of South Carolina recognized the mechanism of exchangeable calcium in liming by 1822, and his 1832 treatise "An Essay on Calcareous Manures" was a widely read and distributed work.

One hundred years later, however, the effects of soil neglect and exploitation were obvious, and in 1928 Hugh Bennett's work "Soil Erosion, a National Menace," as well as work by W. C. Lowdermilk through USDA circulars and education programs, began to lay the groundwork for public support of soil erosion control.

The defining feature of this era was the "Dust Bowl"—a period of massive wind erosion in the central United States caused by the confluence of several events, notably drought, management practices that promoted wind erosion, and low commodity prices, which led to utilization of marginal lands for production.

> Perceptions of soil have changed markedly with time.

> The Dust Bowl epitomizes the dangers of soil erosion.

Development of Soil Conservation Practices/Policy

You could argue whether there was a national policy toward conserving soil resources prior to 1930. There were certainly different periods of thought about conservation, however, that were driven by national demographic changes and technological advances.

In the twentieth century cultivated land has decreased. By 1900 nearly all arable lands except for irrigated land were under tillage. In 1930, for example, 230 million hectares were cultivated, but by 1993 this had

EDMUND RUFFIN—SOIL CHEMIST AND REBEL

Edmund Ruffin (1794–1865) has the distinction of being one of the pioneer agricultural reformers of the south in the early nineteenth century. He also holds the reputation of having fired the symbolic first and last shots of the Civil War. Ruffin was born in Virginia in 1794, and briefly attended the College of William and Mary. He subsequently began performing on-farm experiments in 1818 with fertilizer, rotation, drainage, and tillage that convinced him that soil management could restore declining soil fertility. In particular, he identified the beneficial effects of liming soil with marl (calcareous deposits consisting of shells), which he published in a best-selling book, *An Essay on Calcareous Manures* (1832). Ruffin also published a respected agricultural journal, *The Farmer's Register,* from 1833–1842, and in the 1850s he was president and commissioner of the Virginia Agricultural Society.

By the 1850s Ruffin had become a radical secessionist, and when the Civil War began in April 1861, Ruffin was present in Charleston, South Carolina to fire some of the opening shots at Fort Sumter. On June 17, 1865, sick and despondent at the victory of the Union, and in danger of being imprisoned by Union forces, Ruffin shot himself.

Edmund Ruffin (1794–1865), regarded as the father of U.S. soil chemistry because of his pioneering work with liming agents.

---

decreased to less than 160 million hectares. At the same time farm size has increased while the number of farm families, or full-time farmers, has declined precipitously.

Tillage practices have dramatically changed. The moldboard plow opened vast stretches of native grasslands to cultivation at the turn of the century, but agriculture is increasingly turning to conservation tillage practices, such as no-till, to conserve plant residue on the soil surface that moldboard plowing would bury. Irrigated agricultural land has increased whereas water quality and availability have decreased.

Social and technological advances have affected soil conservation practices.

One of the turning points in conservation thought was in 1901 when President Theodore Roosevelt suggested that conservation (of all sorts, not just soil conservation) was not a question of political management but a scientific endeavor. As you have seen, however, it was the 1930s Dust Bowl that created a sense of urgency about soil conservation (Figure 16-1). It is one thing to talk about soil loss as an abstract idea and quite another to see the soil blowing away around you.

There were some notable voices for soil conservation in this period. In 1939 Hugh Hammond Bennett wrote an essential treatise on soil conservation, and his evangelism for soil conservation led to formation of the Soil Conservation Service (SCS) in 1935 and the Soil and Water Conservation Society in 1946. W. C. Lowdermilk published "Conquest of the Land through 7000 Years" in 1953 to illustrate how land use had led to soil erosion.

(a)

FIGURE 16-1   Scenes from the Dust Bowl era. (a) A great 'roller' moves across the land in Colorado *(Photograph courtesy of the USDA-NRCS);* (b) Farmer and sons walking in the face of a dust storm. Cimarron County, OK *(Photograph courtesy of the USDA-NRCS);* (c) Wind-devastated farmland in Kansas during the Dust Bowl *(Photograph courtesy of the USDA-NRCS).*

(b)

(c)

HUGH HAMMOND BENNETT—FATHER OF U.S. SOIL CONSERVATION

Hugh Hammond Bennett (1881–1960) was born in North Carolina and graduated from the University of North Carolina. He was a soil surveyor for the USDA, and his work nationally and internationally convinced him of the threat that soil erosion posed to the prosperity of U.S. agriculture. This put him at odds with prevailing opinions in the Bureau of Soils that soil was an essentially inexhaustible resource.

Bennett was evangelical in his support of soil conservation, and not above blaming U.S. farmers for their role in soil erosion:

"Americans have been the greatest destroyers of land of any race or people, barbaric or civilized. . ." . (H. H. Bennett)

His 1928 book, *Soil Erosion: A National Menace,* was pivotal in developing support for conservation efforts. Bennett also had the good fortune to testify in front of Congress on Black Sunday, April 14, 1935, when wind-blown soil from the southern Great Plains blackened the sky in much of the Midwest, and also deposited dust as far away as Washington, DC.

The subsequent Soil Conservation Act of April 27, 1935 created the Soil Conservation Service (now NRCS) and Bennett served as its head from 1935 until his retirement in 1953. Part of H. H. Bennett's legacy are the approximately 3000 conservation districts nationwide organized to prevent the erosion he spent his career fighting.

Aldo Leopold, a pioneering conservationist, argued that the soil is not a commodity, and used the analogy that agricultural development, while it could improve the pump, could not improve the well. The urgency to conserve soil ran headlong into the need for food production in support of war efforts in the 1940s and 1950s. In addition, cheap fertilizers and hybrids able to exploit the additional fertilizer made it appear that agricultural practices were minimally affecting soil because crop yields were constantly increasing.

Agricultural productivity based on external inputs like fertilizers showed its limitations, however, and in the 1960s and 1970s environmentalism grew in response to soil and water contamination. One of the culminating events in this period was the passage of the Clean Water Act in 1972. Although the focus of the initial act was point sources of pollution (such as the direct discharge of wastes from industries and sewage treatment plants into water), the focus has since been on nonpoint source pollution (sources of pollution for which direct sources are not obvious or are widespread), for which agricultural activities are often the main culprit.

> The 1972 Clean Water Act was landmark legislation for environmental protection.

National agricultural policy with respect to soil conservation has subsequently become much more direct. In the 1985 and 1990 Food Security Acts, for example, producers with highly erodible land (HEL) were required to follow farm conservation plans if they were to participate in USDA agricultural programs. And in 1997, Idaho courts required establishment of Total Mean Daily Loads (TMDL) limits for 962 water quality impaired streams with a 2002 compliance date. The TMDL is now a standard regulatory practice for agricultural activities that could potentially pollute rivers and streams.

FOCUS ON . . . DEVELOPMENT OF SOIL CONSERVATION ETHICS

1. Would Edmund Ruffin, Hugh Bennett, and Aldo Leopold have agreed on the concept of soil as a commodity?

2. Why does the Dust Bowl era act as a watershed in soil conservation?

3. How can technological advances increase soil erosion and mask soil degradation?

4. What is the distinction between point and nonpoint source pollution?

5. How do changes in national policy affect agricultural producers?

DEVELOPMENT OF NATIONAL POLICIES RELATED TO SOILS

> Five periods in conservation policy can be described.

Did the federal government have a policy before the Dust Bowl era? Five periods in conservation policy have been described (Smith, 1994):

1. Pre-1930: Era of apparent natural abundance.
2. 1930–1960: Era of resource conservation.
3. 1960–1985: Era of diverging agricultural and environmental policy.
4. 1985–1995: Era of consistent production/conservation policy.
5. 1995–Present: Stewardship transition.

Pre-1930: Era of Apparent Natural Abundance

The United States Department of Agriculture (USDA) was established in 1862, but the 1862 Land Grant College Act (Morrill Act) was the seminal moment in the nation's approach to natural resources like soil. The Morrill Act was the first federal legislative act influencing soil and water resource and education programs. The Act encouraged states to establish colleges of agricultural and mechanical arts by granting each state 30,000 acres of public land for every member of congress it had. In 1890 the Morrill Act was extended to include traditionally black colleges (such as Kentucky State University and Tuskegee Institute).

Because of its emphasis on practical and applied arts and sciences, the Morrill Act fostered populist rather than elitist education for a nation that was still largely rural and agricultural. The public lands of the United States were still regarded as boundless. The 1862 Homestead Act subsidized private ownership of federal land to encourage population dispersal and create economic development in the west, and the 1887 Hatch Act established state agricultural experiment stations to assist that development. In the 1902 Reclamation Act, the federal government subsidized water resource development to provide irrigation water at "below cost." The 1914 Smith Lever Act established the Extension Service, which facilitated transfer of research to farmers.

The sum of these acts was essentially to create a better educated rural population with economic and technical assistance. The acts helped utilize abundant land resources of a nation only recently emerging from civil war and eager to exploit opportunities in open territory (views not held by Native Americans in the same territories, who had quite different approaches to land use). A consequence of overexploitation was soil degradation. This opinion from the Bureau of Soils in 1909 probably sums up the commonly held attitudes toward soil at the time:

> *"The soil is the one indestructible, immutable asset that the nation possesses. It is the one resource that cannot be exhausted; that cannot be used up." (http://www.nrcs.usda.gov)*

> The Morrill Act made populist college training in technical and applied sciences possible.

> A commonly held attitude was that soil was an inexhaustible resource.

1930–1960: Era of Resource Conservation

By the 1930s the effects of soil degradation and erosion were obvious, and in 1930 Congress made a $16,000 appropriation for research into the cause and prevention of soil erosion. The Soil Erosion Service was established within the Department of the Interior in 1933. Soil erosion was defined as "a threat to future well-being of society."

In 1935 the Soil Conservation Service (SCS) was established as a permanent agency of the USDA with Hugh Hammond Bennett named as its chief. The SCS consisted of a research division and ten erosion experiment stations. As had been done in the previous era, federal policy first created institutional support systems for farmers, and then enacted policies to influence their activity.

In the 1936 Soil Conservation and Domestic Allotment Act payments were made to farmers willing to adopt less erosive agricultural production systems, and there was economic assistance to shift production. Critics described it as a thinly veiled form of subsidy for the farm community.

The 1940s Agricultural Conservation Program (ACP) consisted of several set-aside programs before and after World War II. Unfortunately, the diversion of cropland from production was made without consideration for the environment—all farmland was eligible—and it allowed retirement of

> The SCS was founded in 1935. H. H. Bennett was its first director.

nonerodible land. Critics of the program argued that erosion control was secondary to commodity support.

For example, the 1956 Soil Bank was a federal program designed to control the supply of agricultural commodities that was justified to taxpayers on the basis of conservation. Payment was made to limit production of specific commodities, but there was no requirement that the land be erodible. Consequently, enrollment occurred on the least productive land and intensification of production occurred on other land, which increased degradation. No provision was made to ensure that retired land was returned to conservation practices.

> The early Soil Bank programs made no provision for permanent conservation practices.

1960–Present: Era of Diverging Agricultural and Environmental Policy and Stewardship

> Public perception about environmental quality has shifted considerably in the last forty years.

Publication of books such as *Silent Spring* by Rachel Carson, the shift of populations from rural to urban settings, and the decrease in the number of individuals who earned their livelihood through farming have been instrumental in causing a distinct shift in public perception of environmental quality during the past forty years. With the 1982 National Conservation Program (NCP), reducing soil erosion became the USDA's highest natural resource conservation priority. The full economic consequences of off-site damage were also increasingly recognized. In 1994, the USDA was reorganized and the SCS was renamed the Natural Resources Conservation Service (NRCS) to reflect its broader role and the broader role of soil conservation in environmental quality.

> The SCS was renamed the NRCS to reflect its broader conservation mission.

Focus on . . . Periods of Conservation Thought

1. How did the Morrill Act shape public policy toward soil as a resource?
2. Why were public policies on land use in the 1940s and 1950s sometimes at cross-purposes with soil conservation?
3. How does changing the name of the SCS to the NRCS reflect a national change in conservation ethics?

Federal Acts Relating to Soil

> Before 1985 most approaches to conservation were voluntary and nonregulatory.

Several specific acts have had a significant influence on the way that soils are managed in the United States and illustrate how public policy toward soils has evolved with time. From 1933 to 1985, for example, the most significant characteristics of public policy were:

❖ voluntary participation
❖ nonregulatory approach
❖ absolute property rights to landowners
❖ research funding for erosion and national soils inventories

A variety of cropland reduction programs occurred from 1933–1985, such as:

- ❖ Conservation Adjustment Program
- ❖ Agricultural Conservation Program
- ❖ Soil Bank
- ❖ Cropland Adjustment Program
- ❖ Acreage Reduction Program

Prior to 1985 there were basically two schools of thought with respect to conservation. In the "education and stewardship" school:

- ❖ Landowners were assumed to be ignorant of adverse effects.
- ❖ The goal was to provide information that would lead to modified practices.
- ❖ Participation was voluntary.

In contrast, the "economic incentive" school held that:

- ❖ Landowners were fully aware of erosive practices.
- ❖ Erosive practices were profitable even in the long term with existing production systems.
- ❖ Economic barriers to adaptation had to be eliminated before conservation would be adopted.

Several federal acts beginning in 1977 with the Surface Mining Control and Reclamation Act (SMCRA) and the Resource Conservation Act (RCA), as well as the Prime Farmland Protection Policy Act (in the 1981 Farm Bill), were indicative of the way policy towards soil conservation was changing in the United States. The 1977 Surface Mining Control and Reclamation Act provided a definition for prime farmland and mandated that states adopt reclamation procedures to restore productivity of mined soil to its "original" state.

In the Resource Conservation Act (1977) the Secretary of Agriculture was directed to conduct a continuing appraisal of soil and water resources on nonfederal land in the United States. Programs for soil and water conservation and resources were mandated to respond to the needs of the country as revealed in the appraisal. Appraisals were conducted in 1981, 1989, and 1997, and policy leaders were asked to quantify potential productivity loss due to soil loss. Attention focused on off-site losses that were assumed to be twice as costly as on-site productivity losses. For example, in the cornbelt, productivity losses to soil erosion were estimated to be $425 million, while off-site losses were estimated to be $1.023 billion. One issue that arose was whether the T value (tolerance value, which is a maximum of 11.2 tons/acre/yr) was too lenient for off-site losses.

In the Prime Farmland Protection Policy Act (contained in the 1981 Farm Bill) federal agencies were asked to identify and account for adverse effects of programs on preservation of prime or unique farmland, and federal agencies were told to consider alternatives to lessen impacts. Federal agencies were also directed to ensure, where possible, that programs were compatible with state, local, and private programs and policies to protect prime and unique farmland.

The 1985 **Food Security Act (FSA)** marked a fundamental change in national policy with respect to soil resources. The 1985 FSA legitimized nonfarm population influences on farm production systems *without the use of incentives.* It introduced cross-compliance (the loss of benefits from one

---

Two schools of thought with respect to conservation were "education and stewardship" and "economic incentives."

Policy changes in the way soil conservation was approached could be seen by 1977.

The 1985 Food Security Act was significant because it regulated farm production without providing incentives.

federal program if the needs of another were not met). Among the types of programs that farmers could lose for noncompliance with conservation practices were:

* ❖ commodity price support
* ❖ commodity loans
* ❖ crop insurance
* ❖ disaster relief
* ❖ CRP—annual rental and half the cost of establishing permanent cover
* ❖ farmer home loans

| The 1985 FSA defined highly erodible land. |

The 1985 FSA also defined highly erodible soil and utilized the "T value" as the goal to be achieved. It included the "Swampbuster" title to protect wetlands and the "Sodbuster" title to protect native prairies. It created the **Crop Reserve Program (CRP)**, a modified version of the Soil Bank, in which land is taken out of production for a ten-year period. An important difference with this legislation, however, is that the land retired from production has to be erodible, and the land returned to production from the CRP has to comply with conservation practices.

| FACTA identified the meaning of sustainable agriculture. |

FACTA, the **Food, Agriculture, Conservation, and Trade Act** (1990) (Section 1603, Title 16) made sustainability an issue in policy. What was meant by *sustainable?* Sustainable agriculture was defined as "an integrated system of plant and animal production practices having a site-specific application that will, over the long term":

1. Satisfy human food and fiber needs.
2. Enhance environmental quality and the natural resource base on which the agriculture economy depends.
3. Make the most efficient use of nonrenewable resources and on-farm resources and integrate, where appropriate, natural biological cycles and controls.
4. Sustain the economic viability of farm operations.
5. Enhance the quality of life of farmers and society.

FACTA also created the Wetlands Reserve Program (WRP) and Water Quality Incentive Program, and restored one million acres of wetlands through wetland easements.

| FAIRA combined several conservation programs under ECARP. |

The **Federal Agriculture Improvement and Reform Act (FAIRA)** was enacted in 1996. It was another major transition period for agricultural production policies and also for conservation policy. Conservation programs became driven not by commodity programs, but by sustainability of the natural resource base. It reauthorized the CRP and WRP and combined them with the Environmental Incentives Program (EQIP). *In totem* these programs constituted the Environmental Conservation Acreage Reserve Program (ECARP), which focused on watersheds, multistate regions, regions of environmental sensitivity, and priority areas. Note how the emphasis of conservation and management has moved with time from the field to the farm to the environment to the watershed.

FAIRA (1996) had some other consequences. The phasing out of commodity support meant penalties for noncompliance were minimal. There was also a shift in CRP focus from windborne soil loss to waterborne loss, from west to east, where rent or lease was higher.

| Society at large began asking agriculture to internalize the costs of pollution control. |

Why did policies change? Cities and industries were being forced to internalize the costs of implementing pollution control systems, so many asked, "Why not agriculture?" Society was also changing. Less than 20 percent of U.S. counties depended on agriculture (that is, relied on agriculture for more than 20 percent of their economic base). Policies based on unique characteristics of agriculture and rural communities also diminished significantly because they now reflected urban and other interests—only 28 percent of farms accounted for 86 percent of all agricultural output.

However, the changes in policy approach have had at least one important consequence that has not been adequately resolved. They pose a monitoring dilemma: How do you measure and attribute nonpoint source pollution fairly and accurately if it comes from widespread agricultural sources?

FOCUS ON . . . FEDERAL POLICY AND SOIL STEWARDSHIP

1. Why did the two schools of thought with respect to conservation widely differ in their approach?

2. What school of thought do national policies with respect to soils seem to follow?

3. What are the most significant features of the FSA, FACTA, and FAIRA?

4. How do you account for soil and water degradation by natural events in a monitoring program?

5. If urban residents benefit from increased investment in conservation practices by rural residents, should urban residents pay more for agricultural products?

SOIL SURVEYS AND POLICY

How can soil surveys be a reflection of policy? Soil surveys have been the cornerstone for land use and management decisions. From a practical perspective, one cannot manage a natural resource without knowing the location and condition of that resource. Surveys are used in the implementation of nearly all government programs in soil and water. Soil surveys were authorized during establishment of the USDA in order to:

". . . acquire and diffuse among the people of the United States useful information on subjects concerned with agriculture, rural development, aquaculture, and human nutrition in the most general and comprehensive sense of those terms."

The Agriculture Appropriation Act of 1896 gave specific authority to the USDA to:

". . . investigate the relation of soils to climate and organic life; for the investigation of the texture and composition of soils in the field and laboratory."

However, it wasn't until 1899 that the first appropriations were received to conduct the surveys and the first report of the Division of Soils was received.

Between 1935 and 1953 two surveys developed. The SCS mapped individual farms to support conservation requests and the Bureau of Plant Industry, Soils, and Agricultural Engineering (BPISAE) conducted countywide

Soil surveys have evolved with time.

mapping. In 1953 the two surveys combined into the National Cooperative Soil Survey (NCSS) under the auspices of the SCS. Soil surveys continue to be updated, and given improvements in aerial and satellite photography and computer technology the various states are revising and digitizing their soil surveys to make the data more accessible.

SOIL EROSION AND POLICY

USLE, RUSLE, and WEPS are key to conservation regulations.

The Universal Soil Loss Equation (USLE) and Revised Universal Soil Loss Equation (RUSLE) (see Chapter 13) are the cornerstones on which government soil conservation programs in humid regions are built. In the western United States the equivalent predictive model is WEPS (Wind Erosion Prediction System). For all its utility, the USLE has always suffered from the flaw that the parameters it uses were developed in only a few (but representative) soils. "Never has such an important policy been based on such a dearth of defensible data" (Keeney and Cruse, 1998; cited in Swanson and Clearfield, 1994). The broad application of the USLE is hard to defend because soils in different environments are unique and can be expected to behave differently than soils in other locations, such as where data for the USLE were generated. Nevertheless, the USLE has been codified into federal and state law. It predicts the tolerable soil loss not leading to productivity loss, and provides the scientific basis for mandated erosion control programs.

The USLE underscores the development and maintenance of research programs on soil erosion. Hugh Hammond Bennett had the conception of normal soil erosion in 1939. D. D. Smith (1941) derived a concept of predicted soil loss based on practices that would lead to no more than 4 tons/acre/year (9 Mg/hectare/year) loss, or 250 years to erode 7 inches (18 cm) of topsoil. This is a much greater tolerance for soil loss than Bennett's replacement ideas, which are only a fraction of that amount. Partly as a result of the work with the USLE, at present there is some form of conservation tillage (30 percent residue cover) on 35 percent of planted acreage, including 26 percent occurring in no-till.

Soil erosion causes problems beyond the loss of soil fertility.

It doesn't hurt to repeat why loss of a natural resource such as soil is a bad thing. Think of willfully destroying, or allowing to decay, something that has taken decades or more to build. When the topsoil (A horizon) erodes, one is left with B horizons that have poor physical and chemical properties. Crop yields decrease unless the replacement of lost nutrients occurs. The B horizons also easily crust and compact, are low in P and other nutrients, impede emergence (because they typically have higher clay content), and have reduced fertility, rooting, and water absorption. As yield decreases, organic matter decreases, and inputs to structure decrease. The producer has reduced net farm return and upward pressure on commodity pricing. Off-site, the costs of soil loss are many:

❖ higher maintenance costs for canal and river dredging
❖ riparian habitat degradation
❖ diversity reduction
❖ water contamination
❖ fisheries impairment
❖ recreational impairment
❖ hydroelectrical wear

Soil Quality and Policy

What is soil quality? Soil quality is defined as:

> *The capacity of a specific kind of soil to function within the natural and managed ecosystem boundaries to sustain plant and animal productivity, maintain or enhance water and air quality, and support human health and habitat.*

How does the idea of soil quality reflect a changing perception of the soil and sustainable agriculture? Current economic systems tend to reward farmers for unsustainable practices. An NRCS report in 1993 determined that the soil quality concept should guide recommendations for use of soil conservation practices and federal programs in resource conservation. One consequence of this approach is that much current policy and research has focused on **C sequestration** (storage). It is estimated that if conservation tillage were universally adopted it would sequester enough C to nullify the greenhouse effect (1.5 billion ha worldwide).

> Soil quality is a driving concept in new regulatory policies.

Current and Potential State Policies Toward Soil Conservation and Environmental Degradation

There has been a shift of conservation policy control from federal government to state and local government, and greater local control of conservation programs by local conservation districts. Some of the lessons that have been learned from this transition are that when delegating implementation of policy to the local level one has to maintain consistency but be flexible; encourage voluntary participation, but continue incentive driven programs; ensure scientific credibility; recognize penalties inherent in conservation; and involve broad interests.

In Iowa, for example, the Iowa Soil Conservation Districts Law (1971) was amended to establish a means by which erosion could be declared a nuisance, and landowners would be required to limit soil loss to tolerable rates. Fertilizers and pesticides were taxed to raise revenue for water quality improvement. In California, Proposition 65 restricted use of some agricultural pesticides that were allowed by the federal government. In Connecticut, liability was imposed on individuals, including farmers, who had contaminated drinking water sources.

Shifting Responsibility for Conservation Policy

One issue is whether states can afford to provide economic incentives for soil conservation or whether their policies will be regulatory (Swanson and Coughenour, 1994). Voluntary programs have not achieved societal expectations for soil and water quality. When conservation subsidies are removed, erosive practices tend to reoccur, sometimes eliminating the contribution of the conservation subsidy. Another issue is whether landowners should be exempt from internalizing the costs of conservation (that is, should they have to eat the cost of conservation on their own?).

Approaches to Conservation Regulation

In 1993 the NRCS suggested what national policy should do:

1. Conserve and enhance soil quality as a first fundamental step to environmental improvement.
2. Increase nutrient, pesticide, and irrigation efficiency.
3. Increase resistance of farming systems to erosion.
4. Make better use of field buffer zones.

Setting state and national standards for conservation to meet these policy goals is a difficult task. The standards may be too stringent if set by individuals without production experience. Will environmentally friendly production systems still keep the United States a net agricultural exporter? Should local communities be encouraged to set stricter standards as necessary (Napier et al. 2000)?

Several approaches to conservation have been tried. Subsidy and compulsory approaches are designed to retire highly erodible land by permanent easements and involve the purchase of cultivation rights. There is a subsidy for use of conservation practices.

The best example of compulsory shift in crop production has been the tobacco buyout experiment in which tobacco producers are paid a fee, based on current allotments, that encourages producers to make the transition to new crops. Sponsored support for tobacco production would, it is hoped, subsequently end. One question with any type of support is whether it can be regarded as a bribe to stop polluting rather than an incentive to practice conservation.

There are also compulsory approaches for soil and water conservation in which the state purchases all property rights—a **takings** in which case the fair market value of the property has to be determined. There can also be state control of nutrient application, enforced through mandatory soil testing. A good example of this is rules for phosphorus application based on soil test P. The state could also have production quotas in which overproduction from inappropriate farming practice would be penalized.

Kentucky is an example of a state that has mandated farming systems. Each farm has to have a farm plan on file specifying exactly how it will comply with the Kentucky Agricultural Water Quality legislation.

> The tobacco buyout program is a good example of a compulsory shift in crop production.

Monitoring and the Cost of Monitoring

An argument against monitoring is that it is too costly. The counterargument is that if the United States can afford to police drug abuse, prostitution, organized crime, illegal crops, moonshine, tax evasion, and other "illegal" behavior, it has the resources to monitor environmental pollution. For example, what if the $2 billion in the current CRP was used to enforce conservation compliance? The money would fund one field monitor at an annual salary of $30,000 for every 32 farmers. Changing demographics has made this more feasible because there are fewer farmers to monitor.

> Monitoring is one of the biggest problems and costs of conservation compliance.

Focus on . . . Policies for Conservation

1. Should Louisiana, Texas, Mississippi, and Alabama sue the cornbelt states for the damage to their shrimp industries caused by the hypereutrophic "Dead Zone" in the Gulf of Mexico?

2. Which is likely to be a more successful approach: conserving soil by taking erodible land out of production or increasing monitoring of current conservation rules?

Local Control of Soil Resources

At the local level how does intervention for conservation occur? It occurs through:

* zoning
* building of highways
* siting of leachfields
* land use planning
* equalization of tax assessments

But there are a host of barriers to implementing soil conservation policy at the local level, such as:

* Determinations that are considered a "takings."
* The NRCS having a partly regulatory role in addition to its role in technical assistance, therefore being a potential threat to local appropriations.
* Desktop plans that show lack of field-based work.
* Reduced water quality, conservation operations, and water resource attention because technical competence is misplaced if it is exclusively devoted to soils.
* Inadequate resources to implement policy.
* Alienation of local districts with different priorities.
* Increased workload and stress on employees due to the requirements of monitoring.
* Loss of other conservation programs such as the Stewardship Incentive Program (SIP) and flood mitigation programs.
* Documentation requirements that slow action.
* Changing conservation standards that are seen as permitting more erosion.
* Local NRCS offices that are not viewed as independent in an appeals process because of their new regulatory role.

❖ No minimum size specified for wetlands.

❖ Indirect effects on commodities and the agricultural economy (the USDA has a policy that no more than 25 percent of country cropland can be in the CRP).

❖ Insufficient communication.

❖ Inadequate education of clientele.

❖ With respect to wetlands, changing ingrained tendencies to drain for better yield and newly emerging concerns about mosquito-borne viral diseases.

❖ Lack of specific skills at local NRCS offices.

❖ Increased structural conservation workloads.

Benefits of Implementing Soil Conservation Policy at the Local Level Under the New Policy Paradigms

What would be the benefits to the federal government of shifting policy development to the local level? Some possible benefits are:

❖ New clients for the NRCS.

❖ Help to significantly reduce erosion.

❖ The largest concerted effort at controlling soil erosion in the shortest period.

❖ NRCS gaining support of nonfarm groups such as Ducks Unlimited and other hunting groups.

❖ Team orientation at the NRCS, FSA, Rural Development (FSA), Corps of Engineers, and Cooperative Extension Service.

❖ Contracting compliance to local districts.

❖ Developing technical expertise in grassland management.

❖ Favorable media.

SUMMARY

In this chapter you looked at soils from the perspective of a natural resource performing essential ecosystem functions. Although renewable, the soil resource forms only slowly with time. As a consequence, the effects of short-term degradation can have long-term consequences. Although early agriculturalists recognized the potential and solutions for soil degradation, for much of the history of the United States soil was considered an exploitable resource. Public policy, from education to economic incentives for farming, served to enhance that degradation.

By the 1930s, the effects of exploitation were obvious and there was a gradual shift of public perception of soil as a resource that could be permanently lost. The formation of the Soil Conservation Service and the first soil surveys were a step to inventory and protect soil resources. Other policies had mixed effects, on the one hand promoting conservation and on the other promoting exploitation.

The realization that soils probably had a maximum tolerable erosive loss combined with the environmental movement and the shift of populations to cities combined to significantly change public policy toward soil conservation. By 1977, active measures were being mandated to protect soil resources and by 1985 a soil conservation ethic was firmly in place. Recent trends have seen a gradual shift of controlling soil conservation policy from the national to the state and local level. The most significant problems with protecting the soil resource concern financing conservation and the mechanics of monitoring nonpoint soil pollution.

END OF CHAPTER QUESTIONS

1. Is soil a renewable resource? Explain your answer.
2. What is "T" and why is it important?
3. What impact has no-till had on Kentucky's agriculture?
4. Create a timeline reflecting key events in soil management and conservation in the United States.
5. How do soil properties affect urban and agricultural development?
6. Identify specific examples of national policies that have promoted soil degradation and explain why they produced this result.
7. Identify specific examples of national policies that have improved soil conservation and explain why they produced this result.
8. How have technological innovations in agriculture spurred or retarded soil erosion?
9. What ecosystem services do soils supply?
10. What is the significance of the 1985 Food Security Act?
11. How does soil resource policy change if it is driven by sustainability and quality issues rather than commodity issues?
12. Provide an example, real or hypothetical, of a "takings" at the local level that will affect soil conservation and land use.
13. How can conservation tillage be related to current and potential national policy with respect to CO_2 emissions?
14. From the latest data in the National Soils Inventory (http://www.usda.nrcs.gov), determine the five leading states in terms of wind erosion and water erosion, respectively.

FURTHER READING

Standard reading for development of the conservation ethic are:

Carson, R. 1962. *Silent spring.* New York: Houghton Mifflin Company.

Leopold, A. 1966. *A Sand County almanac.* Oxford: Oxford University Press, Inc.

Current research on soil conservation as well as updates on changing policies in soil and water conservation can be found in:

Journal of Soil and Water Conservation. Ankeny, IA: Soil and Water Conservation Society.

Much of the background information for this chapter was obtained from the following book:

Swanson, L. E., and F. B. Clearfield (eds). 1994. *Agricultural policy and the environment: Iron fist or open hand?* Ankeny, IA: Soil and Water Conservation Society.

REFERENCES

Daily, G. C., P. A. Matson, and P. M. Vitousek. 1997. Ecosystem services supplied by soil. In G. C. Daily (ed.), *Nature's services.* Washington, DC: Island Press, pp. 113–132.

Napier, T. L., S. M. Napier, and J. Tvrdon (eds.). 2000. *Soil and water conservation policies and programs.* Boca Raton, FL: CRC Press.

Smith, K. R. 1994. Land stewards or polluters: The treatment of farmers in the evolution of environmental and agricultural policy. In L. E. Swanson and F. B. Clearfield (eds.), *Agricultural policy and the environment: Iron fist or open hand?* Ankeny, IA: Soil and Water Conservation Society, pp. 15–28.

Swanson, L. E., and C. M. Coughenour. 1994. The shifting cultural foundation for farming and the environment. In L. E. Swanson and F. B. Clearfield (eds.), *Agricultural policy and the environment: Iron fist or open hand?* Ankeny, IA: Soil and Water Conservation Society, pp. 1–14.

EPILOGUE

"I would rather be tied to the soil as a serf . . . than be king of all these dead and destroyed."

Homer, *Odyssey*

"Nature has endowed the earth with glorious wonders and vast resources that man may use for his own ends. Regardless of our tastes or our way of living, there are none that present more variations to tax our imagination than the soil, and certainly none so important to our ancestors, to ourselves, and to our children."

Charles Kellogg, *The Soils That Support Us*, 1956

In this textbook we have tried to synthesize the most important features of studying soil science. You have examined how soils form, how they are classified, how they are constructed, how they behave in a physical and chemical sense, how they provide an environment for living organisms and the biochemical activities those organisms perform, how soils are managed for agricultural purposes and other activities, and finally how soils are evaluated and treated as a natural resource. We hope that by having finished this text you have developed a greater appreciation for soils and discovered why so many scientists have decided to make soils their choice of study. Even if you have not been turned into a soil scientist, we hope you are prepared in future careers to take soils into consideration in your daily activities. Remember, "The soil is the nation."

Useful Unit Conversion Factors

(Source, American Society of Agronomy)

SI Unit	To Convert to Non-SI Unit Multiply by	Non-SI Unit	To Convert to SI Unit Multiply by
Length			
kilometer	0.621	mile	1.609
meter	1.094	yard	0.914
meter	3.28	foot	0.304
millimeter	3.94×10^{-2}	inch	25.4
nanometer	10	Angstrom	0.1
Area			
hectare	2.47	acre	0.405
square km	247	acre	4.05×10^{-3}
square km	0.386	square mile	2.590
square m	2.47×10^{-4}	acre	4.05×10^{3}
square m	10.76	square foot	9.29×10^{-2}
square mm	1.55×10^{-3}	square inch	645
Volume			
cubic meter	9.73×10^{-3}	acre-inch	102.8
cubic meter	35.3	cubic foot	2.83×10^{-2}
liter	6.10×10^{4}	cubic inch	1.64×10^{-5}
liter	2.84×10^{-2}	bushel	35.24
liter	1.057	quart	0.946
liter	3.53×10^{-2}	cubic foot	28.3
liter	0.265	gallon	3.78
liter	33.78	ounce	2.96×10^{-2}
liter	2.11	pint	0.473
Mass			
gram	2.20×10^{-3}	pound	454
gram	3.552×10^{-2}	ounce	28.4
kilogram	2.205	pound	0.454
kilogram	1.10×10^{-3}	ton	907
megagram	1.102	ton	0.907

(continued)

SI Unit	To Convert to Non-SI Unit Multiply by	Non-SI Unit	To Convert to SI Unit Multiply by
Yield or Rate			
kg ha^{-1}	0.893	lb acre^{-1}	1.12
kg m^{-3}	7.77×10^{-2}	lb bu^{-1}	12.87
kg ha^{-1}	1.49×10^{-2}	bu acre^{-1}	67.19
kg ha^{-1}	1.59×10^{-2}	bu acre^{-1}	62.71
kg ha^{-1}	1.86×10^{-2}	bu acre^{-1}	53.75
L ha^{-1}	0.107	gal acre^{-1}	9.35
Mg ha^{-1}	893	lb acre^{-1}	1.12×10^{-3}
Mg ha^{-1}	0.446	ton acre^{-1}	2.24
m sec^{-1}	2.24	mph	0.447
Pressure			
megapascal	9.90	atmosphere	0.101
megapascal	10	bar	0.10
pascal	2.09×10^{-2}	lb ft^{-2}	47.9
pascal	1.45×10^{-4}	lb in^{-2}	6.90×10^3
Temperature			
Kelvin	1(K-273)	Celsius	1(°C + 273)
Celsius	(9/5 °C) + 32	Fahrenheit	5/9(°F $-$ 32)
Energy, Work, or Heat			
joule	9.52×10^{-4}	Btu	1.05×10^3
joule	0.239	calorie	4.19
joule	107	erg	10^{-7}
joule	0.735	foot-pound	1.36
joule m^{-2}	2.387×10^{-5}	cal cm^{-2}	4.19×10^4
newton (N)	10^5	dyne	10^{-5}
watt (W) m^{-2}	1.43×10^{-3}	cal cm^{-2} min^{-1}	698
Water Measurements			
m^{-3}	9.73×10^{-3}	acre-inches	102.8
m^{-3} h^{-1}	9.81×10^{-3}	ft^3 sec^{-1}	101.9
m^{-3} h^{-1}	4.40	gal min^{-1}	0.227
ha-meter	8.11	acre-feet	0.123
ha-meter	97.28	acre-inch	103×10^{-2}
Water Measurements			
ha-cm	8.1×10^{-2}	acre-feet	12.33
Concentration			
centimole per kg	1	milliequivalent per 100 g	1
g kg^{-1}	0.1	percent	10
mg kg^{-1}	1	ppm	1
Radiation			
becquerel (Bq)	2.77×10^{-11}	curie (Ci)	3.7×10^{10}
Bq kg^{-1}	2.7×10^{-2}	picocurie per g-1	37
gray (Gy)	100	rad	0.01
sievert (Sv)		rem (roentgen equiv. man)	0.01

GLOSSARY

"Soils are developed; they are not merely an accumulation of debris resulting from decay of rock and organic materials. . . . In other words, a soil is an entity—an object in nature which has characteristics that distinguish it from all other objects in nature."

C. E. Millar and L. M. Turk, 1943

Note: A full glossary of soil science terms can be found at the Web site of the Soil Science Society of America (http://www. soils.org).

A

A Horizon (Chapter 3)—The natural surface layer of a mineral soil.

Acellular (Chapter 9)—Organisms lacking structures such as cell nucleus, cell membrane, and cell wall associated with cellular organisms.

Acidity (Chapter 8)—The extent to which a soil has a pH < 7.0.

Acidophilic (Chapter 9)—Preferentially growing in acidic conditions.

Active Acidity (Chapter 8)—The amount of acidity in soil solution directly measured by pH strips or a pH meter.

Active and Passive SOM (Chapter 11)—Two fractions of SOM that are readily available and recalcitrant to decomposition, respectively.

Adsorption (Chapter 14)—The process by which atoms, molecules, or ions are taken up and retained by solid surfaces by physical and chemical mechanisms.

Aeration (Chapter 4)—The extent to which an environment has available oxygen.

Aerobes/Anaerobes/Facultative Anaerobes (Chapter 9)—Aerobes grow in the presence of O_2, anaerobes grow only in its absence, and facultative anaerobes can grow in either the presence or absence of O_2.

Air Dry (Chapter 6)—Soil that has been allowed to dry to the maximum point in ambient conditions through evaporation.

Albedo (Chapter 13)—The ratio of the amount of solar radiation reflected relative to that received expressed as a percent. The albedo of the Earth is approximately 34 percent.

Alfisol (Chapter 3)—One of the soil orders in the U.S. taxonomic system. Alfisols are typically fertile deciduous forest soils that have an accumulation of clay in the B horizon.

Alkali Soil (Chapter 14)—Synonymous with sodic soil, it is a soil with sufficient sodium (Na) to interfere with plant growth.

Alkalinity (Chapter 8)—The extent to which a soil has a pH greater than 7.0.

Alkalophilic (Chapter 9)—Preferentially growing in alkaline environments.

Allelopathic (Chapter 12)—Toxic or inhibitory, specifically, related to the inhibition of one plant by the products of another.

Alluvial (Chapter 2)—Having formed from the deposits of rivers and streams, in general the deposits of any running water.

Alluvial Fan (Chapter 2)—Deposits from a stream as it enters a plain or larger stream.

Ammensalism (Chapter 9)—An interaction in which the activity of one organism is harmful to the growth of another.

Ammonification (Chapter 10)—The mineralization or decomposition of organic N with the release of NH_4^+.

Anaerobic (Chapter 10)—Lacking oxygen or oxygen-free; contrast this with aerobic, which means an environment containing oxygen.

Andisol (Chapter 3)—One of the soil orders in the U.S. taxonomic system. Andisols are fertile soils that form from volacanic deposits.

Anion Exchange Capacity (AEC) (Chapters 8, 14)—The total capacity of the adsorptive components of a soil to attract and exchange anions.

Anthropogenic (Chapters 2, 3)—Of or pertaining to human influence. With respect to soil it is the transformation of soils by human activities such as tillage and earth moving.

Aquicludes (Chapter 13)—An impervious clay and/or rock layer that prevents water movement.

Aquifer (Chapter 13)—An underground layer of permeable material that can store and supply water.

Aquitard (Chapter 13)—A relatively imperious layer in soil that retards water movement.

Archaea (Chapter 9)—A branch of prokaryotic organisms.

Argillic Horizon (Chapter 3)—The B horizon of a soil that contains more clay than the overlying A horizon due to illuviation.

Aridisol (Chapter 3)—One of the soil orders in the U.S. taxonomic system. Aridisols are hot desert soils that have a cambic or argillic horizon. They can be fertile if irrigated.

Aromatic Compounds (Chapter 11)—Compounds that have as their basis a benzene ring.

Assimilation (Chapter 10)—Incorporation of organic and inorganic compounds and/or elements into living organisms.

Atmosphere (Chapter 1)—The gaseous environment surrounding earth and extending into the soil profile.

Autotroph (Chapter 9)—An organism that obtains its carbon for growth from CO_2 or bicarbonate.

B

B Horizon (Chapter 3)—A subsoil horizon underlying either an A or E horizon. B horizons may be exposed if surface erosion is significant.

Bacteriophage (Chapter 9)—Viruses that infect bacteria.

Base Cation Saturation (BCS) (Chapter 8)—The extent to which a soil is saturated with base cations such as Ca^{2+}, Mg^{2+}, and K^+.

Base Cations (Chapter 7)—Commonly refers to the Ca^{2+}, Mg^{2+}, K^+, and Na^+ in soil.

Bedrock (Chapter 2)—The solid rock underlying unconsolidated material.

Biogeochemistry (Chapter 10)—The study of the interaction of chemical and biochemical transformations of elements, minerals, and organic matter in Earth's environment.

Biosphere (Chapter 1)—The living organisms in soil.

Bioturbation (Chapter 9)—The mixing or disturbing of soil by the burrowing or excavating activity of soil organisms.

Buffer Capacity (Chapter 8)—The capacity of a soil to resist change, such as change in pH, and the compounds in soil that contribute to that resistance.

Bulk Density (Chapter 5)—By definition, the mass of soil divided by the volume of soil; the greater the bulk density, the less porous a soil.

C

C Horizon (Chapter 3)—The least developed soil horizon that lies beneath the B horizon.

C Sequestration (Chapter 16)—The storage or retention of carbon in plants, precipitates, and organic matter in soil that makes it temporarily unavailable.

Calcareous (Chapter 14)—High in calcium.

Caliche (Chapter 3)—A type of impermeable layer in soil composed of cemented carbonates.

Cambic Horizon (Chapter 3)—A weakly developed subsoil or B horizon.

Capillarity (Chapter 6)—The attraction of water to itself and to soil particles that allows water to rise through soil pores from above a water table.

Cast (Chapter 9)—The mixed mineral and organic deposits left by earthworms on the surface of soils.

Catena (Chapter 15)—A chain or sequence of soils from the top of a hill to the foot slope. The soils will have formed from the same parent material, and be of approximately the same age, but will differ with respect to such other factors as relief and drainage.

Cation Exchange Capacity (CEC) (Chapters 8, 14)—The capacity of the adsorptive components of a soil to attract and exchange cations.

Chelate (Chapters 10, 11)—From the Greek "claw," this refers to the attraction of reactive constituents of compounds to ions in soil which they bind. Chelating agents bind metals and ions in solutions and keep them in soluble and therefore available form.

Chemodenitrification (Chapter 10)—A chemical process in which N-oxides are lost in gaseous form during N reduction.

Chemotroph (Chapter 9)—An organism that gets its energy from breaking chemical bonds in either organic or inorganic compounds.

Chernozem (Chapter 3)—"Black Earth," a term equivalent to Mollisol used in other soil taxonomic systems.

Chlorosis (Chapter 10)—A lack of chlorophyll in plant tissue that leads to light green or yellowish color and is typically associated with nitrogen or iron deficiency.

Chroma (Chapter 3)—One of the three variables of soil color. Chroma reflects the relative strength or purity of a color.

Clastic (Chapter 2)—Rock formed from mineral or rock fragments or other rocks.

Clay (Chapter 4)—By definition a mineral particle smaller than two micrometers in size. Clay can also refer to specific types of soil minerals such as silicate clays.

Coarse Fragments (Chapter 4)—Mineral soil or material too coarse to pass through a 2-mm sieve.

Coenocytic (Chapter 9)—Fungi that lack internal cell walls in the hyphae.

Colloid (Chapters 7, 14)—Clay or organic particles so small that they remain in suspension in standing water.

Colluvium (Chapter 2)—The soil or rock material deposited at the foot of a slope, primarily through the force of gravity. This includes talus and cliff debris.

Commensalism (Chapter 9)—An interaction in which the activity of one organism is beneficial to the growth of another.

Compaction (Chapter 12)—Reduction in soil volume by mechanical pressure leading to decreased porosity and degraded soil properties.

Competition (Chapter 9)—An interaction in which two organisms have simultaneous demands for the same limiting substrate.

Complex (Chapter 15)—A type of map unit used in soil surveys that reflects the appearance of one or more intermingled soil series.

Conduction (Chapter 6)—Heat transfer brought about by movement of kinetic energy from one body in direct contact with another.

Conservation Tillage (Chapter 12)—One of several methods of tillage that minimizes soil disturbance.

Convection (Chapter 6)—Heat transfer due to the movement of fluid (either liquid or air).

Convective Precipitation (Chapter 13)—Occurs when a moist layer of air near the Earth's surface is heated, rapidly lifted, and cooled. If the uplifting is high enough in the atmosphere ice crystals can result.

CRP (Chapter 16)—Conservation Reserve Program.

Crusting (Chapter 4)—Formation of dense, hard, or brittle layers at the soil surface.

Cryophiles/Psychrophiles (Chapter 9)—Organisms that preferentially grow in cold environments.

Cryptozoans (Chapter 9)—Organisms that preferentially inhabit shaded locations beneath rocks and wood.

D

Denitrification (Chapter 10)—A respiratory process occurring in anaerobic environments in which N-oxides are used as electron acceptors in place of oxygen.

Desertification (Chapter 12)—A process in which arable land is replaced by progressively desertlike conditions.

Diagnostic Soil Horizons (Chapter 3)—The horizons that are most important for soil classification in assigning soils to representative soil orders.

Diffusion (Chapter 14)—The movement, particularly with respect to nutrients, of compounds in soil in response to concentration gradients.

Discharge (Chapter 13)—The amount of water flow in a stream. Discharge (Q) is the product of velocity (V) and cross-sectional area (A) of the stream ($Q = VA$).

Disproportionation (Chapter 10)—A chemical process occurring anaerobically in which S compounds are transformed to yield oxidized and reduced molecules.

Dissolution (Chapter 7)—Another term for *dissolve.*

Divalent Cations or Anions (Chapters 7, 8, 14)—Ions have two charges, either positive (divalent cation, e.g., Ca^{2+}) or negative (divalent anion, e.g., SO_4^{2-}).

Dolomite (Chapter 14)—Calcium-magnesium carbonate, also referred to as limestone.

Drainageways (Chapter 15)—Planned or natural areas through which water drains.

Drilosphere (Chapter 9)—The zone of soil impacted by the burrowing activity of earthworms.

Duripan (Chapter 3)—A relatively impermeable layer in soil composed of cemented silica.

E

Ecosystem (Chapter 16)—A given environment and all its components and those external forces working upon it.

Ectomycorrhizae (Chapter 9)—A type of mycorrhizal association in which the symbiotic fungi forms an extensive mantle around the plant roots.

Eluvial (Chapter 2)—Material that has been removed by suspension or solution; usually in reference to a soil horizon from which material has been removed. Leaching is usually used to describe the loss of material in solution.

Eluviation (Chapter 3)—Removal (loss) of soil material in suspension or solution from a soil layer.

Encyst (Chapter 9)—To enter a state of low metabolic activity protected from the environment until growth conditions improve.

Endomycorrhizae (Chapter 9)—A type of mycorrhizal association that is obligatory.

Entisol (Chapter 3)—One of the soil orders in the U.S. taxonomic system. Entisols are "new" or weakly developed soils.

Eolian (Chapter 2)—Material deposited by wind, such as sand dunes, sand sheets, and particularly loess.

Epipedon (Chapter 3)—The diagnostic surface soil horizon.

Equilibrium (Chapter 8)—A state in which change can occur, but no net change in state is observable.

EQUIP (Chapter 16)—Environmental Quality Incentives Program.

Equivalent Surface Depth (Chapter 6)—The amount of water required to fill a uniform layer of soil completely.

Erodibility (Chapter 4)—The extent to which soil can be eroded or lost.

Erosion (Chapter 12)—The process by which soil is moved, suspended, washed, or blown from one location to another.

Essential Nutrients (Chapter 14)—The elements in the periodic table essential for the life of plants and animals.

Eukaryotes (Chapter 9)—Organisms that possess a cell nucleus.

Eutrophication (Chapters 12, 13)—Excessive growth. Having conditions where nutrients are optimal or excessive for growth. Usually with respect to uncontrolled growth in aquatic environments leading to degraded water quality.

Evapotranspiration (ET) (Chapter 13)—The combined water loss from a given area of evaporation from the soil surface and transpiration by plants.

Exchange Capacity (Chapter 8)—The total ionic charge of the adsorption complex that is active in ion adsorption.

Exchangeable Acidity (Chapter 8)—Acidic cations that can be exchanged from soil colloids by salt solutions.

Exchangeable Bases (Chapter 14)—Typically refers to Ca^{2+}, Mg^{2+}, and K^+ that are adsorbed and desorbed from the CEC.

Expanding Clays (Chapter 14)—2:1 layer silicate clays that shrink and swell in response to water.

F

FACTA (Chapter 16)—Food, Agriculture, Conservation, and Trade Act.

FAIRA (Chapter 16)—Federal Agriculture Improvement and Reform Act.

Fallow (Chapter 12)—Leaving uncropped or untilled.

Family (Chapter 3)—The fifth level of classification in the U.S. system of soil taxonomy. Family names use descriptive terms to describe soil features such as texture, mineralogy, and temperature.

Fauna (Chapter 9)—The animals (macro- and microscopic) present at a site.

Feldspar (Chapter 4)—The most common primary mineral in the Earth's crust.

Fenestration (Chapter 9)—The opening of the leaf epidermis and exposure of tissue to microbial attack; the initial step in the decomposition of leaves.

Ferment (Chapter 9)—Growth by internal cycling of electrons and substrate-level phosphorylation.

Fermentation (Chapter 10)—An oxygen-free process in which a compound acts as both source and sink of electrons, resulting in the release of oxidized and reduced products and the production of high-energy metabolic intermediates.

Fertility (Chapter 1)—The nutrient status of a soil; reflects a soil's capacity to promote growth.

Field Capacity (Chapter 6)—A description of soil water content in which all but the largest freely draining pores are water-filled; typically expressed as the percent of water by weight held by a soil by capillary action after larger pores have drained by the force of gravity.

Flocculate (Chapter 5)—The aggregation of soil particles, particularly in suspension, that causes them to grow in size.

Floodplain (Chapter 2)—The land bordering a stream that has been built up by sedimentation when the stream overflows, and which is subject to periodic inundation.

Flora (Chapter 9)—The plants (macro- and microscopic) present at a site.

Fragipan (Chapter 2)—Dense subsoil layers that are relatively impermeable, brittle when moist and dry but not when wet.

FSA (Chapter 16)—Food Security Act.

Fulvic Acid (Chapter 11)—Yellow to brown material soluble in alkali and acid. Fulvic acid consists of compounds of lower molecular weight and greater oxidation than humic acid.

G

Gelisol (Chapter 3)—One of the soil orders in the U.S. taxonomic system. Gelisols are soils of cold regions typified by permafrost.

Generalized Soil Maps (Chapter 15)—A simplified survey map used at the county level to reflect the major soil associations and their distribution.

Glacial Drift (Chapter 2)—Deposits made by glaciers and their outwash.

Glacial Till (Chapter 2)—Unsorted debris left by a glacier.

Glaciation (Chapter 2)—The process by which large, slowly moving masses of ice transport soil and other debris across a landscape.

Glomalin (Chapter 11)—A glycoprotein produced by mycorrhizal fungi that helps develop soil aggregation.

Granite (Chapter 2)—A light-colored, crystalline igneous rock containing about 25 percent quartz.

Grassed Waterway (Chapter 12)—A grass-covered channel in fields used to carry or divert surface runoff.

Gravimetric Water Content (Chapter 6)—The water content in soil on a mass/mass basis.

Gravitational Potential (Chapter 6)—A mathematical description of the tendency of water to move by the force of gravity.

Great Group (Chapter 3)—The third level of classification in the U.S. system of soil taxonomy. Great group formative elements provide additional information about a soil.

Groundwater (Chapter 13)—Water beneath the Earth's surface in saturated soil or porous rock.

Gully Erosion (Chapter 12)—A type of soil erosion, along with ephemeral gully erosion, in which a permanent channel caused by extreme soil loss occurs.

H

Hardpan (Chapter 3)—A soil layer that acts as a barrier to root and water movement.

Heterotroph (Chapter 9)—An organism that obtains its energy and C for growth from organic matter.

Heterotrophic Nitrification (Chapter 10)—A process carried out by heterotrophic bacteria and fungi in which reduced organic N is oxidized to yield NO_2^- and NO_3^-.

Highly Erodible Land (Chapter 15)—Land susceptible to erosion. By definition, based on the RUSLE, these lands have an erodibility index greater than 8.

Histosol (Chapter 3)—One of the soil orders in the U.S. taxonomic system. Histosols are typified by accumulation of organic material.

Horizon (Chapters 1, 2, 3)—A natural layer in soil in either the surface or subsurface that forms parallel to the land surface during the process of soil formation. Horizons differ from one another by physical, chemical, and biological properties, texture consistency, acidity, and color, among other properties.

Hue (Chapter 3)—One of the three variables of soil color. Hue reflects the light of certain wavelengths, or perceived color.

Humic Acid (Chapter 11)—brown to black organic material extractable from soil by alkali solution and insoluble in acid.

Humification (Chapter 9)—The formation of soil humus through digestion, fragmentation, and metabolism.

Humin (Chapter 11)—The fraction of SOM that cannot be extracted from soil with an alkali solution and that consists of high-molecular-weight compounds resistant to decomposition.

Humus (Chapter 11)—Brown to black substances of relatively high molecular weight formed during random synthesis reactions in soil. The stable fraction of the organic matter in soil that persists once readily decomposable material is gone.

Hydration (Chapter 7)—The adding of water to form new chemical bonds.

Hydrolysis (Chapter 2)—A chemical reaction in which hydrogen and hydroxyl from a water molecule react with a soil mineral to form an acid and base.

Hydrometer (Chapter 4)—A device used to measure soil texture based on the density of a suspension.

Hydrophytic (Chapter 13)—Water loving, particularly with respect to plant growth.

Hydrosphere (Chapter 1)—The water environment of Earth consisting of seas, lakes, and streams, but also extending to the water films surrounding soil particles.

Hygroscopic Coefficient (Chapter 6)—The water content at which no additional water can be lost through evaporation.

Hysteresis (Chapter 6)—The phenomenon in soil in which water content during wetting and drying cycles does not coincide.

I

Igneous (Chapter 2)—Rock formed by the cooling and solidification of molten parts (e.g., lava) of the lithosphere. Igneous rock is relatively unchanged since its formation.

Illite (Chapter 7)—A type of silicate clay material composed of hydrous mica.

Illuviation (Chapter 3)—The process of gaining or depositing material removed by one soil horizon or layer from another that typically lies above it.

Immobilization (Chapter 10)—Synonymous with assimilation, usually used in the sense of making elements unavailable for use by other organisms in soil.

Inceptisol (Chapter 3)—One of the soil orders in the U.S. taxonomic system. Inceptisols are moderately developed soils.

Inclusions (Chapter 15)—Locations of different soil series within a larger soil series that do not constitute a large enough area to be mapped separately.

Infiltration (Chapter 4)—Downward entry of water into soil.

Inorganic (Chapter 10)—Elemental or simple molecular forms of compounds.

Interveinal (Chapter 14)—Between veins.

Ion Exchange (Chapter 7)—The exchange between soil solution and soil solids of cations and anions.

Isomorphous Substitution (Chapters 7, 14)—The exchange of atoms of equal size but lesser charge for those of greater charge in silicate clays that leads to a permanent negative charge.

K

Kaolinite (Chapter 7)—A type of silicate clay. Kaolinite is a 1:1 layer clay with the general formula $Al_2Si_2O_5(OH)_4$.

L

Lacustrine (Chapter 2)—Referring to lakes or lake water. Lacustrine deposits were deposited in lake water and then exposed either through drainage of the lake or elevation of the land.

Land Capability Class (Chapter 15)—A parallel system of classification to taxonomic classification that assigns values based on potential productivity and use.

Landform (Chapters 2, 3)—A natural feature of the Earth's surface.

Landscape Position (Chapter 15)—Where a soil exists within a landscape, a critical feature in soil genesis.

Leaching (Chapter 2)—One type of elluvial activity. It is the removal of materials (solutes) in solution typically by downward movement through soil.

Levee (Chapter 12)—A natural or constructed barrier for flood protection.

Lignin (Chapter 11)—A complex polymer of phenyl propanoid subunits that forms a major part of the structural integrity of plants and is a significant soil organic matter precursor.

Lime (Chapter 14)—Ground limestone, either calcite or dolomite.

Liming Agent (Chapter 14)—A material that raises the pH of a soil.

Lithosphere (Chapter 1)—The rocks and minerals in the Earth's surface.

Lithotroph (Chapter 9)—An organism that obtains its energy from inorganic compounds.

Loam (Chapter 4)—A classification of soil texture reflecting a mixture of < 52 percent sand, 28–50 percent silt, and 7–27 percent clay-sized particles.

Loess (Chapter 2)—A type of eolian deposit. Loess is fine material (silt-sized particles) transported by wind and deposited elsewhere.

Luxury Consumption (Chapter 14)—Uptake of nutrients in excess of a plant's growth needs.

M

Macrofauna (Chapter 9)—Large, visible organisms such as insects.

Macronutrients (Chapter 14)—Elements required in the greatest amount by plants and animals.

Macropores (Chapter 5)—Large soil pores > 100 μm in diameter.

Manure—The excrement of animals.

Mapping Unit (Chapters 3, 15)—A conceptual unit in a soil survey of one to many delineations that represent one or more soils either individually or as a mixture.

Mass Action (Chapter 14)—Mass flow, usually with respect to nutrients, in response to water movement.

Mass Flow (Chapter 14)—Flow of water in response to gradients.

Mass Wasting (Chapter 12)—Downslope movement of large volumes of soil or rock due to gravity.

Matric Potential (Chapter 6)—A mathematical description of the force or attraction of water to solid particles in soil.

Mesofauna (Chapter 9)—Multicellular eukaryotic animals such as mites and nematodes.

Mesophiles (Chapter 9)—Organisms that preferentially grow in moderate temperature regimes.

Mesopores (Chapter 5)—Medium-sized pores 30–100 μm in diameter.

Metamorphic (Chapter 2)—Rock that has been formed by recrystallization of primary and secondary materials (igneous or sedimentary rock) under conditions of high temperature and pressure or chemical reactions.

Methemoglobnemia (Chapter 13)—A condition, particularly in infants, in which excess NO_3^- consumption causes hemoglobin to lose its ability to effectively transport oxygen.

Micelle (Chapter 14)—A negatively charged colloid particle.

Microfauna (Chapter 9)—Microscopic eukaryotic animals such as protozoa.

Micronutrients (Chapter 14)—Elements required in small, sometimes microgram, amounts by plants and animals.

Microorganisms (Chapter 9)—Small acellular, single-celled, and multicellular organisms typically invisible to the naked eye.

Micropores (Chapter 5)—Small pores < 30 μm in diameter.

Mineralization (Chapter 10)—Decomposition or transformation of organic compounds into inorganic molecules and carbon dioxide.

Mites (Chapter 9)—Macro- to microscopic arthropods in soil.

Mollisol (Chapter 3)—One of the soil orders in the U.S. taxonomic system. Mollisols are deep, fertile soils that typically develop in grassland.

Moraine (Chapter 2)—A blanket or ridge of unconsolidated material left by a glacier.

Mottling (Chapter 3)—Spotted areas of color in soil reflecting periodic saturation.

Mulch (Chapter 12)—A layer of material, either organic or inorganic, covering the soil surface that acts as a barrier to evaporation.

Mutualism (Chapter 9)—An interaction in which the activities of two organisms are mutually beneficial.

Mycorrhiza (pl. mycorrhizae) (Chapter 9)—A generic term for fungi that form symbiotic associations with plant roots.

N

Neutralism (Chapter 9)—Lack of interaction between two organisms.

Nitrification (Chapter 10)—The oxidation of reduced inorganic and organic N to ultimately release NO_3^-.

Nitrogen Fixation (Chapters 9, 10)—Biological (bacterial) conversion of molecular dinitrogen gas (N_2) to ammonia and subsequently organic nitrogen.

Nitrogenase (Chapter 10)—The enzyme complex responsible for biological nitrogen fixation.

Nonexchangeable Ions (Chapter 14)—Ions so tightly adsorbed to exchange surfaces that they do not participate in cation or anion exchange reactions.

Nonhumic Compounds (Chapter 11)—Organic compounds in soil that retain some recognizable chemical identity.

No-Tillage (Chapter 12)—A conservation tillage practice in which the soil is not disturbed except for a small slit for seed placement.

O

O Horizon (Chapter 3)—A surface organic layer overlying the mineral layer.

Ochric Epipedon (Chapter 3)—A surface soil that is pale throughout or consists of a thin dark layer over a pale layer of soil.

Octahedral (Chapter 7)—Refers to the overall arrangement or pattern of the aluminum, oxygen, and hydroxyls in silicate clay layers.

Organic (Chapter 10)—Containing C and/or other elements such as N and S and synthesized by living organisms.

Orographic Precipitation (Chapter 13)—Orographic precipitation occurs when warm air is lifted and cooled as air masses go over mountains. The lifting and cooling results in precipitation commonly occurring in the upper elevations of the windward side of the mountain (where the air mass is coming from) and little to no precipitation on the leeward side of the mountain (the opposite side of the mountain). This is commonly called the rain shadow effect.

Orthophosphate (Chapter 10)—The soluble form of inorganic P, at typical soil pH it is usually a mixture of HPO_4^{2-} and $H_2PO_4^-$.

Osmosis (Chapter 6)—Usually used in terms of diffusion of water through differentially permeable membranes.

Osmotic Potential (Chapter 6)—A mathematical description of the force by which water is attracted to soil solutes.

Oxidation (Chapter 10)—Loss of electrons from an element or compound.

Oxisol (Chapter 3)—One of the soil orders in the U.S. taxonomic system. Oxisols are weathered soils that have oxic horizons.

P

Pan (Chapter 3)—Horizons or layers in soil that are compacted.

Parasitism (Chapter 9)—An interaction in which one organism obtains its sustenance from another organism without necessarily killing it except in the long term.

Parent Material (Chapter 2)—The consolidated or unconsolidated material from which a soil forms.

Pedon (Chapter 3)—The smallest volume of soil that can be called a soil. It is a three-dimensional body of soil with dimensions large enough to permit the study of horizon shapes and relations.

Pedogenesis (Chapter 2)—Ped generation. Synonymous with the process of soil formation.

Pedosphere (Chapter 1)—The environment of the soil profile.

Peds (Chapter 1)—Units of soil structure in a soil horizon. They may be blocky, platey, or angular. *Soil aggregate* and *soil ped* are synonymous.

Percent Base Saturation (Chapter 14)—The extent to which the total CEC of a soil is saturated with respect to base cations such as Ca^{2+}, Mg^{2+}, and K^+.

Percolation (Chapter 4)—The downward movement of water into soil.

Permeability (Chapter 15)—The ease with which gases, liquids, and plant roots pass through the bulk mass of soil or a soil layer.

Pesticide (Chapter 9)—Any chemical used to kill unwanted microorganisms, plants, or animals.

pH (Chapter 8)—A measure of acidity, the negative log of the hydrogen ion (or hydronium) concentration in soil.

Phosphatases (Chapter 10)—Enzymes that cleave orthophosphate from organic P compounds.

Photosynthesis (Chapter 10)—The use of light energy to convert inorganic C into organic C.

Phototroph (Chapter 9)—An organism that generates its energy by photophosphorylation.

Piezometers (Chapter 13)—Hollow tubes with incisions inserted into soil and used to sample fluctuating groundwater tables.

Plow Layer (Chapter 2)—That layer of soil, typically at the soil surface, that is subject to frequent tillage.

Podzol (Chapter 3)—A term used in some taxonomic systems that is equivalent to Spodosol.

Polypedon (Chapter 3)—A group of contiguous, similar pedons.

Pore Size Distribution (Chapters 4, 5)—The percentage of pores of different size classification in a given soil.

Porosity (Chapters 4, 5)—The volume percentage of the total bulk of soil not occupied by solids.

Precipitation (Chapters 2, 13)—Rainfall or snowfall, or the deposition of dissolved solutes.

Predation (Chapter 9)—An interaction in which one organism consumes another.

Primary Minerals (Chapters 4, 7)—Minerals formed naturally by the crystallization of molten rock.

Proctor Density Test (Chapter 12)—An engineering test used to determine the maximum compaction of soils at different water contents.

Productivity (Chapter 1)—The ability to promote or sustain growth.

Profile (Chapter 2)—A two-dimensional, vertical cross-section of a soil through all horizons.

Prokaryotes (Chapter 9)—Organisms lacking a cell nucleus, usually equated with bacteria.

Proteobacteria (Chapter 9)—A taxonomic division of bacteria in soil; the dominant group of bacteria in soil.

Q

Quartz (Chapter 4)—A resistant crystalline mineral composed of silicon and oxygen.

R

Redox Potential (Chapter 9)—A measure of the extent to which an environment is oxidized or reduced; often equated with the aeration state of an environment.

Redox Reactions (Chapter 10)—Oxidation reduction reactions in which electrons are lost and gained by various compounds in soil.

Reduction (Chapter 10)—Gain of electrons by an element or compound.

Regolith (Chapter 2)—Loose earth material over solid bedrock.

Relief (Chapter 2)—The elevation or differences in elevation in a landscape.

Residual Acidity (Chapter 8)—Soil acidity that is neutralized by lime or other alkaline materials, but that cannot be replaced by an unbuffered salt solution.

Residual Soil (Chapter 2)—A soil that develops in place from the weathering of rock in a location.

Respiration (Chapter 10)—An energy-yielding process in which a compound is oxidized and its electrons used to generate ATP during transport through the cell membrane.

Respire (Chapter 9)—The process of generating energy by electron transport across membranes.

Rhizosphere (Chapter 9)—The zone of soil around the plant root that is influenced by the plant root and its metabolism.

Rill Erosion (Chapter 12)—A type of soil erosion caused by water removing soil from preferential flow channels. The channels can be easily obscured by tillage, which conceals the problem.

Riparian Areas (Chapter 13)—The land surrounding streams or other freshwater bodies.

Runoff (Chapter 12)—That part of precipitation that does not infiltrate the soil but flows over land.

S

Salinization (Chapter 12)—The process by which salt content in soils increases to levels detrimental to plant growth and soil function.

Saltation (Chapter 12)—An erosion process in which windborne sands strike the ground and disperse silt and clay particles into suspension. The soil particles leap and jump along the soil surface during high winds.

Sand (Chapter 4)—By definition, soil particles smaller than 2 mm and larger than 0.05 mm.

Sandstone (Chapter 2)—A type of sedimentary rock, usually dominated by quartz, bound together by cementing agents such as silica or iron oxide.

Saprophytic (Chapter 10)—Growing on dead and decayed organic compounds.

Saturated (Chapter 6)—A soil condition in which all pore space is occupied by water.

Saturated Flow (Chapter 6)—Water flow in soil in which all pores are water-filled.

Saturation (Chapter 6)—The soil condition when all pores are filled with water or all exchange sites are filled by one or more ions.

Secondary Minerals (Chapters 4, 7)—Minerals formed by the weathering of primary minerals. Examples are kaolinite and all carbonates.

Sedimentary (Chapter 2)—Referring to material that is deposited by water, wind, ice, or gravity.

Sesquioxide (Chapter 14)—A general term for oxides and hydroxides of iron and aluminum.

Shale (Chapter 2)—A type of sedimentary rock made of clay, silt, and very fine sand.

Sheet Erosion (Chapter 12)—A type of erosion in which soil is more or less removed uniformly from an exposed surface.

Side-Dress (Chapter 14)—Application of an amendment to the side of a growing plant.

Siderophores (Chapter 10)—Low-molecular-weight organic compounds with a high affinity for iron.

Silicate Clay (Chapter 14)—A secondary mineral composed of crystalline layers of silica and alumina.

Silt (Chapter 4)—By definition, soil particle smaller than 0.05 mm and larger than 0.002 mm.

Slickenside (Chapters 2, 14)—Polished surfaces on blocks of soil formed by the blocks sliding past one another as occurs in Vertisols.

Smectite (Chapter 7)—A type of 2:1 silicate clay with high cation exchange capacity; montmorillonite is the best known.

Soil Association (Chapter 15)—A kind of map unit used in soil surveys comprised of delineations of landscape units in which two or more soil series appear in a fairly repetitive and describable pattern.

Soil Fertility (Chapter 14)—The status of a soil with respect to its ability to supply the nutrients essential to plant growth.

Soil Management (Chapters 12, 14)—The total of all operations, agricultural or otherwise, that are used to manipulate a soil.

Soil Moisture (Chapter 6)—Water contained in soil.

Soil Order (Chapter 3)—The first level of classification in the U.S. system of soil taxonomy. Soil orders represent the most basic grouping of soils in terms of the soil-forming factors.

Soil Organic Matter (SOM) (Chapter 11)—Usually understood to be the nonliving, organic fraction of soil consisting of decayed and partially decayed plant and animal remains, and unrecognizable humic substances.

Soil Separates (Chapter 4)—Mineral particles < 2 mm in diameter that range between specified size limits.

Soil Series (Chapter 3)—The sixth and final level of soil classification used in the U.S. system of soil taxonomy. Soil series are named after adjacent cities and towns and represent soil types with very similar properties.

Soil Solution (Chapter 14)—The water, including all dissolved solutes, in soil.

Soil Structure (Chapter 5)—The three-dimensional array of solids and pore space.

Soil Survey (Chapter 15)—The identification, classification, mapping, examination, and evaluation of the soils in a landscape.

Soil Taxonomy (Chapter 3)—The book specifying the soil classification system used in the United States.

Soil Textural Classes (Chapter 4)—The terms used to describe the relative proportions of various separates in a soil, such as clay, loamy clay, etc.

Solum (Chapter 2)—The A and B horizons together or either one separately above the C horizon; the true soil.

Specific Surface Area (Chapter 7)—The soil particle surface area divided by the solid particle mass, expressed as $m^2\ kg^{-1}$.

Splash Erosion (Chapter 12)—The translocation of soil particles by the impact of falling rain.

Spodosol (Chapter 3)—One of the soil orders in the U.S. taxonomic system. Spodosols have a pronounced spodic horizon where iron, aluminum, and organic matter have accumulated.

Stemflow (Chapter 13)—Precipitation interacting with vegetation that flows down stems and trunks thereby picking up and transporting loose deposits from those locations.

Stream Gauging (Chapter 13)—The process whereby the height and flow of water past a site is monitored.

Subgroup (Chapter 3)—After "great group," the fourth level of classification in the U.S. system of soil taxonomy. Subgroup names provide information about special features of a soil.

Suborder (Chapter 3)—After "order," the second level of classification in the U.S. system of soil taxonomy. Suborder formative elements provide information about climate and parent material.

Subsidence (Chapter 11)—Generally refers to a specific phenomenon in which the decomposition of high–organic matter soils leads to the rapid lowering of the soil surface.

Subsoil (Chapter 2)—In a fully formed soil it would be that formed below an A horizon. Subsoils can be exposed through erosion and become surface soils, however.

Substrate-Level Phosphorylation (Chapters 9, 10)—A metabolic process in which high-energy phosphate bonds are created on intermediate compounds from which the phosphate is later cleaved to yield energy for cellular metabolism.

Surface Creep (Chapter 12)—The process of wind-driven erosion and movement of soil across a landscape.

Symbiosis/Symbiotic (Chapters 9, 10)—An interaction, sometimes obligatory, in which two different organisms grow together for their mutual benefit; contrast with *asymbiotic* in which no cooperation between two organisms occurs.

Synergism (Chapter 9)—An interaction of two organisms in which the activity of each is enhanced in combination compared to when they are separate.

T

Takings (Chapter 16)—Confiscation of private property by government, usually through the process of eminent domain, in which the property owner is compensated for the loss, or loss of certain uses of their property.

Tensiometers (Chapter 13)—Measure the force or tension that results from a water-filled column in contact with the soil through a porous ceramic cup. They measure the soil water matric potential.

Terrace (Chapter 2)—A bench-like landform occurring on the border of rivers, lakes, and oceans that results from the deposition of material by flooding or surf activity from subsequently receded water. It can also refer to an artificial structure made by earth movement for the purpose of erosion control.

Tetrahedral (Chapter 7)—Refers to the overall arrangement or pattern of the silicon, oxygen, and hydroxyls in silicate clay layers.

Textural Triangle (Chapter 4)—A graphing device used to determine the textural class of a soil based on the percent of sand-, silt-, and clay-sized particles.

Thermophiles (Chapter 9)—Organisms that grow preferentially at elevated temperatures.

Throughfall (Chapter 13)—Precipitation that passes through a forest canopy, but interacts with leaves.

Tile Drain (Chapters 2, 6, 12, 14)—Ceramic or plastic pipe placed at suitable depths and intervals to artificially lower water tables.

Tillage (Chapter 12)—The process of moving, turning, or stirring soil. Mechanical manipulation of the soil for any purpose.

Tilth (Chapter 15)—The physical condition of a soil in relation to the ease of its tillage and fitness as a seedbed for crop growth.

Top Dress (Chapter 14)—Application of an amendment to the soil surface.

Topography (Chapter 2)—The lay of the land, its levelness or hilliness.

Toposequence (Chapter 15)—A sequence of related soils that differ primarily because of topography as the soil-forming factor.

Topsoil (Chapter 2)—Most often refers to the soil immediately involved in cultivation. Often used synonymously with the term *A horizon.*

Tortuosity (Chapters 5, 6)—Refers to the twisted nature of soil pores.

Trace Element (Chapter 14)—Synonymous with micronutrient, an essential element required in very small amounts.

Transpiration (Chapter 13)—Loss of water to the atmosphere through the leaves of plants.

Truncated (Chapter 3)—Having lost all or part of a soil horizon, or all or part of a metabolic pathway.

Tundra (Chapters 3, 11)—Treeless plains characteristic of arctic environments.

U

Ultisol (Chapter 3)—One of the soil orders in the U.S. taxonomic system. Ultisols develop in warm, humid forests. They are typically more weathered and less fertile than Alfisols.

Undifferentiated Soils (Chapter 15)—A kind of map unit used in soil surveys to reflect locations in which different soil series are so intermixed as to make individual distinctions irrelevant.

Universal Soil Loss Equation (USLE) (Chapter 12)—A predictive model for soil erosion.

Unsaturated—The condition in which only some pore space is filled with water.

Unsaturated Flow (Chapter 6)—Water movement in soil by capillarity through soil in which the largest pores are air-filled.

V

Vadose Zone (Chapter 6)—The unsaturated zone of soil extending from the top of the water table to the soil surface.

Value (Chapter 3)—One of the three variables of soil color. Value reflects the relative lightness of intensity of color.

Vermiculite (Chapter 7)—A 2:1 type silicate clay mineral formed from mica. It displays an adsorption preference for potassium and cesium.

Vertisol (Chapter 3)—One of the soil orders in the U.S. taxonomic system. Vertisols are characterized by high content of shrinking and swelling clay that causes soil heaving.

Volumetric Water Content (Chapter 6)—Soil water content expressed as volume of water relative to volume of bulk soil (water content on a vol/vol basis).

W

Water Erosion Prediction Project (WEPP) (Chapter 12)—A process-driven computer model for soil erosion prediction.

Water-Holding Capacity (Chapter 4)—The total amount of water that a soil can hold once all freely draining pores have emptied by the force of gravity.

Water Potential (Chapters 6, 9)—A mathematical description of water availability or the potential of water to move and do work in response to free energy gradients; a measure of the availability of water.

Watershed (Chapter 13)—An area draining ultimately to a particular body of water such as a lake or river.

Water-Stable Aggregate—A ped or aggregate that retains its structure despite rain impact or agitation in water.

Water Table (Chapters 6, 13)—The surface of the groundwater.

Weathering (Chapters 1, 2, 7)—All physical, biological, and chemical changes produced at or near the Earth's surface that result in the disintegration and decomposition of rocks and minerals.

Wetland (Chapters 3, 15, 16)—An area of land characterized by hydric soils and hydrophytic vegetation.

Wilting Point (Chapter 6)—The water content or potential at which a plant is unable to extract water from soil; water availability in soil for plant growth becomes insufficient to supply adequate plant turgor or rigidity.

Wind Erosion Equation (WEQ) (Chapter 12)—The computer-based equivalent of WEPS.

Wind Erosion Prediction System (WEPS) (Chapter 12)—The equivalent of USLE for predicting wind erosion.

Windbreak (Chapter 12)—An artificial or natural plant barrier designed to protect crops and soil from strong winds.

X

Xeric (Chapter 3)—A soil moisture regime common to Mediterranean climates that is typified by moist, cool winters and warm, dry summers. Soil moisture is usually not present at times for optimal plant growth.

INDEX

A horizon, 41, 42
Acid phosphatases, 220
Acid rain, 167, 279
Acidification, 28–29
Acidity, 162–166, 167–168
ACP, 367
Actinomycetes, 179, 185
Active acidity, 164
Active SOM pool, 242
Actual evapotranspiration (AET), 286, 287
Additions, 39
AEC, 158
Aeration, 104, 187
Aerobes, 188
AET, 286, 287
Aggregate size, 102
Aggregates, 101
Agricultural activity, 37
Agricultural Conservation Program (ACP), 367
Agriculture, 256–257, 296
Air dry, 117
Albic horizon, 53
Albolls, 70
Alfisols, 59–62
Algae, 179
Alkaline soils, 327–331, 330
Allelopathic, 256
Allophane, 22
Alluvial fans, 20
Alluvial soils, 19–20
Alluvium, 32
Aluminosilicates, 134
Aluminum, 165
Aluminum-magnesium octahedral sheet, 143, 144
Ammensalism, 198
Ammonification, 214
Amorphous minerals (noncrystalline) minerals, 140
Andisols, 62–63
Anecic earthworms, 176
Angular blocky, 101, 102
Anion exchange capacity (AEC), 158
Anion exchange reactions, 157
Anions, 314
Anthrepts, 68
Anthropic epipedon, 52
Anthropogenesis, 36–37
Ants, 175
A_p horizon, 42

Aqualfs, 62
Aquands, 62
Aquents, 65
Aquepts, 68
Aquerts, 74
Aquic soils, 54
Aquiclude, 291
Aquitard, 290, 291
Aquods, 73
Aquolls, 70
Aquox, 71
Aquults, 73
Arachnids, 175
Archaea, 179
Arents, 66
Argids, 64
Argillic horizon, 53
Aridic environments, 55
Aridsols, 63–64
Artesian well, 290
Artificial potting and rooting media, 90
Aspect, 31, 246
Assimilation, 207
   carbon, 211
   iron, 226–227
   manganese, 226–227
   nitrogen, 214
   phosphorus, 220
   sulfur, 224
Atacama Desert, 283
Atmosphere, 5
Autotrophs, 182, 183
Available water, 116–118

B horizon, 41, 42
Bacillus, 180
Back slope, 32
Bacteria, 179–180, 182
Bacterial colonies, 194
Bacteriophage, 181, 193
Baerman funnel, 193
Banded iron formations, 227
Base cation saturation (BCS), 160, 165, 166
Baseflow, 292
BCS, 160, 165, 166
Bedrock, 258
Beetles, 175, 176
Bennett, Hugh Hammond, 255, 363–365, 367, 372
Berlese funnel, 193
Berry, Wendell, 3, 361

Best management practices (BMPs), 296
Biogeochemical cycles. *See* Soil biogeochemical cycles
Biogeochemical transformations, 206–209
Biogeochemistry, 206
Biological activity. *See* Soil organisms
Biological weathering, 28–30
Biosphere, 5
Bioturbation, 35, 195
Black alkali soils, 331
Blake, William, 83
Blocky structure, 101, 102
Blue baby syndrome, 279
BMPs, 296
Boreal forest, 34
BPISAE, 371
Broadcast fertilizer, 320
Bt horizon, 95
Buffer capacity, 166–167
Buffering, 316
Bulk-blend fertilizers, 319
Bulk density, 106–110, 353
Bureau of Plant Industry, Soils, and Agricultural Engineering (BPISAE), 371
Buried horizon, 42

C, 210
C horizon, 41, 42
C sequestration, 373
Calcareous, 330
Calcids, 64
Calcite weathering, 137
Caliche, 63
Cambic horizon, 53
Cambids, 64
Capillary equation, 122
Capillary flow, 122–123
Capillary fringe, 290
Carbon (C), 210
Carbon cycle, 210–212
Carbon dioxide, 138
Carbonation, 27
Catena, 348
Cation exchange, 156
Cation exchange capacity (CEC), 158–161, 166
Cations, 314
CEC, 158–161, 166
CENTURY model, 243

395